Recent Advances in Georesources, Intelligent Materials, and Environmental Science

Elhoucine Essefi
https://orcid.org/0000-0002-0887-9940
University of Sfax, Tunisia

Wissem Ben Soltan
University of Gabes, Tunisia

Jamel Khériji
University of Gabes, Tunisia

Lazhar Labiadh
Huaqiao University, Tunisia

Salah Ammar
University of Gabes, Tunisia

IGI Global Scientific Publishing
Publishing Tomorrow's Research Today

Vice President of Editorial	Melissa Wagner
Director of Acquisitions	Mikaela Felty
Director of Book Development	Jocelynn Hessler
Production Manager	Mike Brehm
Cover Design	Jose Rosado

Published in the United States of America by
IGI Global Scientific Publishing
701 East Chocolate Avenue
Hershey, PA, 17033, USA
Tel: 717-533-8845 | Fax: 717-533-7115
Website: https://www.igi-global.com E-mail: cust@igi-global.com

Library of Congress Cataloging-in-Publication Data

Names: Essefi, Elhoucine, 1976- editor | Ben Soltan, Wissem, 1983- editor | Kheriji, Jamel, 1983- editor | Labiadh, Lazhar, 1988- editor | Ammar, Salah, 1958- editor

Title: Recent advances in georesources, intelligent materials, and environmental science / edited by Elhoucine Essefi, Wissem Ben Soltan, Jamel Kheriji, Lazhar Labiadh, Salah Ammar.

Description: Hershey, PA : IGI Global Scientific Publishing, 2026. | Includes bibliographical references and index. | Summary: "This book aims to provide a comprehensive overview of the latest advancements in georesources, intelligent materials, and environmental science"-- Provided by publisher.

Identifiers: LCCN 2025022509 (print) | LCCN 2025022510 (ebook) | ISBN 9798337335063 hardcover | ISBN 9798337335070 paperback | ISBN 9798337335087 ebook

Subjects: LCSH: Environmental monitoring--Technological innovations | Climate change mitigation--Technological innovations | Environmental sciences--Technological innovations

Classification: LCC QH541.15.M64 R435 2026 (print) | LCC QH541.15.M64 (ebook)

LC record available at https://lccn.loc.gov/2025022509

LC ebook record available at https://lccn.loc.gov/2025022510

British Cataloguing in Publication Data
A Cataloguing in Publication record for this book is available from the British Library.

Table of Contents

Detailed Table of Contents

calcined kaolinitic clay and activated using sodium hydroxide (NaOH) solutions. The resulting geopolymers were characterized by XRD, XRF, compressive strength measurements, FTIR spectroscopy, and leaching tests to assess their chemical composition, structural development, and capacity for heavy metal immobilization. The influence of NaOH concentration (8, 10, and 12 M), mine waste content (0, 10, 20, 50, 80, and 100 wt.%), and curing duration (7, 14, 21, and 28 days) was systematically evaluated. The optimal formulation was achieved using 10 M NaOH and curing for 28 days, resulting in materials with high compressive strength, a dense and cohesive matrix, and efficient immobilization of potentially toxic elements (Pb, Zn, As, and Cd). These results highlight the potential of Lakhouat mine waste as a raw material for sustainable geopolymer production, offering an environmentally responsible approach to mine waste management and the development of eco-friendly construction materials.

Hajer Haj Ali, Water, Energy and Environment Laboratory (L3E), Faculty of Sciences of Sfax, University of Sfax, Sfax, Tunisia
Jamel Touir, Water, Energy and Environment Laboratory (L3E), Faculty of Sciences of Sfax, University of Sfax, Sfax, Tunisia

This study investigates the role of dolomitization on porosity in the Middle Turonian Bireno Member in Jebel M'rhila, Central Tunisia. Through an integrated approach combining field study and petrography scanning electron microscopy (SEM) and geochemistry and mineralogy the present work aims to delineate the mineralogical variations and diagenetic processes that have controlled the porosity development in this dolomitic succession. The results reveal that dolomitization occurred during multiple stages, influenced by marine and meteoric fluids, and was accompanied by processes such as recrystallization, cementation, dissolution, silicification, and fracturing. These diagenetic events exerted a complex and interdependent control on the development of different types of porosity (intercrystalline, intracrystalline, vuggy, and fracture porosity). The spatial distribution of dolomite textures and geochemical signatures further supports the role of meteoric diagenesis, particularly along a fracture networks, in enhancing or occluding porosity. A detailed diagenetic sequence is proposed in this work, illustrating the temporal and spatial evolution of pore systems. This study provides key insights into the diagenetic controls on reservoir quality in dolomitized carbonate platforms, with implications for hydrocarbon exploration and reservoir characterization.

worry regarding the environmental quality of the studied locations because PM10 and elements values were high. The findings of this investigation may be utilized to raise awareness of air pollution in Nigeria.

The chapter gives a detailed review of self-healing materials (SHMs) in terms of their mechanisms, classifications, and smart applications in various fields. It also emphasizes on the intrinsic and extrinsic healing systems which are activated by stimuli like heat, light, pH and electrical or mechanical forces. Important design concepts such as dynamic bonding, environment responsiveness and biomimicry such as skin, bone and spider silk are discussed. Some of the latest developments in the multifunctional SHMs, Nano composites, environmentally friendly materials, and self-governing structures are also analyzed in the chapter. Applications are described in construction, electronics, biomedical, energy storage and environmental remediation. The problems of scalability, material fatigue, and regulatory problems are covered, as well as the future direction of the use of AI-driven design, Industry 4.0 integration, and circular economy strategies. SHMs find themselves at the centre of sustainable and intelligent material systems.

The Earth's climate is rapidly changing, and it has been projected that the global climate will continue to change even beyond this century. The magnitude of change directly relies on the quantity of greenhouse gases emitted by the mankind and the earth's sensitivity towards the emissions. The climate hazards will continue

Intelligent materials, also known as smart materials, are emerging as transformative tools in the pursuit of sustainable environmental solutions. These materials possess the ability to respond adaptively to external stimuli such as temperature, pH, light, or pollutants, enabling real-time monitoring, self-repair, and energy-efficient operations. As environmental degradation and climate change intensify, the integration of intelligent materials into green technologies presents a promising path toward sustainable development. Innovations in areas such as photocatalytic degradation of pollutants, self-healing construction materials, responsive membranes for water purification, and energy-harvesting systems demonstrate the potential of these materials to address environmental challenges effectively. performance.This Chapter explores the latest advancements in intelligent materials, critically examines the challenges facing their widespread adoption, and outlines future prospects for their role in achieving sustainable environmental goals.

This study presents a comparative analysis of two contemporary coastal sedimentary systems located in the Bekalta region esat of Tunisia: El Jazira barrier island and the Echraf dune cordon. The objective is to better understand their geomorphological development, sedimentary dynamics, and environmental evolution. Using an integrated approach that combines geomorphological mapping, sedimentological analysis, and interpretation of hydrodynamic data (waves, winds, and climatic variables), the study identifies the main morphological units of each system while detailing the spatial patterns and ecological roles of vegetation cover. Special attention is given to the interaction between plant colonization and sediment stabilization. Temporal changes in the morphology and extent of these systems are assessed through a diachronic analysis of aerial photographs from the years 1962 and 1996. The results highlight the critical role of vegetation in enhancing sediment stability and buffering against erosional processes, particularly in response to seasonal climatic variations. This

Preface

UNVEILING INTERDISCIPLINARY FRONTIERS IN GEORESOURCES, INTELLIGENT MATERIALS, AND ENVIRONMENTAL SCIENCE

In an era defined by unprecedented environmental shifts, escalating resource demands, and the relentless march of technological innovation, the need for integrated scientific approaches has never been more critical. The volume, *Recent Advances in Georesources, Intelligent Materials, and Environmental Science*, curated by Essefi et al. stands as a testament to this growing interdisciplinarity. It brings together a diverse collection of cutting-edge studies that collectively illuminate the complex, symbiotic relationships between our planet's natural systems, the materials we engineer, and the advanced technologies we develop to understand and manage them. As articulated by Essefi et al. (this volume), the intensifying global challenges—ranging from the finite nature of our resources and pervasive environmental pollution to the unpredictable variability of climate and the urgent demand for sustainable materials—mandate approaches that are not only integrated and data-driven but also fundamentally ecologically responsible.

This compilation serves as a beacon, highlighting how researchers at the confluence of geosciences, materials engineering, environmental monitoring, and intelligent technologies are profoundly redefining our understanding and interaction with natural systems. The chapters contained within delve into a spectrum of innovative strategies, from the valorization of waste materials that once posed environmental burdens to gaining intricate insights into the diagenetic processes governing subsurface carbonate reservoirs. They explore the transformative potential unlocked by sensor-driven geological frameworks, offering unprecedented real-time data for decision-making. Furthermore, the book examines advancements in critical areas such as thermal protection systems, vital for extreme engineering applications, and provides rigorous atmospheric pollution assessments, directly addressing public health concerns. A significant portion is dedicated to the exciting realm of emerging smart materials, showcasing their remarkable capabilities for self-healing and environmental responsiveness—qualities that promise a new generation of resilient and sustainable infrastructure. Complementing these thematic pillars, several chapters leverage the power of artificial intelligence (AI) and sophisticated hybrid computational models to monitor and predict environmental change with enhanced accuracy. Finally, the volume offers detailed analyses of coastal evolution and resilience, crucial for understanding and adapting to dynamic interfaces between land and sea.

Collectively, these contributions underscore a shared, imperative vision: to develop sustainable systems, enhance resource efficiency across all sectors, and foster the creation of adaptive technologies inherently capable of responding with agility and efficacy to the rapidly changing environmental conditions that characterize our epoch. As the editors Essefi, Ben Soltan, Kheriji, Labiadh, and Ammar emphasize, this book aims beyond merely disseminating novel scientific knowledge. Its profound ambition is to ignite further research endeavors, stimulate cross-disciplinary collaborations, and inspire innovation that will collectively shape the future trajectory of environmental stewardship and materials science globally.

CHAPTER OVERVIEW: A DEEP DIVE INTO INTERDISCIPLINARY INNOVATION

The chapters in this volume are meticulously organized to provide a comprehensive overview of the latest advancements, each contributing a unique perspective to the overarching themes. The following sections provide a detailed exposition of each chapter's focus, meticulously linking its content to the book's broader objectives and citing the principal authors.

Chapter 1: Valorization of Abandoned Lakhouat Mine Waste in Metakaolin-Based Geopolymers, Mechanical Properties and Immobilization of Toxic Elements: Valorization of Abandoned Lakhouat Mine Waste

This seminal contribution by Mefteh, Romdhani, Lazaar, Haj Ali, Andrejkovicová, Novais, Labrincha, Sebei, Medhioub, and Rocha stands as a powerful example of addressing the twin challenges of resource depletion and environmental pollution through innovative material science. Mefteh et al. specifically investigate the valorization of abandoned Lakhouat mine waste (LKH) by incorporating it into metakaolin-based geopolymers. The sheer scale of mine waste globally presents an enormous environmental burden, often containing potentially toxic elements (PTEs) that can leach into ecosystems. This research tackles this problem head-on by transforming a waste product into a valuable resource. The authors meticulously characterize the LKH material, noting its primary composition of carbonates along with minor but significant amounts of metal-bearing minerals such as sphalerite (ZnS), galena (PbS), and marcasite (FeS_2). Their methodology involves blending this waste with calcined kaolinitic clay and activating the mixture using sodium hydroxide (NaOH) solutions. A critical aspect of their investigation is the systematic evaluation of key parameters: NaOH concentration (testing 8, 10, and 12 M), mine waste content (ranging from 0 to a remarkable 100 wt.%), and curing duration (7, 14, 21, and 28 days). The findings are highly promising, demonstrating that an optimal formulation—achieved using 10 M NaOH and a 28-day curing period—resulted in geopolymers exhibiting superior mechanical strength and a dense, cohesive matrix. Crucially, this optimal material also proved exceptionally efficient at immobilizing potentially toxic elements like lead (Pb), zinc (Zn), arsenic (As), and cadmium (Cd). The work by Mefteh et al. thus profoundly highlights the potential of Lakhouat mine waste as a viable raw material for sustainable geopolymer production. This offers a dual benefit: an environmentally responsible approach to mine waste management and the concurrent development of eco-friendly, high-performance construction materials, thereby contributing significantly to the demand for sustainable materials and enhancing resource efficiency.

Chapter 2: Effects of Dolomitization on Porosity During Various Diagenesis Stages in Carbonate Fractured Reservoirs; A Case Study from Central Tunisia: Diagenesis Stage in Bireno Member

Delving deep into the foundational principles of geosciences and our understanding of natural systems, Haj Ali and Touir present a compelling study on the role of dolomitization in controlling porosity within carbonate reservoirs. Their research focuses on the Bireno Formation in Jebel M'rhila, Central Tunisia, a region of significant geological interest. Haj Ali and Touir employ an integrated approach to delineate the complex mineralogical variations and intricate diagenetic processes that have shaped the porosity development within this dolomitic succession. Their findings reveal that dolomitization—the process where limestone is converted to dolomite—did not occur in a single event but rather through multiple stages, profoundly influenced by interactions with both marine and meteoric fluids. This multi-stage process was further complicated by a suite of other diagenetic events including recrystallization, cementation, dissolution, silicification, and fracturing. The authors meticulously demonstrate how these interlinked diagenetic processes exerted a complex and interdependent control on the formation and evolution of various types of porosity. The spatial distribution of dolomite textures observed in their study provides further robust evidence for the pivotal role of meteoric diagenesis, particularly when channeled along fracture networks, in either enhancing or occluding porosity. The chapter culminates in the proposal of a detailed diagenetic sequence, illustrating the temporal and spatial evolution of pore systems within the Bireno Formation. This work by Haj Ali and Touir provides key insights into the diagenetic controls on reservoir quality in dolomitized carbonate platforms, directly enhancing our understanding of natural systems and offering critical knowledge for georesource management.

Chapter 3: Thermal Barrier Coatings - Conventional, New Materials and Methods, Properties and its Applications

Shifting focus to materials engineering and the demand for sustainable materials, the chapter by Prakasha, B.U., Malladad, and Shetty delves into the critical area of Thermal Barrier Coatings (TBCs). These specialized coatings are indispensable in high-temperature engineering applications, particularly for structural members in aircraft engines, gas turbines, and combustion engines, where extreme temperatures and corrosive environments are common. Prakasha et al. provide a comprehensive examination of TBCs, detailing how different coating methods are employed to provide essential protection against corrosion, wear, erosion, and to offer thermal insulation. Among these, TBCs stand out due to their complex multi-layered structure, which is crucial for their performance. The authors explain that TBCs typically utilize either ceramic or metallic multi-layers to form a protective barrier. This coating acts as an effective insulator against hot gas or steam streams to which components like turbine blades, casings, and auxiliary equipment are constantly exposed. By mitigating thermal stress and degradation, TBCs significantly improve the operational efficiency and extend the durability of these high-performance systems. To further enhance the properties and longevity of TBCs, Prakasha et al. emphasize the crucial need to understand the structural and property changes that occur when these materials operate under extreme temperature and pressure conditions. This deep dive into TBCs directly contributes to the development of sustainable materials and adaptive technologies, crucial for performance and longevity in demanding environments.

Chapter 4: Evaluation of PM10 Air Quality in Akure, Nigeria

Directly addressing the pervasive issue of environmental pollution and public health, Abulude's chapter focuses on assessing PM10 air quality in Akure, Nigeria. This study is a crucial contribution to environmental monitoring and understanding the localized impacts of global environmental challenges. Abulude's primary goals were to determine the elemental components of PM10 (particulate matter with an aerodynamic diameter of 10 micrometers or less) and to identify their sources within the urban environment of Akure. The research involved selecting three distinct sites: Oba-Ile, the National Museum and Monuments, and the Federal University of Technology, Akure (FUTA), to capture a representative cross-section of the urban landscape. Samples were meticulously collected using a Nano sampler for eight hours daily over a ten-month period. Elemental composition was identified using Energy Dispersive X-ray Fluorescence (EDXRF), while PM10 content was gravimetrically estimated. The study also examined the influence of meteorological parameters on PM10 concentrations. The findings were stark and concerning: the Museum site recorded the highest average PM10 results (125.20 $\mu g/m^3$), which the author attributes, in part, to significant human activity in that area. Critically, Abulude's study revealed that Akure's PM10 air quality was unequivocally unhealthy when compared to international Air Quality Index standards, and significantly exceeded the World Health Organization (WHO) recommended limits. Principal Component Analysis (PCA) was employed to deduce the sources of PM10, identifying both natural and anthropogenic activities as contributors. In summary, Abulude's investigation raises serious concerns regarding the environmental quality of the studied locations due to high PM10 and elemental values. The findings of this investigation are invaluable for raising awareness of air pollution in Nigeria and providing foundational data for informed policy and intervention strategies to mitigate environmental pollution.

Chapter 5: Emerging Trends in Self-Healing Materials for Smart and Sustainable Applications

Boopathy, Bose, Srinivasan, and Suresh contribute a comprehensive review that firmly places the spotlight on emerging smart materials capable of self-healing and environmental responsiveness, a key theme of this volume. Their chapter provides an in-depth examination of self-healing materials (SHMs), meticulously detailing their underlying mechanisms, diverse classifications, and a wide array of smart applications across various fields. The authors specifically emphasize the distinction between intrinsic and extrinsic healing systems, both of which are activated by specific stimuli such as heat, light, pH changes, or electrical and mechanical forces. They delve into important design concepts that underpin SHM development, including dynamic bonding, environment responsiveness, and biomimicry—drawing inspiration from natural systems like skin, bone, and spider silk which possess inherent self-repair capabilities. The chapter analyzes some of the latest developments in multifunctional SHMs, exploring advancements in nanocomposites, environmentally friendly materials, and self-governing structures. Applications of SHMs are thoroughly described across critical sectors including construction, electronics, biomedical engineering, energy storage, and environmental remediation, demonstrating their vast potential for developing sustainable systems. Boopathy et al. also address the inherent problems and challenges associated with SHM widespread adoption, such as scalability, material fatigue, and regulatory hurdles. They conclude by outlining future directions, including the integration of AI-driven design, Industry 4.0 principles, and circular economy strategies, positioning SHMs at the very center of future sustainable and intelligent material systems.

Chapter 6: An Intelligent Framework to Monitor and Predict Land Temperatures Using Learning Algorithms

This chapter by Sharanya, Rajalakshmi, Pavithra, Vijayalakshmi, and Radha exemplifies the profound impact of leveraging artificial intelligence and hybrid computational models to monitor environmental change. Their work addresses one of the most pressing global challenges: rapidly changing Earth's climate and its projected continuous alteration beyond this century. Sharanya et al. recognize that climate hazards will increasingly trigger interactive and accumulated impacts on natural systems, thus making mitigation of adverse effects quintessential. The authors argue that an improved understanding of the ecosystem can be achieved by accurately predicting global warming temperatures and land surface warming. To this end, their work focuses on building an effective framework that harnesses the power of Gated Recurrent Unit (GRU) cells, a type of recurrent neural network particularly adept at processing sequential data, for predicting the mean temperature of a region. The proposed work develops a comprehensive AI system designed to predict temperature using extensive previous and historical data. This system promises significant benefits to the community by offering accurate predictions of land temperature, which is critically important for a wide variety of human activities, including agriculture, urban planning, disaster preparedness, and resource management. The intelligent framework presented by Sharanya et al. directly contributes to our capacity to understand and anticipate climate-related changes, a core component of creating adaptive technologies capable of responding to rapidly changing environmental conditions.

Chapter 7: A Hybrid Modeling Approach with AdaGrad Optimization for Geotechnology-Based Environmental Change Monitoring

Further expanding on the theme of leveraging artificial intelligence and hybrid computational models to monitor environmental change, Kidambi Raju, Bakthavachalam, Varadarajan, Sundaramoorthy, and El-kenawy present a sophisticated study on using advanced technologies to measure environmental changes. Recognizing the emergence of a new generation of computational models, Kidambi Raju et al. evaluate the performance of several powerful hybrid machine learning models, including XGBoost+LightGBM (a voting classifier), GNN+CatBoost, AdaBoost+ANN, and SVM+BiLSTM. These models were assessed using four global metrics relevant to environmental monitoring. Among the initial comparisons, the GNN+CatBoost model demonstrated superior performance, achieving a training accuracy of 97.37% and a test accuracy of 90.91%. To further enhance this promising model, Kidambi Raju et al. integrated additional optimizations. The most significant improvement was realized with their proposed model: the GNN+CatBoost model optimized by AdaGrad. This optimized version achieved an even higher training accuracy of 98.53%, while maintaining an impressive test accuracy of 90.91%. Crucially, the proposed model successfully passed rigorous goodness-of-fit tests, yielding robust statistical results. The authors report a T-statistic of 61.44, an F-statistic from ANOVA of 0.35, and a p-value of 0.5688, all of which outperformed other models in their comparative analysis. This rigorous statistical validation suggests that the Generalized Neural Networks (GNNs) + CatBoost model with AdaGrad optimization (GCAO) is a powerful and highly accurate environmental monitoring tool. This work by Kidambi Raju et al. directly supports the imperative to enhance resource efficiency and create adaptive technologies capable of responding to rapidly changing environmental conditions through superior predictive reliability.

Chapter 8: Intelligent Materials for Sustainable Environmental Solutions: Innovations, Challenges, and Future Prospects

Jaiswal and Bahadur contribute a vital chapter that directly explores the role of intelligent materials for environmental sustainability, a cornerstone of the volume's focus. Their work positions intelligent materials, also known as smart materials, as transformative tools in the pursuit of truly sustainable environmental solutions. Jaiswal and Bahadur elucidate the defining characteristic of these materials: their remarkable ability to respond adaptively to external stimuli such as temperature, pH, light, or pollutants. This responsiveness enables critical functions like real-time monitoring, self-repair mechanisms, and energy-efficient operations. As environmental degradation and climate change continue to intensify globally, the authors argue that the integration of intelligent materials into green technologies presents an exceptionally promising pathway toward sustainable development. They highlight key innovations in areas such as the photocatalytic degradation of pollutants, which offers cleaner methods for environmental remediation; the development of self-healing construction materials, promising more durable and resilient infrastructure; responsive membranes for advanced water purification, critical for addressing water scarcity; and energy-harvesting systems, which contribute to cleaner energy production. The chapter thoroughly explores the latest advancements in these intelligent materials, critically examines the challenges that currently hinder their widespread adoption (such as cost, scalability, and long-term performance), and outlines compelling future prospects for their integral role in achieving sustainable environmental goals. This contribution by Jaiswal and Bahadur directly articulates how these materials embody the essence of creating adaptive technologies to address pressing environmental and societal challenges.

Chapter 9: Comparative Study of Coastal Sedimentary Systems, Environmental Insights from El Jazira Barrier Island and the Echraf Dune Cordon, Tunisia: Comparative Study of Coastal Sedimentary Systems

The final chapter by Haj Ali, Touir, and Ben Ali provides a crucial detailed analysis of coastal evolution and resilience, a highly relevant topic in the context of climate variability and environmental change. This study presents a comparative analysis of two contemporary coastal sedimentary systems situated in the Bekalta region, east of Tunisia: El Jazira barrier island and the Echraf dune cordon. The primary objective of Haj Ali et al. was to gain a deeper understanding of their geomorphological development, sedimentary dynamics, and environmental evolution. They employed an integrated approach that combined rigorous geomorphological mapping, detailed sedimentological analysis, and insightful interpretation of hydrodynamic data. Through this comprehensive methodology, the study successfully identified the main morphological units of each system and provided detailed insights into the spatial patterns and crucial ecological roles of their respective vegetation covers. The results unequivocally highlight the critical role that vegetation plays in enhancing sediment stability and providing a vital buffer against erosional processes, particularly in response to seasonal climatic variations. This interaction creates a dynamic feedback mechanism that significantly contributes to the inherent resilience of these coastal systems. The findings of this investigation provide invaluable insights for integrated coastal zone management and informed conservation planning, especially in regions facing escalating pressures from climate change-induced sea-level rise and intensified human activities. The work by Haj Ali et al. directly informs our collective imperative to develop sustainable systems and enhance resource efficiency by understanding and protecting our vulnerable coastal environments.

CONCLUSION

In conclusion, this meticulously assembled collection of chapters, guided by the editorial vision of Essefi, Ben Soltan, Kheriji, Labiadh, and Ammar, robustly demonstrates the accelerating convergence between geosciences, intelligent materials, and advanced environmental technologies. Across the diverse studies presented—from the innovative waste valorization techniques pioneered by Mefteh et al., the intricate diagenetic insights provided by Haj Ali and Touir, and the revolutionary IoGT framework proposed by Essefi, to the advanced material science of TBCs by Prakasha et al., the critical air quality assessments by Abulude, the groundbreaking self-healing materials reviewed by Boopathy et al., the AI-driven climate predictions by Sharanya et al., the optimized environmental monitoring models by Kidambi Raju et al., the comprehensive view of intelligent materials by Jaiswal and Bahadur, and the essential coastal resilience studies by Haj Ali, Touir, and Ben Ali—a powerful and shared theme emerges. This theme is the urgent need for innovative, sustainable solutions, solutions that are firmly grounded in rigorous scientific inquiry and robustly supported by modern analytical and computational tools. The profound insights gathered within this volume not only deepen our fundamental understanding of natural systems but also illuminate practical, actionable pathways for effectively addressing the most pressing environmental and societal challenges of our time. It is the sincere hope of Essefi et al. that this comprehensive work will not only inspire but actively catalyze continued exploration, foster dynamic interdisciplinary collaboration, and drive transformative research in the years ahead, ultimately shaping a more sustainable and resilient future for all.

Elhoucine Essefi
University of Sfax, Tunisia

Wissem Ben Soltan
University of Gabes, Tunisia

Jamel Kheriji
University of Gabes, Tunisia

Lazhar Labiadh
University of Gabes, Tunisia

Salah Ammar
University of Gabes, Tunisia

Chapter 1
Valorization of Abandoned Lakhouat Mine Waste in Metakaolin-Based Geopolymers:
Mechanical Properties and Immobilization of Toxic Elements

Samir Mefteh
https://orcid.org/0009-0002-1454-8520
Laboratory of Spectroscopic Characterization and Optical Materials, Faculty of Sciences of Sfax, University of Sfax, Sfax, Tunisia

Yamina Romdhani
https://orcid.org/0009-0008-8382-4035
Laboratory of Spectroscopic Characterization and Optical Materials, Faculty of Sciences of Sfax, University of Sfax, Sfax, Tunisia

Khalil Lazaar
https://orcid.org/0000-0002-4230-8834
Laboratory of Spectroscopic Characterization and Optical Materials, Faculty of Sciences of Sfax, University of Sfax, Sfax, Tunisia

Hajer Haj Ali
https://orcid.org/0000-0002-6254-3778
Laboratory of Spectroscopic Characterization and Optical Materials, Faculty of Sciences of Sfax, University of Sfax, Sfax, Tunisia

Slavka Andrejkovicová
Department of Geosciences, University of Aveiro, Aveiro, Portugal

Rui M. Novais
Department of Materials and Ceramic Engineering, University of Aveiro, Portugal & Universitário de Santiago, Aveiro, Portugal

DOI: 10.4018/979-8-3373-3506-3.ch001

João A. Al Labrincha
Department of Materials and Ceramic Engineering, University of Aveiro, Portugal & Universitário de Santiago, Aveiro, Portugal

Mounir Medhioub
Laboratory of Spectroscopic Characterization and Optical Materials, Faculty of Sciences, University of Sfax, Sfax, Tunisia

Abdelaziz Sebei
Department of Geology, Faculty of Sciences, University of Tunis El Manar, Tunisia

Fernando Rocha
https://orcid.org/0000-0002-3636-3933
Department of Geosciences, University of Aveiro, Aveiro, Portugal

ABSTRACT

This study investigates the valorization of abandoned Lakhouat mine waste (LKH) through its incorporation into metakaolin-based geopolymers. The waste material, mainly composed of carbonates with minor amounts of metal-bearing minerals such as sphalerite (ZnS), galena (PbS), and marcasite (FeS_2), was blended with calcined kaolinitic clay and activated using sodium hydroxide (NaOH) solutions. The resulting geopolymers were characterized by XRD, XRF, compressive strength measurements, FTIR spectroscopy, and leaching tests to assess their chemical composition, structural development, and capacity for heavy metal immobilization. The influence of NaOH concentration (8, 10, and 12 M), mine waste content (0, 10, 20, 50, 80, and 100 wt.%), and curing duration (7, 14, 21, and 28 days) was systematically evaluated. The optimal formulation was achieved using 10 M NaOH and curing for 28 days, resulting in materials with high compressive strength, a dense and cohesive matrix, and efficient immobilization of potentially toxic elements (Pb, Zn, As, and Cd). These results highlight the potential of Lakhouat mine waste as a raw material for sustainable geopolymer production, offering an environmentally responsible approach to mine waste management and the development of eco-friendly construction materials.

1. INTRODUCTION

Tunisia possesses a rich geological and mining heritage, with mineral exploitation dating back to Antiquity and continuing into the present day (Mobbs, 2005). Mining activities have primarily targeted various ores, notably iron, lead, zinc, and barite, which are mainly concentrated in the northwestern region of the country (Jemmali et al. 2014; Ayari et al. 2025). While these activities provide valuable resources, they also generate substantial amounts of waste, posing significant environmental and socio-economic challenges. Former mining operations often leave large volumes of residues in open-air dumps, which may contain potentially toxic elements (PTEs) such as lead, zinc, cadmium, and arsenic. These contaminants can threaten soils, groundwater, and surface water, presenting serious risks to ecosystems and human health. (Surenbaatar et al. 2023; Huang et al. 2024; Zineb et al. 2024; Parlak et al. 2024; Khalil et al. 2025; Camacho-Alcantar et al. 2025).

In response to these challenges, there is increasing interest in the valorization of mining waste, which can be seen as a secondary resource rather than conventional waste (Kekez et al. 2025; Lazorenko et al. 2025). The mineralogical and chemical composition of these residues, rich in aluminosilicates and metal oxides, offers opportunities for producing sustainable alternative materials (Mokhtari et al. 2025; Morales Aranibar et al. 2025). Among these, geopolymers are particularly promising (Carrillo Beltran et al. 2025a). They are obtained through the alkaline activation of silica- and alumina-rich materials, resulting in a solid and cohesive matrix (Carrillo Beltran et al. 2025b; Siyal et al. 2024). Beyond their strong mechanical and chemical properties, geopolymers can immobilize heavy metals, thereby limiting the mobility and environmental impact of toxic elements contained in the residues (Maqbool et al. 2023; Palma et al. 2024; Genua et al. 2025).

In this context, the present study investigates mining wastes from the abandoned Lakhouat mine, located in northwestern Tunisia. The primary aim is to characterize these wastes in terms of their mineralogical, chemical, and physic-chemical properties and to evaluate their potential for the formulation of high-performance geopolymers. Subsequently, the research focuses on the characterization of these geopolymers, with particular emphasis on their mechanical strength, matrix cohesion, and ability to immobilize heavy metals.

2. MATERIALS AND METHODS

2.1. Materials

Geopolymers were prepared using a mixture of metakaolin (MK) and mine waste from the abandoned Lakhouat mine (LKH). The Lakhouat mine is located in the Siliana governorate, northwestern Tunisia, specifically in the Gaâfour delegation, approximately 28 km northwest of the city of Siliana (Fig. 1). Sampling was performed at the tailings embankments of the mine, collecting multiple samples from different points to ensure adequate representativeness of the material. The collected samples were homogenized to obtain a representative sample for characterization analyses The samples were then dried at 50 °C for 48 h, ground, and stored for further use.

Figure 1. Geological map showing the location of the Lakhouat mining site (A and B), tailings and former mining installations (C and D).

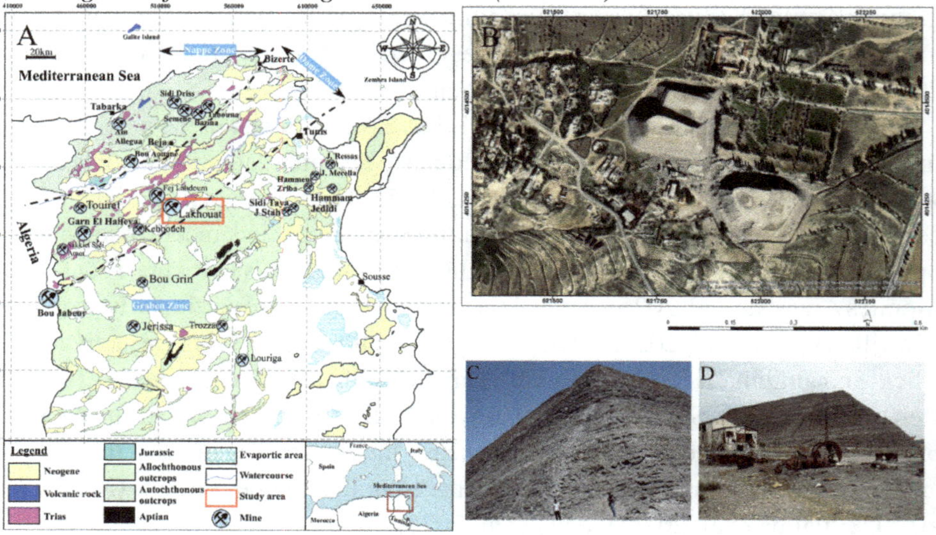

The kaolinitic clay (KA) used for geopolymer preparation was provided by the Tunisian company SANIMED®, which employs it in the production of sanitary materials. Metakaolin (MK) was produced by calcining this clay at 700 °C for 3 h. Sodium hydroxide (NaOH) solutions of various concentrations (8 M, 10 M, and 12 M) were used for activation of the raw materials.

The mineralogical composition of the raw materials and geopolymer formulations was determined by X-ray diffraction (XRD) using a Philips®/Panalytical X'Pert-Pro MPD instrument with Cu Kα radiation ($\lambda = 1.5405$ Å) and a step size of 0.02° 2θ s^{-1} in the 4–65° 2θ range. Data processing was performed with X'Pert HighScore Plus software. Semi-quantification of clay minerals and associated phases was carried out by the classical peak area method, corrected according to the chemical composition (Galindo et al. 1996).

The chemical composition (major and trace elements) was analyzed by X-ray fluorescence (XRF) using a Niton FXL Thermo Scientific FM-XRF spectrometer. Loss on ignition (LOI) was determined by calcination at 1000 °C. Particle size distribution of the raw materials was measured by laser diffraction using a Malvern Mastersizer 3000 Hydro MV laser granulometer.

For the identification of organic and inorganic functional groups, Fourier Transform Infrared (FTIR) spectroscopy was conducted using a PerkinElmer Spectrum Two spectrometer in transmission mode over the range of 4000–450 cm^{-1}.

To measure the compressive strength (RC) of the synthesized geopolymer samples, a LLOYD LR50K Plus testing machine connected to NEXYGEN Plus data analysis software was used, at a displacement rate of 2 mm/min.

2.2. Methods

Geopolymers were synthesized from Lakhouat mine tailings in combination with metakaolin. Several experimental parameters were systematically evaluated, including the concentration of the sodium hydroxide (NaOH) activator solution (8 M, 10 M, and 12 M), the proportion of tailings incorporated (0 %, 10 %, 20 %, 50 %, 80 %, and 100 %), and the curing duration (7, 14, 21, and 28 days).

The NaOH-based activation solution was used as the primary geopolymerization agent. Three concentrations (8, 10, and 12 M) were tested to assess their influence on the dissolution of aluminosilicate phases and the reactivity of the mixture. Drying and curing of the samples were performed in an oven at 60 °C, a condition considered optimal for ensuring proper polycondensation reactions.

The evolution of the mechanical properties of the formulated geopolymers, particularly their compressive strength, was monitored, along with their fire resistance behavior. These tests aim to identify the most suitable formulation conditions to optimize both the mechanical performance and durability of the materials, promoting the sustainable valorization of mine tailings as eco-friendly geomaterials.

3. RESULTS AND DISCUSSION

3.1. Raw materials characterization

3.1.1. Mineralogical analysis

The XRD pattern of the representative sample of the Lakhouat mine tailings, obtained after data processing (Fig. 2), revealed the presence of the following mineral species: calcite, dolomite, quartz, gypsum, kaolinite, sphalerite, marcasite, and galena. Calcite is the dominant mineral (63%), followed by gypsum (14%), quartz (11%), and dolomite (7%). Phyllosilicates are present in minor amounts, not exceeding 2%. Metal-bearing minerals, occurring as sphalerite (ZnS), galena (PbS), and marcasite (FeS_2), were detected in low proportions, with an average content of about 1%. Nevertheless, these phases represent the main carriers of lead (Pb) and zinc (Zn) in the tailings.

Figure 2. X-ray diffractogram of bulk rock powder from a representative sample of Lakhouat mine tailings (LKH).

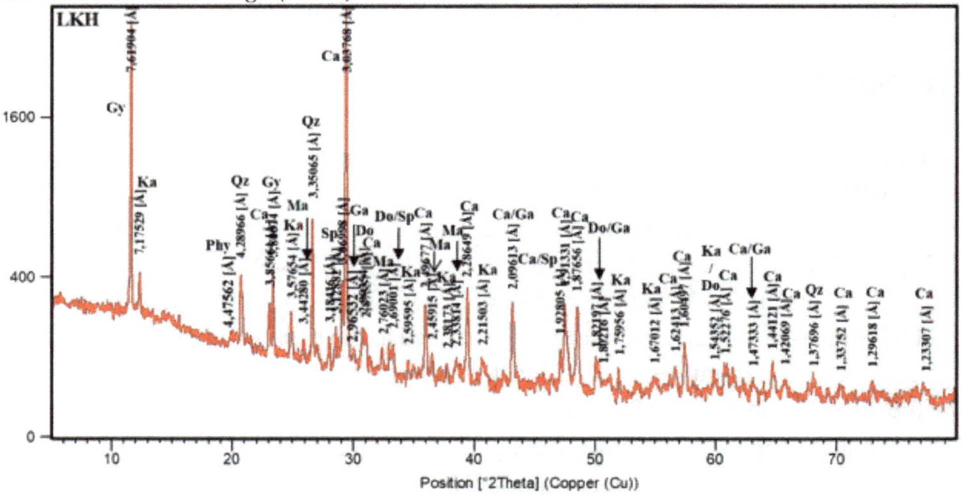

For the kaolinitic clay (KA), the diffractogram (Fig. 3) indicates that kaolinite is the predominant mineral, identified by its characteristic peaks at 7.12 Å and 3.57 Å. Quartz is also present, with peaks at 4.26 Å and 3.33 Å, along with minor amounts of illite and alunite, which is a potassium aluminum sulfate mineral. It contains amorphous or partially crystalline aluminum, which can act as an additional source of aluminum during geopolymer formation after thermal treatment (Yildizay et al. 2017)

Figure 3. X-ray diffractograms of kaolinitic clay (KA) and metakaolin (MK) obtained after calcination at 700 °C. The thermal transformation of kaolinite into an amorphous phase is indicated by the appearance of a broad diffuse peak in the 2θ = 20–30° region, characteristic of metakaolin.

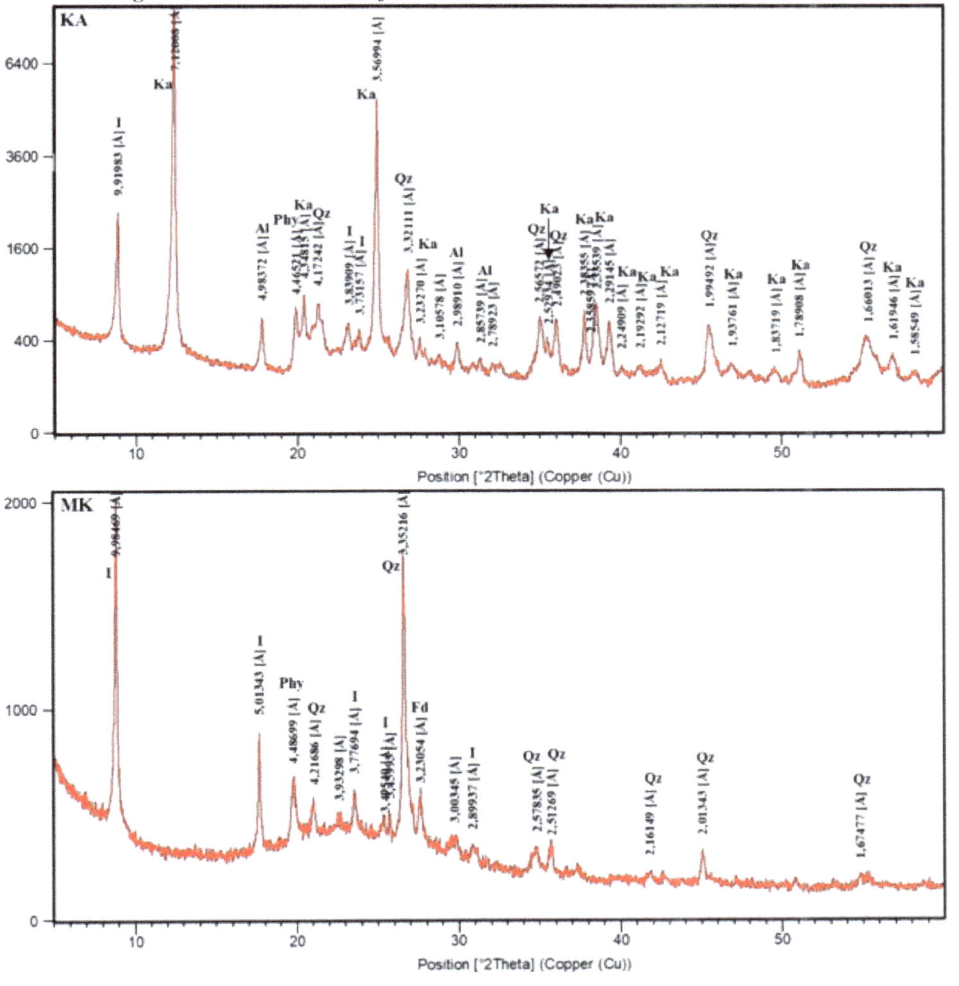

To obtain metakaolin, the kaolinitic clay was calcined at 700 °C/3h, which is the optimal temperature for the dehydroxylation and structural breakdown of kaolinite. The XRD pattern displays the appearance of a broad diffuse hump between 20° and 30° 2θ, characteristic of an amorphous phase (Fig. 3), thereby confirming the formation of metakaolin. This amorphous structure, being highly disordered, exposes numerous reactive aluminum and silicon sites, which significantly enhances its reactivity in alkaline solutions (Scherb et al. 2020; Kassa et al. 2022).

3.1.2. Particle size distribution

The particle size distribution of the Lakhouat mine tailings is bimodal (Fig. 4), highlighting the predominance of medium-sized grains and a significant fine fraction, which is favorable for chemical reactivity and the formation of the aluminosilicate gel. The fine fraction (<63 μm) represents 35.71 % of the material, while the fraction between 63 μm and 2 mm accounts for 64.29 %, with no particles exceeding 2 mm.

Figure 4. Particle size distribution and fraction percentages of Lakhouat mine waste.

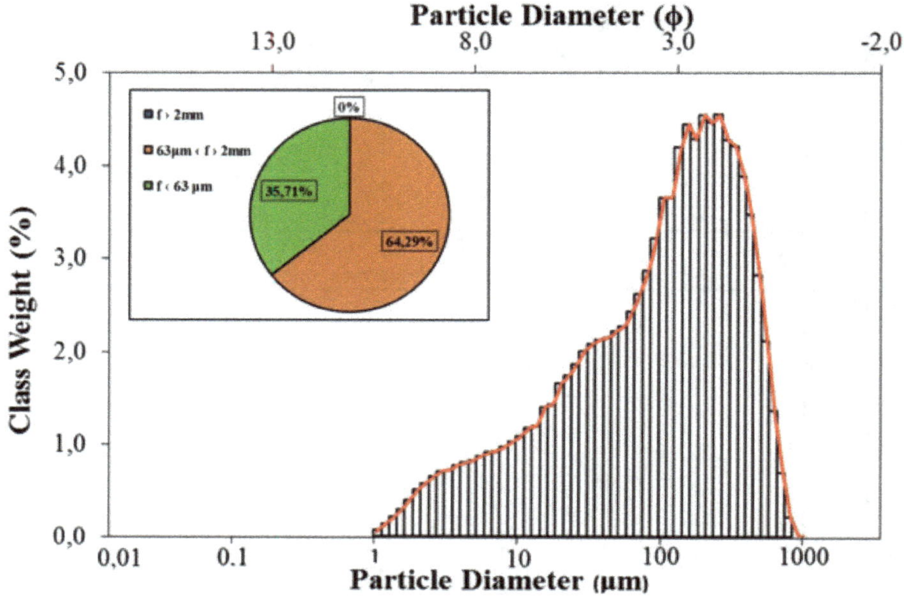

These tailings exhibit an essentially sandy-silty texture, dominated by high sand content and supplemented by smaller proportions of clay. The medium grains, mainly corresponding to sand, contribute to the mechanical stability and density of the matrix. The absence of particles larger than 2 mm ensures a homogeneous mixture and efficient incorporation of the tailings.

Overall, the predominance of sand, a moderate clay fraction, and an abundant fine fraction, this abundance of fines also makes the tailings particularly susceptible to wind and water erosion; however, it is an important factor to consider in their management and utilization, and it is favorable for geopolymer synthesis (Rao et al. 2015; Zhang et al.2023)

3.1.3. Chemical analysis

The chemical analysis of major oxides, presented in Table. 1, reveals the composition of the Lakhouat mine tailings (LKH), kaolinitic clay (KA), and metakaolin (MK).

Table 1. Major elements composition (wt. %) of the representative Lakhouat mine tailings (LKH), Kaolinitic clay (KA), and Metakaolin (MK).

Raw Materials	Major Elements (wt.%)											
	SiO_2	Al_2O_3	MgO	CaO	K_2O	MnO	Na_2O	P_2O_5	SO_3	TiO_2	Fe_2O_3	LOI
LKH	17.343	7.444	0.978	31.585	0.253	0.137	0.191	0.145	10.092	0.341	5.697	22.02
Kaolin (KA)	48.568	34.914	0.544	0.05	3.086	0.014	0.079	0.075	0.019	0.122	0.991	11.21
Metakaolin (MK)	53.321	39.601	0.496	0.052	3.004	0.013	0.08	0.123	0.039	0.129	1.579	1.15

In the Lakhouat tailings, silica (SiO_2) accounts for approximately 17 %, reflecting the notable abundance of quartz, consistent with the mineralogical analysis. Alumina (Al_2O_3) reaches 7.44 %, confirming the presence of aluminosilicate minerals, particularly phyllosilicates. The Fe_2O_3 content (5.69%) indicates the oxidation of sulfide minerals such as marcasite and/or the presence of secondary iron oxides. Calcium oxide (CaO) is the major component, exceeding 30 %, highlighting the carbonated nature of the mineralized gangue. Magnesium oxide (MgO), at ~1 %, is likely associated with dolomite. The loss on ignition (LOI) is very high (~22 %), attributed to the dehydration of sulfates and decarbonation of carbonates. The tailings contain low concentrations of Na_2O, TiO_2, and K_2O, with Na_2O at 0.19 % and TiO_2 and K_2O present in minor amounts.

The kaolinitic clay (KA) is primarily composed of SiO_2 (48.57 %) and Al_2O_3 (34.91 %), with low contents of alkali oxides (K_2O 3.09 %, Na_2O 0.079 %) and alkaline-earth oxides (CaO 0.05 %, MgO 0.544 %). Minor oxides such as Fe_2O_3, TiO_2, and MnO are present in trace amounts, indicating the relative purity of this clay. The LOI is high (11.21 %), reflecting interlayer water and hydroxyl groups in kaolinite. After calcination at 700 °C, metakaolin (MK) exhibits increased SiO_2 (53.32 %) and Al_2O_3 (39.60 %) due to water loss and the formation of a highly reactive amorphous phase. Alkali and alkaline-earth oxides remain nearly stable, ensuring their availability for alkaline activation during geopolymerization (Diffo et al. 2015; Al-Duais et al. 2024). Minor oxides show minimal variation, although Fe_2O_3 increases slightly from 0.991 % to 1.579 %, reflecting component concentration after dehydration. The LOI decreases markedly to 1.15 %, confirming the removal of interlayer water and hydroxyl groups from the kaolinite structure.In summary, calcination transforms kaolinite into amorphous metakaolin, rich in reactive aluminum and silicon sites, with low LOI and stable oxides. These characteristics make metakaolin an ideal precursor for geopolymer synthesis (Jindal et al. 2023; Butle et al. 2025).

Regarding trace elements (Table. 2), the analysis of the representative sample from the abandoned Lakhouat mine reveals high concentrations of several metallic elements. In particular, lead (4,520 ppm), zinc (18,870 ppm), strontium (9,770 ppm), and barium (4,200 ppm) show significant levels. Arsenic (100 ppm) and cadmium (60 ppm) also display non-negligible concentrations. These results reflect the mineralogical nature and the complex composition of the Lakhouat mine wastes. Such high concentrations are mainly attributed to the limited efficiency of former ore extraction and processing techniques, leading to the accumulation of residual metals in the tailings. These observations are consistent with previous studies conducted on the site (Boussen, 2010; Sahraoui et al. 2016; Ayari et al. 2023). For the kaolinitic clay before calcination and the metakaolin after calcination at 700 °C, calcination leads to a redistribution of trace elements, with a decrease in some elements (V, Cu, As, W, U) and an increase in others (F, Rb, Sr, Ba, Ga, Nb, Sn, Bi). These variations are mainly related to the loss of volatile matter and structural water, as well as the formation of a reactive amorphous phase (Diffo et al. 2015; Yin et al. 2024).

Table 2. Trace elements content (ppm) of the Lakhouat mine waste sample (LKH). the kaolinitic clay sample before calcination (KA). and after calcination at 700 °C (MK).

Trace Elements (ppm)	Raw Materials		
	LKH	Kaolin (KA)	Metakaolin (MK)
Pb	4520	20	30
As	100	70	20
Cd	60	-	-
Zn	18870	30	30
Ba	4200	160	260
V	90	20	10
Cr	70	7	7
Co	10	-	-
Ni	40	6	6
Sr	9770	110	150
Y	20	20	20
Ga	30	40	50
Rb	30	390	430
Ag	10	-	-

continued on following page

Table 2. Continued

Trace Elements (ppm)	Raw Materials		
	LKH	**Kaolin (KA)**	**Metakaolin (MK)**
Bi	8	-	-
Ce	50	60	60
La	30	30	30
Nd	20	30	30
Tl	20	3	3
Th	150	20	20
Cl	190	-	-
Sc	0.002	-	4.3
Se	7.3	-	-
Br	3.9	1.1	-
Mo	5.9	-	-
U	22.3	11.6	7.6
Cu	-	60	40
Nb	-	14.6	19.5
Sm	-	8.4	9.1
W	-	21.6	11.5
Bi	-	5.9	7.8
F	-	1850	2620

Concerning potentially toxic elements (PTEs), among those detected (As, Pb, Cu, U, Sb), the concentrations remain very low. For instance, arsenic decreases from 70 ppm to 20 ppm, lead from 20 ppm to 30 ppm, and antimony disappears completely. These low concentrations have no significant impact on the safety or reactivity of metakaolin for geopolymer synthesis (Tian et al. 2022). Other chemical elements, including Zn, Ni, Ce, La, Nd, and Th, remain stable during calcination, thereby contributing to the overall chemical stability of the material. Elements such as F, Rb, Sr, Ba, Nb, and Ga exhibit slight enrichment after calcination; however, this enrichment does not adversely affect the suitability of the metakaolin for subsequent applications.

3.2. Characterization of geopolymer specimens

3.2.1. General appearance of geopolymers after curing

After demolding, the geopolymer specimens exhibit a generally stable morphology; however, certain areas reveal surface heterogeneities (Fig. 5). These appear as slight roughness or localized foaming, probably related to variations in the particle size distribution of the mixture components and to alkaline activation reactions, despite homogeneous mixing.

As the curing time progresses, the specimens tend to stabilize and maintain a regular cylindrical shape without any signs of cracking or deformation. This observation indicates good internal cohesion and effective geopolymerization, leading to a compact and durable structure.

Figure 5. Photographs showing the general appearance of geopolymers after 14 days of curing, formulated with different concentrations of NaOH activating solution (8 M, 10 M, 12 M).

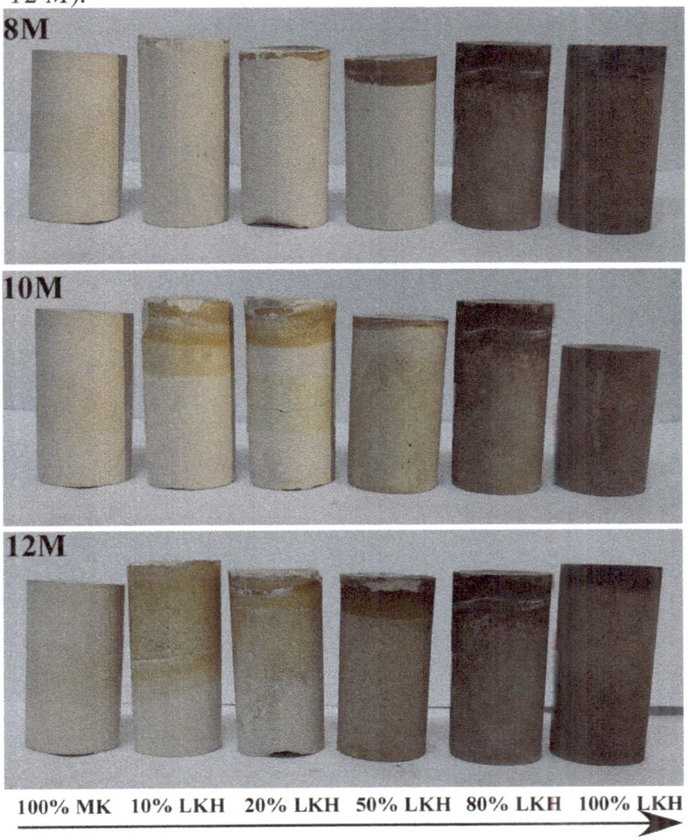

3.2.2. Compressive strength

The histograms (Fig. 6) illustrate the evolution of compressive strength as a function of three activator concentrations (8 M, 10 M, and 12 M), various proportions of incorporated mine waste (0, 10, 20, 50, 80, and 100 wt .%), and different curing durations (7, 14, 21, and 28 days).

Figure 6. Compressive strength (CS) histograms of geopolymers as a function of Lakhouat tailings content and curing time.

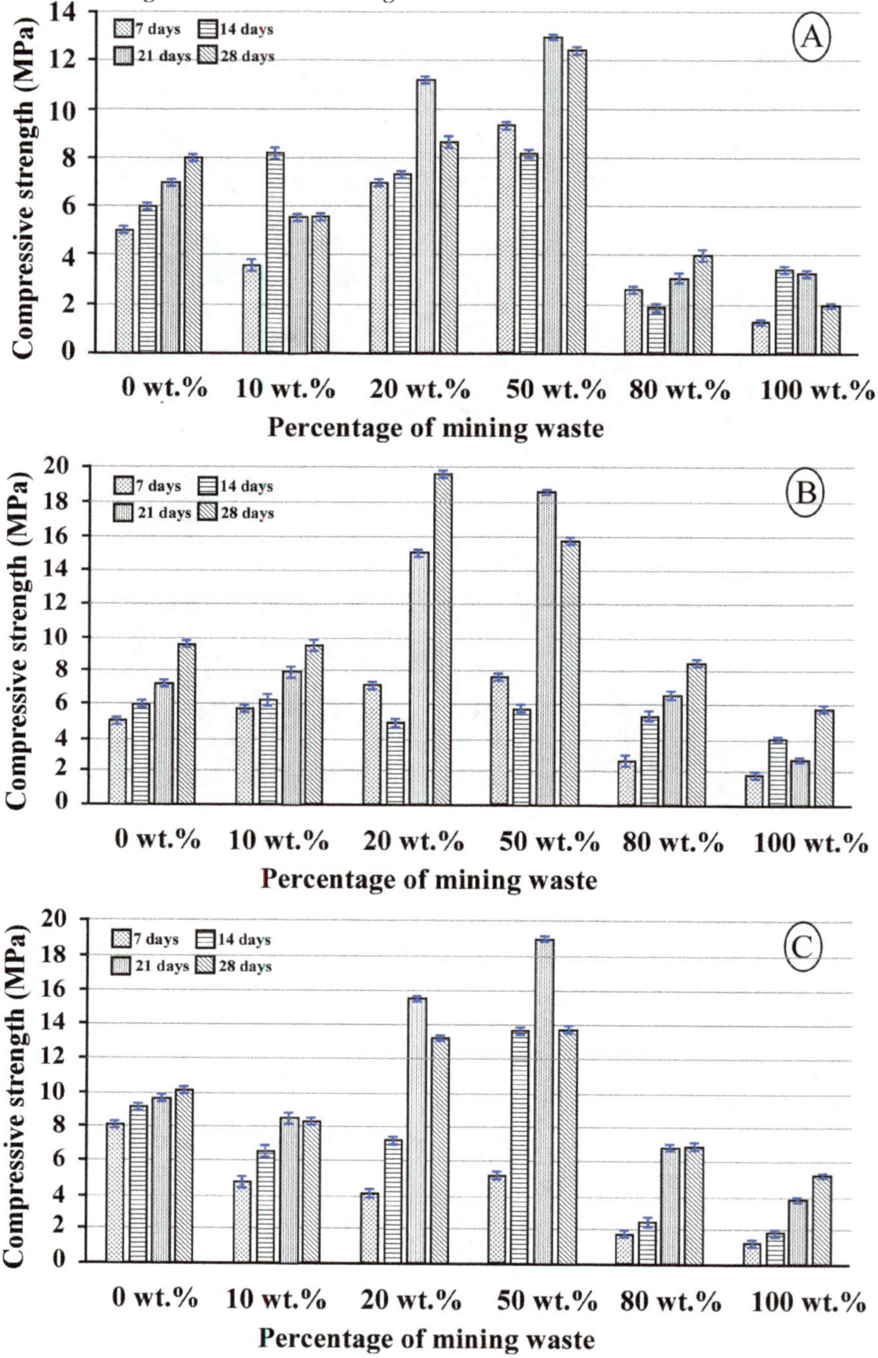

For the 8 M concentration, compressive strength remains relatively low, reaching a maximum of 12.5 MPa. A slight increase is observed with longer curing time, with the optimum recorded after 21 days at 50% tailings. In contrast, the 10 M activator concentration demonstrates superior performance, with optimal strength values ranging between 18.5 and 19 MPa, particularly after 21 days and for a 20 wt.% mine waste. The 12 M solution exhibits intermediate strength values, between 14.7 and 18.4 MPa, following a trend similar to that of the 10 M formulation. Overall, the 10 M and 12 M formulations achieve the highest compressive strengths, frequently exceeding 12 MPa and reaching up to 18 MPa after 28 days, even with high tailings contents (50 wt.%).

In general, compressive strength increases progressively with curing time from 7 to 28 days, regardless of the activator concentration. These results highlight the crucial role of activator concentration and curing duration in optimizing the mechanical properties of geopolymers (Tuyan et al. 2018; Poojalakshmi et al. 2025), demonstrating that the 10 M and 12 M formulations provide the best performance, while the 8 M concentration remains significantly less effective.

3.2.4. Leaching test

Leaching tests were conducted on selected geopolymer specimens (10, 20, 50, 80, and 100 wt.%) prepared with a 10 M activator solution after 28 days of curing (Table. 3).

Table 3. Concentrations (ppm) of leached elements in 10 M NaOH geopolymers after 28 days of curing, as a function of the percentage of Lakhouat mine waste.

Geopolymers	Element (ppm)										
	S	K	Ca	V	Fe	Cu	Zn	As	Rb	Sr	W
10 wt.% LKH ; 90 wt.% MK	-	6.624	0.341	0.193	0.185	0.013	0.060	0.441	0.377	-	0.307
20 wt.% LKH ; 80 wt.% MK	-	5.688	0.620	0.148	0.299	0.021	0.245	1.4	0.229	0.014	0.149
50 wt.% LKH; 50 wt.% MK	97.04	9.515	0.685	0.110	0.037	0.030	0.048	1.4	0.232	0.020	0.079
80 wt.% LKH; 20 wt.% MK	205.6	13.17	7.46	0.202	3.099	0.072	0.270	1.911	0.152	6.546	0.403
100 wt.% LKH	178.18	0.744	4.660	0.089	0.052	0.057	0.180	3.220	0.020	8.992	0.086

For low LKH contents (10 -20 wt.%), leaching was very limited, with concentrations generally below 7 ppm for major elements such as potassium (K), calcium (Ca), and sulfur (S), and negligible levels for trace metals including iron (Fe), copper (Cu), zinc (Zn), arsenic (As), rubidium (Rb), and strontium (Sr), indicating that the geopolymer matrix is stable and that minor incorporation of mine tailings does not compromise element retention. For intermediate proportions (50 wt.% LKH), increased leaching was observed for certain elements, notably sulfur (S, 97 ppm) and potassium (K, 9.5 ppm), while other elements remained low, reflecting a slight reduction in retention capacity with increasing tailings content. At higher proportions (80–100 wt.% LKH), leaching of major elements, particularly sulfur (S), potassium (K), and calcium (Ca), increased significantly, and enhanced leaching of some trace metals such as arsenic (As) and strontium (Sr) was also observed, highlighting the importance of controlling the tailings fraction to ensure environmental stability. Overall, the 10 M leaching tests demonstrate that formulations with low tailings incorporation ($\leq$ 20 wt.%) provide optimal chemical stability, whereas higher LKH contents require careful consideration to limit the leaching of both major and trace elements. In summary, potentially toxic elements present in the tailings, including lead (Pb, 4,520 ppm), cadmium (Cd, 60 ppm), arsenic (As, 100 ppm), zinc (Zn, 18,870 ppm), strontium (Sr, 9,770 ppm), and barium (Ba, 4,200 ppm), were either undetectable or present at very low levels in the leachates, confirming their effective immobilization within the geopolymer matrix and demonstrating the material's capacity to fix these elements and limit their release into the environment, even after incorporation of mine tailings. These findings are consistent with studies by Hu et al. (2020) and Zhang et al. (2022), which reported similar immobilization efficiencies for heavy metals in geopolymer matrices.

3.2.5. FTIR characterization of geopolymers

FTIR analysis was conducted on several geopolymer formulations containing varying percentages of mine tailings, prepared with a 10 M; NaOH activator solution, and cured for 28 days (Fig. 7).

Figure 7. FTIR spectra of geopolymers with different percentages of Lakhouat mine tailings using a 10 M NaOH activating solution

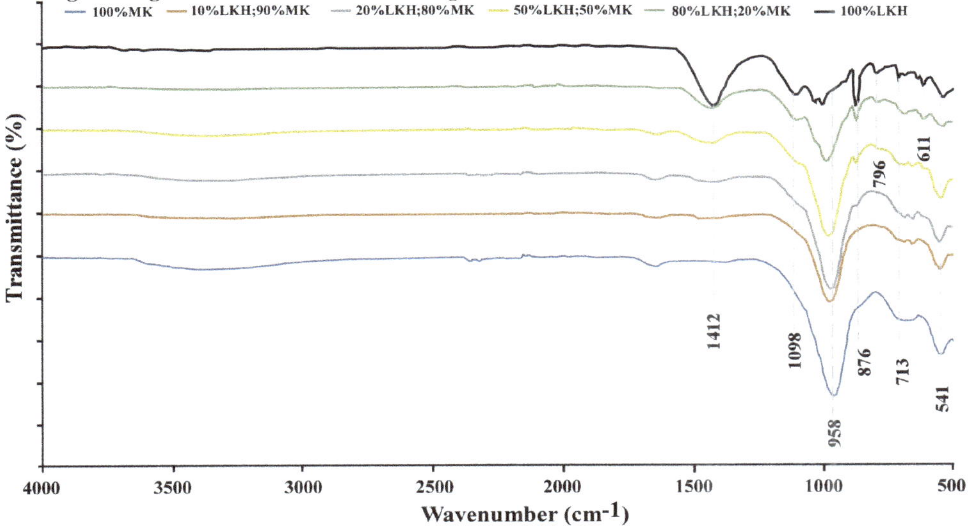

The spectra confirm that geopolymerization effectively occurred, as evidenced by the shift of the characteristic metakaolin band (1098 cm^{-1}), attributed to Si–O–Si vibrations, towards lower wavenumbers (~958 cm^{-1}). This shift indicates the formation of three-dimensional networks, reflecting modifications in the microstructure of the mine waste and the development of new aluminosilicate geopolymer structures (Li et al. 2019).

The band at 958 cm^{-1} corresponds to the symmetric and asymmetric stretching vibrations of Si–O–Si, Si–O–Al, and Si–O–Fe bonds (Lee et al. 2003). Comparison of the spectra from different geopolymer formulations reveals variable band shifts, suggesting that the extent of geopolymerization and the density of the resulting three-dimensional networks differ depending on the formulation (Zhang et al. 2012).

The C–O stretching vibration of carbonate groups appears at 1412 cm^{-1}, indicating minor carbonate impurities associated with higher proportions of Lakhouat mine tailings (El Azizi et al. 2025). Bands with peaks at 876 cm^{-1} and 796 cm^{-1} are assigned to the symmetric stretching of Si–O–Si groups. Additionally, the band at 541 cm^{-1}, observed in geopolymers containing mine tailings, is attributed to Fe–O bonds (Santos et al. 2024).

Alkaline activation of the mine tailings using NaOH solution induces a shift of the characteristic metakaolin band (796 cm^{-1}) and the mine tailings band (876 cm^{-1}) towards 713 cm^{-1}. This spectral shift reflects the formation of silicon-rich silicate species in which Na^+ ions are incorporated into the structural network, indicating a chemical modification associated with the geopolymerization process (Li et al. 2019).

4. CONCLUSIONS

This study focused on developing a geopolymer formulation based on metakaolin (MK) incorporating mining waste from the abandoned Lakhouat mine (LKH). In these wastes, carbonates are the dominant minerals, whereas metal-bearing minerals such as sphalerite (ZnS), galena (PbS), and marcasite (FeS_2) are present in low proportions, with an average content of around 1%. Chemical analyses revealed high concentrations of major oxides (CaO, SiO_2, Al_2O_3), along with notable levels of potentially toxic elements (PTEs), particularly lead, zinc, arsenic and cadmium. Based on this context, valorization tests were conducted by combining Lakhouat mine waste with calcined kaolinitic clay and activating the mixture with NaOH solutions. Several experimental parameters were systematically investigated, including NaOH molarity (8 M, 10 M, and 12 M), the proportion of mine waste incorporated (0 wt.%, 10 wt.%, 20 wt.%, 50 wt.%, 80 wt.%, and 100 wt.%), and curing time (7, 14, 21, and 28 days).

A 10 M NaOH activating solution proved to be the optimal condition for producing stable and high-performance geopolymers. Curing period of 28 days was particularly effective in achieving higher compressive strength, better matrix cohesion, and the formation of an abundant geopolymeric gel. These conditions also enhanced the immobilization of heavy metals, thereby reducing their mobility and potential environmental impact.

This work therefore provides a dual contribution. First, it proposes a sustainable approach for the management and rehabilitation of Lakhouat mining waste. Second, it opens up promising perspectives for the development of environmentally friendly and cost-effective alternative materials based on geopolymers.Nevertheless, further research is required to optimize the formulations, evaluate the long-term behavior of the developed materials, and explore their integration into industrial applications such as construction, infrastructure, or safe waste storage.

ACKNOWLEDGEMENTS

This work was supported by the TUNISO-MAURITANIAN project (1004/23). We would also like to express our sincere gratitude to the Department of Geosciences and the Department of Materials and Ceramic Engineering at the University of Aveiro, Portugal, for providing the materials necessary to carry out this study.

REFERENCES

Al-Duais, I. N., Ahmad, S., Al-Osta, M. A., Maslehuddin, M., & Saleh, T. A. (2024). Properties of alkali-activated concrete made using the optimum combinations of precursor materials and activation parameters. *Journal of Materials in Civil Engineering*, *36*(4), 04024042. DOI: 10.1061/JMCEE7.MTENG-16944

Ayari, J., Barbieri, M., Barhoumi, A., Boschetti, T., Braham, A., Dhaha, F., & Charef, A. (2023). Trace metal element pollution in media from the abandoned Pb and Zn mine of Lakhouat, Northern Tunisia. *Journal of Geochemical Exploration*, *247*, 107180. DOI: 10.1016/j.gexplo.2023.107180

Ayari, J., Barbieri, M., Boschetti, T., Sellami, A., Ballirano, P., & Charef, A. (2025). Geological, Mineralogical, and Alteration Insights of the Intermediate-Sulfidation Epithermal Mineralization in the Sidi Aissa District, Northern Tunisia. *Geosciences*, *15*(7), 269. DOI: 10.3390/geosciences15070269

Boussen, S. (2010). Evolution de haldes plombo-zincifères dans le nord de la Tunisie: l'exemple d'un contexte carbonaté. Ph.D. Thesis, Faculté des Sciences de Tunis, Université de Tunis El Manar, Tunis, Tunisia

Butle, V. V., Srinivasan, V., Reddy, D. S., & Gouda, J. (2025). Comparative assessment of limestone-calcined clay blend and fly ash as binders in roller-compacted concrete pavement: Evaluation of fresh and hardened properties. *Road Materials and Pavement Design*, •••, 1–36. DOI: 10.1080/14680629.2025.2486526

Camacho-Alcantar, M., González-Méndez, B., Loredo-Portales, R., Romo-Leon, J. R., & Molina-Freaner, F. (2025). Evaluating potentially toxic element accumulation in crops near abandoned mine tailings in northwestern Mexico: A One Health perspective. *Environmental Monitoring and Assessment*, *197*(10), 1–18. DOI: 10.1007/s10661-025-14587-1 PMID: 40973812

Carrillo Beltran, R., Martinez, A., Lopez, J., & Gonzalez, F. (2025a). Geopolymers manufactured by the alkali activation of mining, ceramics, and olive industry by-products. *Materials (Basel)*, *18*(3), 688. DOI: 10.3390/ma18030688 PMID: 39942354

Carrillo Beltran, R., Picazo Camilo, E., Perea Toledo, G., & Corpas Iglesias, F. A. (2025b). Study of Novel Geopolymer Concrete Prepared with Slate Stone Cutting Sludge, Chamotte, Steel Slag and Activated with Olive Stone Bottom Ash. *Materials (Basel), 18*(9), 1974. DOI: 10.3390/ma18091974 PMID: 40363481

Diffo, B. K., Elimbi, A., Cyr, M., Manga, J. D., & Kouamo, H. T. (2015). Effect of the rate of calcination of kaolin on the properties of metakaolin-based geopolymers. *Journal of Asian Ceramic Societies, 3*(1), 130–138. DOI: 10.1016/j.jascer.2014.12.003

Galindo, A. L., Ruiz, J. T., & Lopez, J. G. (1996). Mineral quantification in sepiolite-palygorskite deposits using X-ray diffraction and chemical data. *Clay Minerals, 31*(2), 217–224. DOI: 10.1180/claymin.1996.031.2.07

Genua, F., Lancellotti, I., & Leonelli, C. (2025). Geopolymer-based stabilization of heavy metals, the role of chemical agents in encapsulation and adsorption. *Polymers, 17*(5), 670. DOI: 10.3390/polym17050670 PMID: 40076162

Hu, S., Zhong, L., Yang, X., Bai, H., Ren, B., Zhao, Y., Zhang, W., Ju, X., Wen, H., Mao, S., Tao, R., & Li, C. (2020). Synthesis of rare earth tailing-based geopolymer for efficiently immobilizing heavy metals. *Construction & Building Materials, 254*, 119273. DOI: 10.1016/j.conbuildmat.2020.119273

Huang, J. L., Li, Z. Y., Mao, J. Y., Chen, Z. M., Liu, H. L., Liang, G. Y., Zhang, D.-B., Wen, P.-J., Mo, Z.-Y., & Jiang, Y. M. (2024). Contamination and health risks brought by arsenic, lead and cadmium in a water-soil-plant system nearby a non-ferrous metal mining area. *Ecotoxicology and Environmental Safety, 270*, 115873. DOI: 10.1016/j.ecoenv.2023.115873 PMID: 38150749

Jemmali, N., Souissi, F., Carranza, E. J. M., Vennemann, T. W., & Bogdanov, K. (2014). Geochemical constraints on the genesis of the Pb–Zn deposit of Jalta (northern Tunisia): Implications for timing of mineralization, sources of metals and relationship to the Neogene volcanism. *Chemie der Erde, 74*(4), 601–613. DOI: 10.1016/j.chemer.2014.01.002

Jindal, B. B., Alomayri, T., Hasan, A., & Kaze, C. R. (2023). Geopolymer concrete with metakaolin for sustainability: A comprehensive review on raw material's properties, synthesis, performance, and potential application. *Environmental Science and Pollution Research International, 30*(10), 25299–25324. DOI: 10.1007/s11356-021-17849-w PMID: 35000173

Kassa, A. E., Shibeshi, N. T., & Tizazu, B. Z. (2022). Kinetic analysis of dehydroxylation of Ethiopian kaolinite during calcination. *Journal of Thermal Analysis and Calorimetry, 147*(22), 12837–12853. DOI: 10.1007/s10973-022-11452-y

Kekez, S., Novakova, I., Lach, M., Setlak, K., Fiala, L., Přikryl, J., Furtos, G., Petean, I., Alexandersson, K. F., Baronins, J., & Hager, I. (2025). Waste valorization for fabrication of geopolymers. *Case Studies in Construction Materials*, *22*, e04388. DOI: 10.1016/j.cscm.2025.e04388

Khalil, A., Ait-Khouia, Y., Beniddar, H., El Ghorfi, M., Hakkou, R., Taha, Y., & Benzaazoua, M. (2025). Sustainable reprocessing of Pb–Zn mine tailings through froth flotation for resource recovery and environmental remediation in abandoned mining regions. *Minerals Engineering*, *222*, 109132. DOI: 10.1016/j.mineng.2024.109132

Lazorenko, G., Ivanov, S., Petrova, T., & Dimitrov, V. (2025). Towards coal mining waste valorization: Gangue as a resource for geopolymers and related alkali-activated materials. *Journal of Hazardous Materials*, *420*(3), 126–138. DOI: 10.1016/j.grets.2025.100205

Lee, W. K. W., & Van Deventer, J. S. J. (2003). Use of infrared spectroscopy to study geopolymerization of heterogeneous amorphous aluminosilicates. *Langmuir*, *19*(21), 8726–8734. DOI: 10.1021/la026127e

Li, Z., Zhang, S., Zuo, Y., Chen, W., & Ye, G. (2019). Chemical deformation of metakaolin based geopolymer. *Cement and Concrete Research*, *120*, 108–118. DOI: 10.1016/j.cemconres.2019.03.017

Maqbool, Q., Mobili, A., Blasi, E., Sabbatini, S., Ruello, M. L., & Tittarelli, F. (2023). Sustainable Alkali-Activated Mortars for the Immobilization of Heavy Metals from Copper Mine Tailings. *ACS Sustainable Resource Management*, *1*(1), 154–164. DOI: 10.1021/acssusresmgt.3c00002

Mobbs, P. M. (2005). The mineral industry of Tunisia. Minerals Yearbook: Area Reports, International. *Africa and the Middle East.*, *III*, 37.

Mokhtari, P., Rahimi, H., & Hosseini, R. (2025). Valorization of mine tailings for sustainable geopolymer production. *International Journal of Ceramic Engineering & Science*, *7*(4), e70014. DOI: 10.1002/ces2.70014

Morales Aranibar, C. G., Torres, J., & Ramirez, L. (2025). Reuse of mine tailings through geopolymerization and 3D printing: A sustainable approach. *Sustainability*, *17*(6), 2617. DOI: 10.3390/su17062617

Palma, G., Bolaños, H., Huamani, R., Clements, C., & Hedayat, A. (2024). Optimization of Geopolymers for Sustainable Management of Mine Tailings: Impact on Mechanical, Microstructural, and Toxicological Properties. *Minerals (Basel)*, *14*(10), 997. DOI: 10.3390/min14100997

Parlak, M., Tunçay, T., & Parlak, A. Ö. (2024). Heavy metals in tailings and soils in the Pb-Zn mining areas of North-west Türkiye and health risk evaluations. *International Journal of Agriculture Environment and Food Sciences*, *8*(1), 131–148. DOI: 10.31015/jaefs.2024.1.14

Poojalakshmi, E. S., Nagarajan, P., Sudhakumar, J., & Thomas, B. S. (2025). Impact of alkaline activator concentration on mechanical properties and microstructure of a ternary blended one-part geopolymer cement. *Scientific Reports*, *15*(1), 33808. DOI: 10.1038/s41598-025-01610-1 PMID: 41028007

Rao, F., & Liu, Q. (2015). Geopolymerization and its potential application in mine tailings consolidation: A review. *Mineral Processing and Extractive Metallurgy Review*, *36*(6), 399–409. DOI: 10.1080/08827508.2015.1055625

Sahraoui, H., Attia, R., Hamrouni, H., & Hachicha, M. (2016). Assessment of trace elements contamination of agricultural topsoil around Lakhouat mine area, Tunisia. [WJST]. *Walailak Journal of Science and Technology*, *13*(11), 965–976.

Santos, A., Andrejkovičová, S., Perna, I., Almeida, F., & Rocha, F. (2024). Mechanical and thermal properties of geopolymers derived from metakaolin with iron mine waste. *Applied Clay Science*, *258*, 107452. DOI: 10.1016/j.clay.2024.107452

Scherb, S., Köberl, M., Beuntner, N., Thienel, K. C., & Neubauer, J. (2020). Reactivity of metakaolin in alkaline environment: Correlation of results from dissolution experiments with XRD quantifications. *Materials (Basel)*, *13*(10), 2214. DOI: 10.3390/ma13102214 PMID: 32408524

Siyal, A. A., Mohamed, R. M. S. R., Shamsuddin, R., & Ridzuan, M. B. (2024). A comprehensive review of synthesis kinetics and formation mechanism of geopolymers. *RSC Advances*, *14*(1), 446–462. DOI: 10.1039/D3RA06205H PMID: 38173565

Surenbaatar, U., Lee, S., Kwon, J.-Y., Lim, H., Kim, J.-J., Kim, Y.-H., & Hong, Y.-S. (2023). Bioaccumulation of Lead, Cadmium, and Arsenic in a Mining Area and Its Associated Health Effects. *Toxics*, *11*(6), 519. DOI: 10.3390/toxics11060519 PMID: 37368619

Tian, Q., Bai, Y., Pan, Y., Chen, C., Yao, S., Sasaki, K., & Zhang, H. (2022). Application of geopolymer in stabilization/solidification of hazardous pollutants: A review. *Molecules (Basel, Switzerland)*, *27*(14), 4570. DOI: 10.3390/molecules27144570 PMID: 35889449

Tuyan, M., Andiç-Çakir, Ö., & Ramyar, K. (2018). Effect of alkali activator concentration and curing condition on strength and microstructure of waste clay brick powder-based geopolymer. *Composites. Part B, Engineering*, *135*, 242–252. DOI: 10.1016/j.compositesb.2017.10.013

Yildizay, H., & Goren, R. (2017). The utilization of Şaphane alunite in geopolymer production. *Journal of the Australian Ceramic Society*, *53*(2), 265–270. DOI: 10.1007/s41779-017-0032-z

Yin, H., Liu, H., Bu, Y., Chen, W., Ding, F., Lin, W., & Zhang, Y. (2024). Regulation of reactivity of calcined ultrafine kaolin products. *Processes (Basel, Switzerland)*, *12*(10), 2268. DOI: 10.3390/pr12102268

Zhang, B., Yu, T., Deng, L., Li, Y., Guo, H., Zhou, J., Li, L., & Peng, Y. (2022). Ion-adsorption type rare earth tailings for preparation of alkali-based geopolymer with capacity for heavy metals immobilization. *Cement and Concrete Composites*, *134*, 104768. DOI: 10.1016/j.cemconcomp.2022.104768

Zhang, N., Hedayat, A., Figueroa, L., Steirer, K. X., Li, L., & Sosa, H. G. B. (2023). Physical, mechanical, cracking, and damage properties of mine tailings-based geopolymer: Experimental and numerical investigations. *Journal of Building Engineering*, *75*, 107075. DOI: 10.1016/j.jobe.2023.107075

Zhang, Z., Wang, H., & Provis, J. L. (2012). Quantitative study of the reactivity of fly ash in geopolymerization by FTIR. *Journal of Sustainable Cement-Based Materials*, *1*(4), 154–166. DOI: 10.1080/21650373.2012.752620

Zineb, M., Abdelhak, B., Bilal, B., & Fadila, A. (2024). Characterization and pollution assessment of potentially toxic metals (PTMs) in the sulfide tailings from Boudoukha abandoned mine (Zn–Pb–Cu), NE Algeria. *International Journal of Environmental Research*, *18*(4), 70. DOI: 10.1007/s41742-024-00620-7

Chapter 2
Effects of Dolomitization on Porosity During Various Diagenesis Stages in Carbonate Fractured Reservoirs:
A Case Study From Central Tunisia

Hajer Haj Ali
https://orcid.org/0000-0002-6254-3778
Water, Energy and Environment Laboratory (L3E), Faculty of Sciences of Sfax, University of Sfax, Sfax, Tunisia

Jamel Touir
Water, Energy and Environment Laboratory (L3E), Faculty of Sciences of Sfax, University of Sfax, Sfax, Tunisia

ABSTRACT

This study investigates the role of dolomitization on porosity in the Middle Turonian Bireno Member in Jebel M'rhila, Central Tunisia. Through an integrated approach combining field study and petrography scanning electron microscopy (SEM) and geochemistry and mineralogy the present work aims to delineate the mineralogical variations and diagenetic processes that have controlled the porosity development in this dolomitic succession. The results reveal that dolomitization occurred during multiple stages, influenced by marine and meteoric fluids and was accompanied by processes such as recrystallization, cementation, dissolution, silicification, and fracturing. These diagenetic events exerted a complex and interdependent control on the development of different types of porosity (intercrystalline, intracrystalline,

DOI: 10.4018/979-8-3373-3506-3.ch002

vuggy, and fracture porosity). The spatial distribution of dolomite textures and geochemical signatures further supports the role of meteoric diagenesis, particularly along a fracture networks, in enhancing or occluding porosity. A detailed diagenetic sequence is proposed in this work, illustrating the temporal and spatial evolution of pore systems. This study provides key insights into the diagenetic controls on reservoir quality in dolomitized carbonate platforms, with implications for hydrocarbon exploration and reservoir characterization.

1- INTRODUCTION

The development of porosity in carbonate reservoirs (Choquette and Pray, 1970 ; Purser, 1978 ; Troudi et al., 2000 ; Hood et al., 2004) is a fundamental factor controlling their potential as hydrocarbon reservoirs. Unlike siliciclastic systems, where porosity is primarily determined by depositional and compactional processes, carbonate systems are significantly modified by post-depositional diagenetic alterations. Among these, dolomitization plays a crucial role in reshaping both the mineralogical fabric and the pore system of the host rock (Sibley, and Gregg, 1987 ; Wilson et al., 1990 ; Warren, 2000 ; Flügel, 2004 ; Yang et al., 2020 ; Hu et al., 2020).

The Bireno Member, a Middle Turonian dolomitic succession in Central Tunisia, offers an ideal case study to examine the complex interplay between dolomitization and other diagenesis processes and porosity evolution. This succession, part of a rudist-bearing shallow marine carbonate platform (Touir and Soussi, 2003 ; Zagrarni, 1999) has been subjected to multiple early and burial diagenetic events, including recrystallization, cementation, dissolution, silicification, and fracturing. These processes are intricately linked to changes in fluid chemistry, particularly the influx of meteoric waters during the shallowing-up of the marine environmemts and the subaerial exposure phases, as well as to tectonic activity that generated fracture networks serving as fluid conduits (Haj Ali et al., 2016b).

While previous studies have documented the regional stratigraphy and general depositional framework of the Bireno Member (Fournié, 1978 ; Touir et al., 1989 ; Zagrarni, 1999), detailed investigations into the textural, geochemical, and microstructural characteristics of the dolomites, and their implications for porosity development, remain limited. This paper aims to fill that gap by providing a comprehensive analysis of the diagenetic transformations affecting the Bireno carbonates and assessing their impact on reservoir potentiality. By integrating polarizing microscopy and scanning electron microscopy (SEM), petrography, and geochemical data, we reconstruct the diagenetic history of this Formation and elucidate the spatial and temporal relationships between dolomitization and porosity development.

2- PALEOGEOGRAPHY

The Bireno carbonate succession outcrops in several Atlas anticlinal structures across Central Tunisia, including Jebel M'rhila our study area (Fig. 1). The succession is composed with three carbonate units, the middle one of which is predominantly including dolomitic unit experienced a subaerial exposure linked to the global sea-level fall known as SB 90.5 Ma according to the global eustatic chart of Haq et al. (1987) (Abdallah et al., 2000; Touir et al., 2009) (Fig. 2). The dolomitic unit is overlain with a transgressive oolitic and bioclastic limestone unit. The Bireno Member represents a large carbonate platform 60-80 km wide that developed on a structural block bounded by two major NW-SE trending faults ; the M'rhila Fault to the North and the Kasserine Fault to the South (Touir and Soussi, 2003). Its total extent exceeds 300 km, stretching from the Algerian border in the West to the N-S megastructure in the East (M'Rabet et al., 1996; Troudi et al., 2002). Following Wright and Burchette's (1996) classification, the Middle Turonian Bireno Member represents a shoal-rimmed carbonate platform (Touir & Soussi, 2003). Along its northern margin, a belt of hippuritid rudist lithosomes (Gili et al., 1995) forms an E-W-oriented organic shoal, separating an inner shelf to the South from an outer shelf to the North of the platform. The Jebel M'rhila is located in the platform's edge, where this rudist shoal developed.

Figure 1. Location map of the Jebel M´rhila.

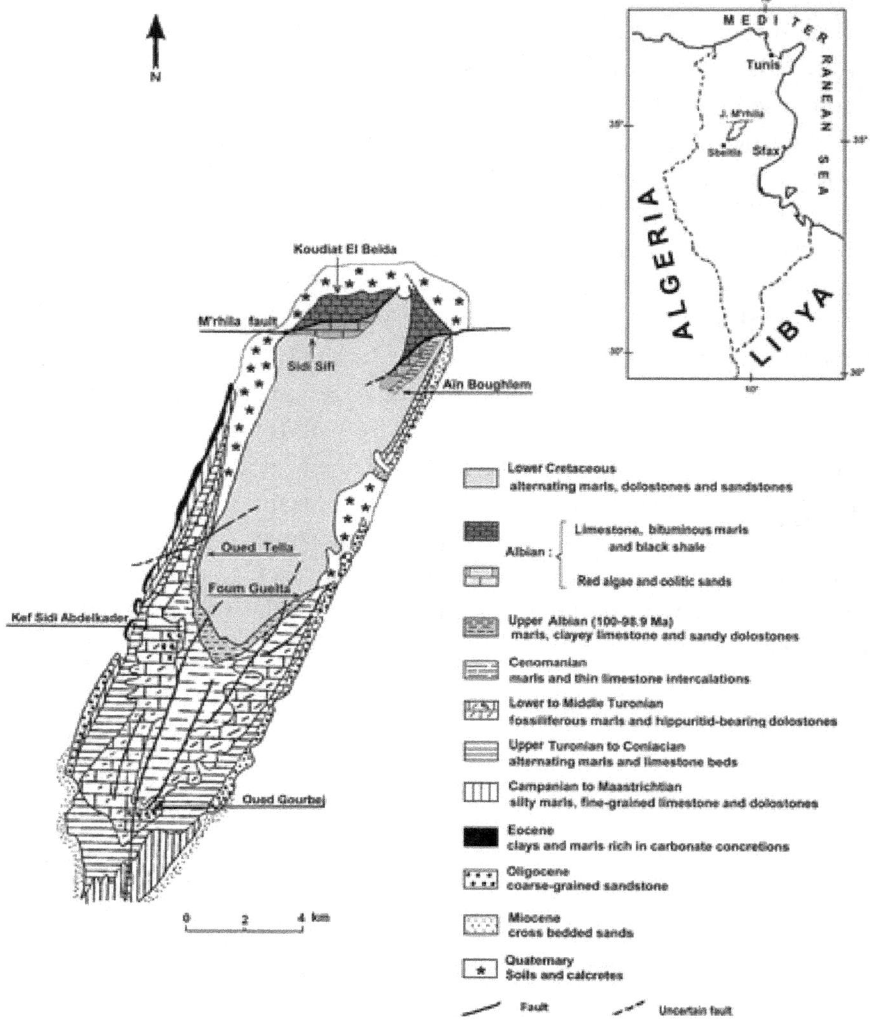

Figure 2. Stratigraphy and sequence stratigraphy of the Bireno platform, showing the dolomitic middle unit (Haj Ali et al., 2016 a).

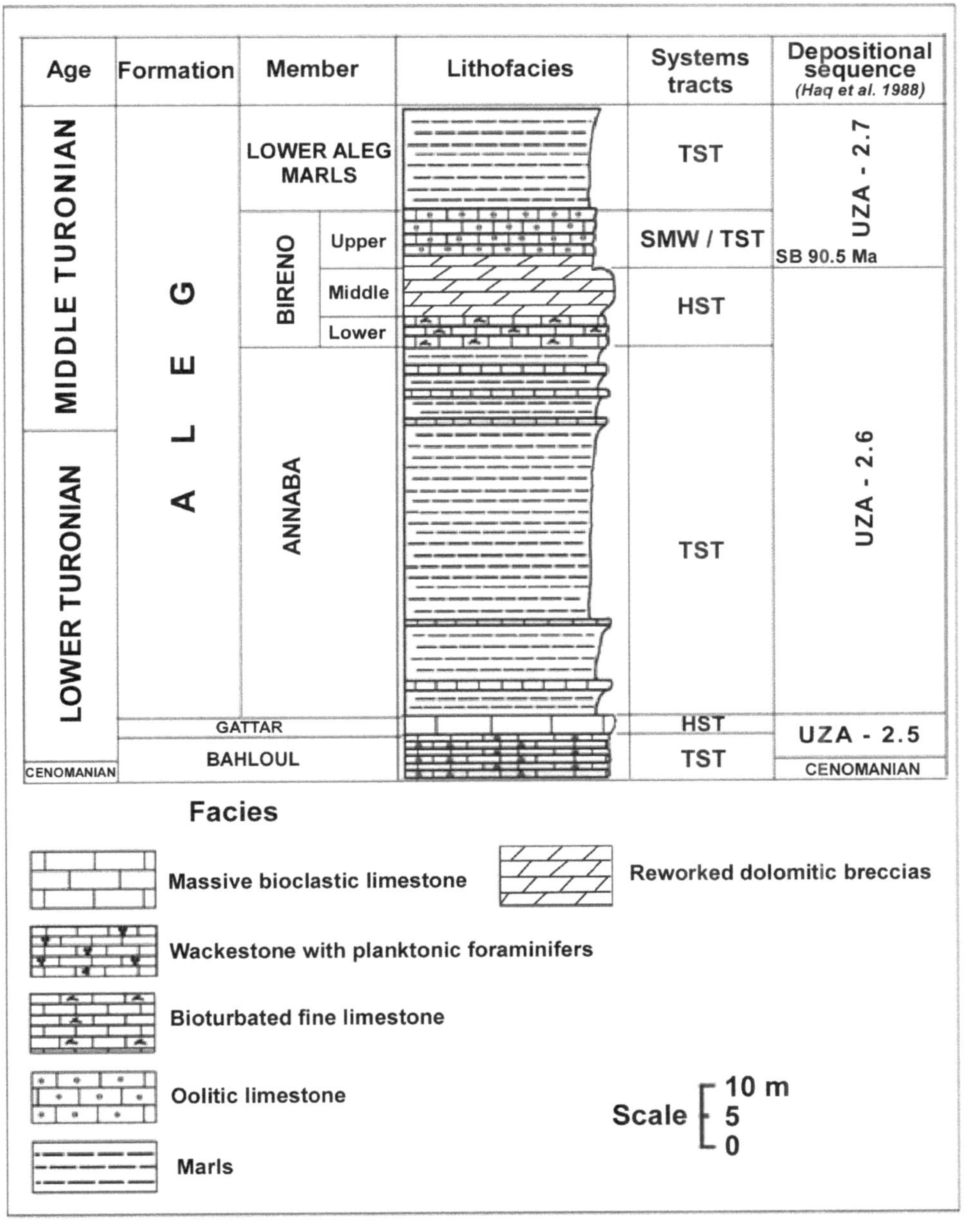

3- STRUCTURAL SETTING

The Bireno platform is intersected by a synsedimentary NE-SW fault system that was reactivated during the Middle Turonian (Touir et al., 1987; Ben Ayed et al., 1991) (Fig. 3). This deformation occurred within a regional transtensional tectonic setting (Chihi et al., 1984; Bouaziz et al., 2002). In the Jebel M'rhila region, the dolomitic middle unit exhibits multiple fracture networks.

Figure 3. Sketch map showing the Middle Turonian Tunisia paleogeography, the Bireno facies distribution and the location of the studied sections (in Troudi et al., 2002, slightly modified).

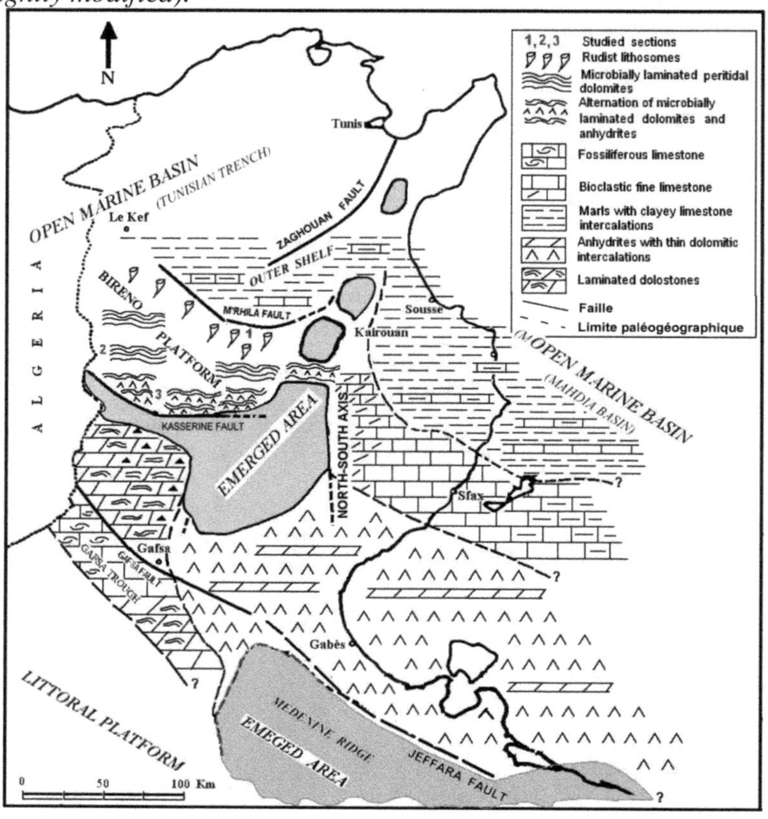

The most extensive and well-preserved fractures within the Bireno Member are observed near Kef Sidi Abdelkader and Oued Gourbej localities (Fig. 4). These fractures are typically infilled either by limestone from the Upper unit of Bireno Member or, in some cases, by marls from the overlying Aleg Formation. The presence of such marl-filled fractures (Haj Ali et al., 2016b) confirms their Middle

Turonian age. Furthermore, certain fractures and microfractures display dolomite cement coatings along their walls (Haj Ali et al., 2016b), demonstrating that fracturing preceded dolomitization.

Figure 4. Facies map of the middle unit of the Bireno platform in Jebel M'rhila, indicating the main localities of Sidi Abdelkader and Oued Gourbej where the Bireno dolomites and the associated fractures are the most preserved (Haj Ali et al., 2016 b).

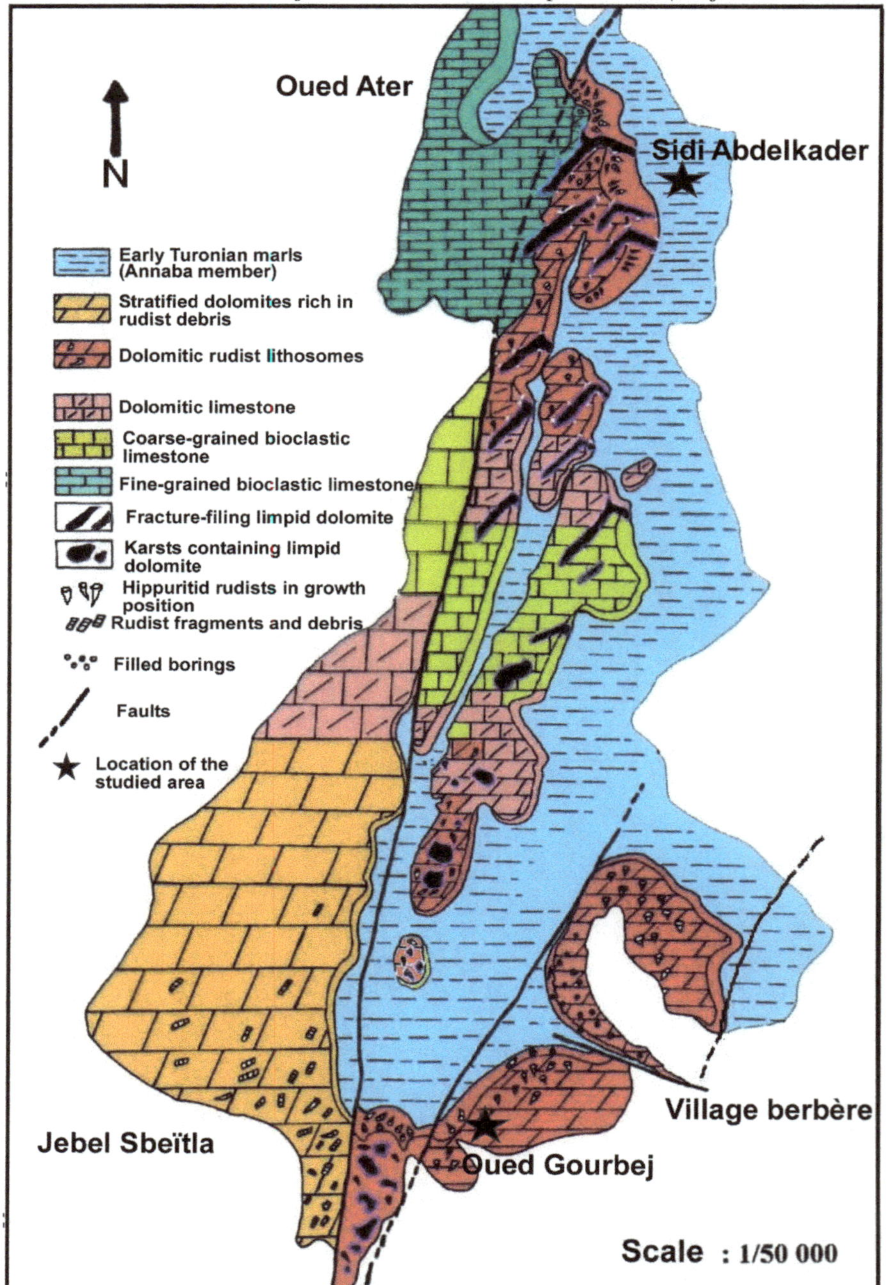

At Sidi Abdelkader locality the fractures are multi-directional, but domminated by N-S and E-W trends. These fractures penetrate deeply into the Bireno's lower unit, with secondary sets oriented between N60 and N130. Similarly, at Oued Gourbej locality, fractures predominantly align along N-S and NE-SW directions (Touir et al., 2009, 2012, Haj Ali et al., 2016a,2016b).

4- METHODOLOGY

To investigate the role of dolomitization on porosity evolution and reconstruct the diagenetic history and its impact on reservoir properties within the Middle Turonian Bireno Member in Central Tunisia, a multidisciplinary approach was followed. This involved detailed fieldwork, petrographic analysis, mineralogical and geochemical profiling, and scanning electron microscopy (SEM).

Fieldwork and Sampling

Field investigations were conducted through the Jebel M'rhila outcrops, which provides an excellent exposure of the Bireno Member. Stratigraphic sections were measured, and detailed lithological descriptions were recorded. A total of 35 representative rock samples were collected from different stratigraphic levels and facies types, with particular attention to dolomitized horizons and zones with visible porosity or fracturing.

Petrographic Analysis

Selected samples were examined under a polarizing microscope to determine their petrographic characteristics, including textural features, and the nature of diagenetic alterations. In addition, Scanning Electron Microscopy (SEM) analysis was conducted to characterize microtextural attributes, mineralogical assemblages, and pore network architecture at high magnification. SEM imaging enabled a refined examination of pore morphology, size distribution, and interconnectivity, as well as the crystallographic and textural relationships between diagenetic and detrital phases. The petrographic and microstextural observations were integrated to decipher diagenetic pathways, identify successive diagenetic overprints, and reconstruct the paragenetic sequence, thereby elucidating the evolution of porosity and its controlling diagenetic mechanisms within the studied unit.

Mineralogical and Geochemical Analyses

We Selected a few samples for X-ray diffraction (XRD) analysis to quantify mineralogical compositions and detect the presence of detrital or authigenic phases (e.g., quartz, anhydrite). The analyses were carried out using a Philips X'Pert diffractometer in the Chemistry Department of the Faculty of Sciences of Sfax.

To trace fluid origins and diagenetic pathways, geochemical analyses were performed on powdered samples that were dissolved after a tri-acid attack and transformed into solution, so major and trace element concentrations (Ca, Mg, Sr, Fe, Mn) were determined using atomic absorption method (ZEEnit 700 spectrome try, Analytik Jena) in the LARSEN laboratory (ENIS, Sfax).

5- RESULTS

5.1- Stratigraphy of Bireno Member

The stratigraphy of the Bireno Member was studied by several workers (Matmati et al., 1990; Troudi, 1990, 1997; Saïdi et al., 1997; Touir et al., 1989). The most detailed stratigraphy of the Late Cretaceous in Central Tunisia was carried out by Touir et al. (1989), who established the biostratigraphy of the Late Cretaceous at Jebel M'rhila, including the Bireno Member. In the absence of biostratigraphic fossils, these workers attributed this Formation to the Middle Turonian, based on its stratigraphic setting between the Annaba Formation (Early and Middle Turonian) at the base and the marls of the Aleg Formation (Middle and Late Turonian) at the top (Fig. 5). Our field observations identified three main lithological units: a lower unit of limestone and marls rich in marine fossils and oolites, a middle dolomitic unit often bioconstructed with hippuritid rudists, and an upper unit of oolitic and bioclastic limestone. The median unit shows significant lateral facies variability such as the hippuritid buildups passe to the South into oncolithic limestones and to the North into bioclastic and oolitic limestones. This facies change is clearly illustrated by the Late Cretaceous facies map of Jebel M'rhila, confirming the predominance of dolomitic facies in the study areas, thus justifying their selection for this study. According to this map, the dolostone are the best exposed in two localities, Kef Sidi Abdelkader and Oued Gourbej (Fig. 4).

Figure 5. Field views illustrating the Turonian stratigraphic framework in Jebel M'rhila (A) Vertical lithostratigraphic succession from the Fahdene to Bireno Members (B) Stratigraphic relationships between the Annaba, Bireno, and overlying Aleg Formations. (C) Internal architecture of the Bireno Member showing SB 90.5 Ma between median and upper terms.

Kef Sidi Abdelkader

In this locality (Fig. 1), the Bireno Member corresponds to a 10 to 20 m-thick dolomitic bar with a lenticular structure, forming a dominant cliff (Kef) in the topography (Fig. 4). This buildup is bounded to the NW by the synsedimentary normal fault N030, before transitioning to marls with marine fauna (Annaba Formation). Towards the SE, this carbonate body gradually thins to a thickness of 5 m, while becoming more stratified. The lenticular carbonate bar of the Bireno Member at Kef Sidi Abdelkader includes of three distinct terms:

- The lower unit, approximately 4 m thick, is arranged into interbedded limestone and marly or clayey limestone.
- The middle unit, 12 m thick, is a rudist-rich lenticular bar, the rudists represent the genus Hippurites. Thus bioherm extends from SE to NW on over approximately 200 m, its thickness varies from 2 to 12 m, before abruptly ending to the NW against the N030 synsedimentary normal fault. The surface of these buildups is irregularly gullied, leaving small hollows filled with erosion breccia or, locally, with marls containing marine fauna. This surface shows locally epikarsts, it represents a subaerial exposure surface (Touir et al., 2009).

Towards the SE, a bar, highly ferruginous surface occurs, with perforations filled with fine dolomitic and ferruginous detritus. Locally, the Hippuritidae individuals are grouped into 'clusters' representing some patch-reefs (Philip, 1972). It is worth noting that this dolomitic unit is affected by a network of small normal and reverse faults, directed N020 to N060, sealed and associated with slip figures and oblique stratifications.

- The upper unit is a 5 m thick bar of bedded limestone overlying unconformably the eroded and brecciated surface of the middle unit and sealing the synsedimentary N030 fault. It consists of oo-bioclastic limestone with green algae (dasycladaceae) and bryozoans. Towards the NW of the Jebel M´rhila, the dolomites are relayed with a fossiliferous limestone bar 5 to 20 m in thickness containing various broken mollusques and echinoderms, suggesting post-mortem bioaccumulations.

Moving away from Kef Sidi Abdelkader towards the SE, the Bireno dolomitic body becomes thinner (5 m) and clearly stratified and less dolomitic. Dolomites remain only close to the N030 oriented fractures or in the form of patches covering the surface of the limestones beds.

Oued Gourbej

The Oued Gourbej is located approximately 12 km south of Kef Sidi Abdelkader. Here, the dolomitic unit of Bireno Member is arranged into a series of 5-8 m thick lenticular buildups rich in hippuritids. These are abrumpltly boundedto the NW by the N020-030 fault (Fig. 4). The surface of these bodies is overlain by bioclastic limestone with ooids and rudist debris. The dolomitic bodies are separated from each other by a normal fault with fractures oriented N020-030. These lenticular bodies are themeselves affected by a network of multidirectional fractures. The associated rudists belong to the genus Hippurites (Pervinquière, 1903) and they are often fossilized in a vertical or oblique live position. The surface of the dolomitic unit shows locally epikarsts. Further east, in the Berber village locality, the Bireno dolomites preserve genuine 'covered karsts', which are currently occupied by collapse breccias. To the NW of the N20-30 fault, bordering the dolomitic lenticular bodies, the middle unit of Bireno Member pass abruptly into stratiform dolomites in Jebel Sbeïtla locality.

5.2- Mineralogy

The Bireno dolomitic unit is predominantly composed of dolomite. However, near fractures affecting this unit, in one hand, and from bottom to top of the unit, in another hand, in favor of the increase of calcite the proportion of dolomite decreases (Fig. 6) (Merino, 2006 ; Touir et al., 2009 ; Haj Ali et al.,2016b ; Bachari et al., 2022). Toward the top of the dolomitic unit, dolomite is locally associated with silica content not exceeding 18%.

Figure 6. Carbonate powder diffractograms showing the mineralogy of the carbonate samples respectively coming from site Sidi Abdelkader locality

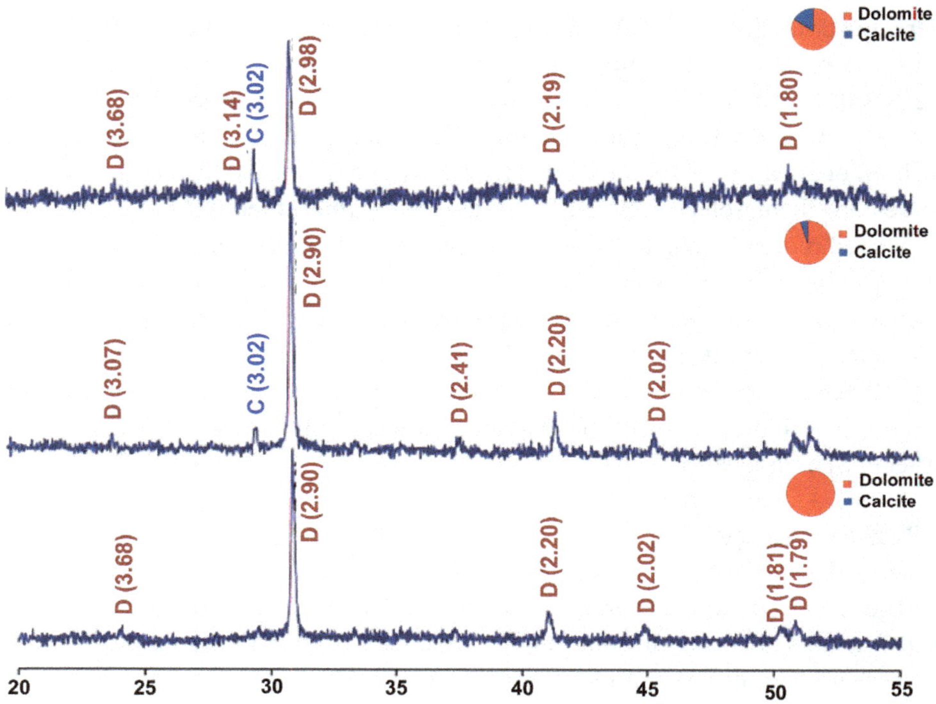

This spatial distribution of minerals within the dolomitic body reflects the influence of meteoric waters, which diluted interstitial marine fluids, reducing the Mg^{2+}/Ca^{2+} ratio and promoting calcite precipitation (Folk and Siedelcka, 1974; Folk and Land 1975; Ward and Halley 1985). These meteoric waters likely circulated through early fractures in the Bireno unit (Haj Ali et al., 2016b).

The silica associated with dolomite at the top of the Bireno unit is of continental origin. It likely resulted from leaching of continental profiles by meteoric water, followed by dissolved silicon (Si) transport via meteoric recharge and authigenic silica precipitation (SiO_2) within the residual pore network of the studied dolomites. The meteoric water influence was linked to the shallowing up and following subaerial exposure of the Bireno platform during the Middle Turonian (Touir and Soussi. 2003 ; Touir et al., 2009)

5-3. Geochemistry

The dolostones of the Bireno Member exhibit a compositional spectrum ranging from sub-stoichiometric dolomite to calcitic dolomite, displaying two distinct geochemical patterns (Table 1). First, progressive calcium enrichment occurs in fracture-proximal zones, while vertical profiles show localized Ca-enrichment trends from the unit's top to bottom. This spatial distribution suggests significant meteoric water influence (Hanshaw et al., 1971 ; Badiozamani, 1973), with infiltration occurring both vertically through the emergent platform surface and laterally along fracture networks. The resulting dilution of marine connate waters reduced Mg/Ca ratios, facilitating the formation of calcium-rich (hypercalcic) dolomite (Bourrouilh, 1973 ; Reeder et Prosky, 1986). This interpretation is supported by systematic depletion of marine indicators (Na and Sr) near fractures, while Mn and Fe distributions show consistently low concentrations with random spatial patterns that preclude meaningful diagenetic-environment interpretation. The combined geochemical evidence points to meteoric diagenesis as the dominant control on dolomite composition within the Bireno Member.

Table 1. Mineralogy and geochemistry of the Middle dolomitic Bireno term in Jebel M'rhila area.

Locality	Sample	Mineralogy	CaCO3 (moles%)	MgCO3 (moles%)	FeCO3 (moles%)	Sr (ppm)	Na (ppm)
Kef Sidi Abdelader	S0	Dol Sub stoec	49	23	0,6	3562	16,34
	S1	Dol hyper cal	63	24	0,37	3567	15,99
	S2	Dol hyper cal	79	25	0,13	1,882	13,01
	S3	Dol hyper cal	75	17	0,11	1,797	12,15
	S4	Dol hyper cal	69	19	0,41	2713	12,53
	S5	Dol hyper cal	74	27	0,79	2,531	11,725
	S6	Dolom calciq	61	28	0,41	3,896	25,28
Oued Gourbej	OG0	Dol hyper ca	65	33	0,51	2,173	10,486
	OG1	Dol hyper ca	61	32	0,54	0,7789	8,452
	OG2	Dol hyper ca	66	37	0,189	0,8175	7,89
	OG3	Dol hyper ca	75	34	1,58	1,779	9,98
	OG4	Dol hyper ca	73	28	1,85	4,935	10,18
	OG5	Dol hyper ca	69	29	0,73	1,472	13,945
	OG6	Dol hyper ca	61	26	0,34	1,087	10,795

5.4- Petrographic Examination

In this study, we conducted a detailed petrographic analysis of selected samples under a polarizing microscope to identify textural relationships, and pore distribution within the dolomitic unit. This approach enabled us to recognize evidence of recrystallization, cementation, dissolution, and other textural modifications related to dolomitization. We then complemented these observations with Scanning Electron Microscopy (SEM) examination, which provided higher magnification insights into pore morphology, size distribution, interconnectivity, and crystallographic relationships between dolomite crystals and associated diagenetic phases. By integrating petrographic and SEM data, we sought to characterize the full range of diagenetic phenomena, with particular emphasis on dolomitization, in order to reconstruct the diagenetic history of the dolomitic unit and evaluate its impact on porosity evolution in the reservoir Bireno.

5.4.1- Microscopic Dolomite Textures

The polarized light microscopy with SEM observations of dolomite samples reveal a textural variation. According to the Gregg and Sibley's (1984) classification, the textures of the observed dolomite range from anhedral to euhedral, showing clear microstratigraphic and fracture-related distribution patterns. Fracture walls are lined with straight-edged, euhedral zoned dolomite cement with fringes arranged into rhombohedral to hemirhombohedral crystals, often separated by dissolution voids (***Figure** 7*; H). Close to fractures, dolomite displays subhedral to euhedral textures with typically a single cement fringe (***Figure** 7* ; A-B). Moving away from fractures, crystals change to non-zoned, subhedral to anhedral crystals with turbid dolomicritic cores (***Figure** 7* ; F-G). Such textures change highlights the role of fracture-controlled fluid circulation in dolomitization resulting in cement precipitation of dolomite concentrated along fluid pathways.

Figure 7. Photomicrographs of different dolomitic facies seen in the polarizing microscope: A- euhedral dolomitic with fringes ; B- intercrystalline and intracrystalline dissolution ; C- subhedral dolomite with compaction ; D- subhedral to euhedral dolosparite with zoned rhombohedra ; E- intercrystalline porosity impregnated with iron oxide ; F- fractured anhedral dolomite ; G- dolosparite to dolomicrosparite subhedral ; H- fracture filled with calcite, dolomite and silica

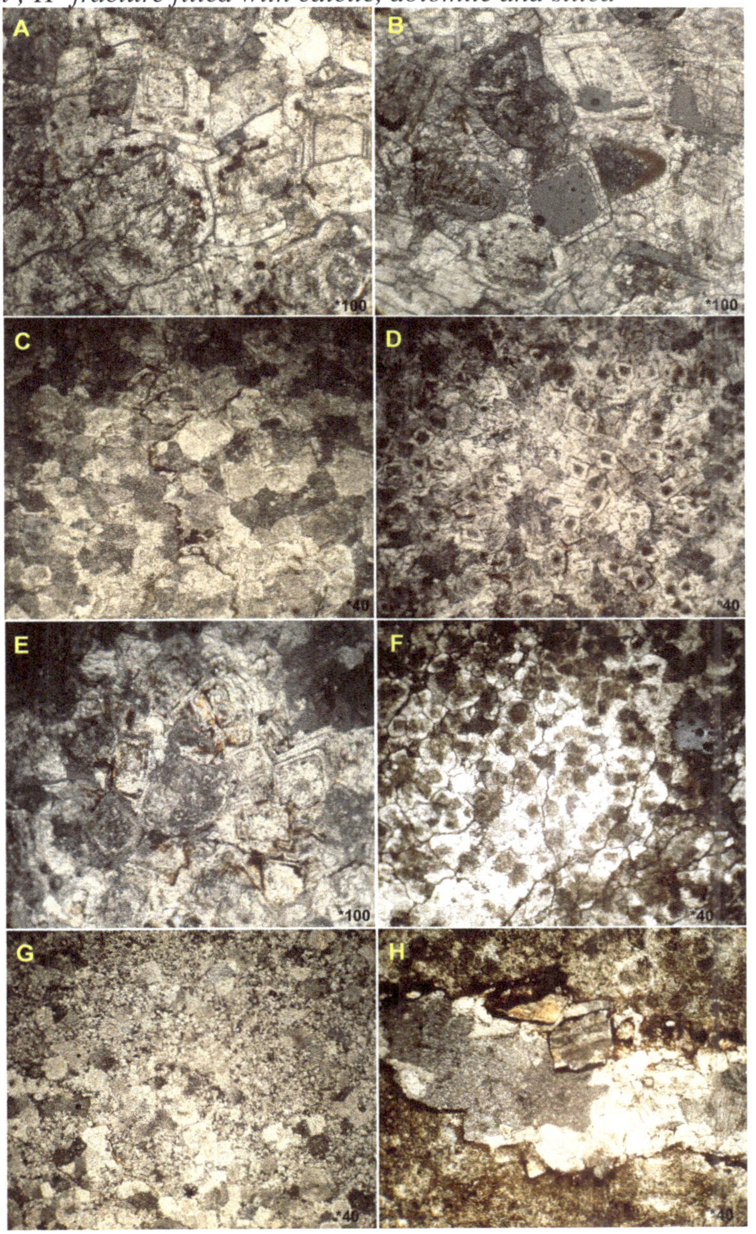

5.4.2- SEM Observation of Porosity

Scanning Electron Microscopy (SEM) observations allowed a detailed characterization of the different types and geometries of porosity that were firstly identified under the polarizing microscope. These include basically intra-crystalline, inter-crystalline, vuggy, and fracture porosity. It appears that dissolution, in association with fracturing, generated a significant amount of secondary porosity, resulting in a reservoir with high potentiality.

- **Inter-crystalline porosity**

This porosity corresponds to voids left between dolosparite crystals (**Figure 8** ; A). These voids, absent in the original dolomicrite, were created during its recrystallization into dolosparite and the resulting interlocking arrangement of crystals. Inter-crystalline porosity is also present between dolomite cement crystals within microfractures. This type of porosity was further enhanced by partial dissolution of dolomite crystals, as evidenced by the occurrence of rounded crystal edges.

- **Intra-crystalline porosity**

This type of porosity is relatively uncommon in the Bireno dolomites and occurs in two forms (**Figure 8** ; B):

- Completely hollow rhombohedra resulting from total dissolution, or rhombohedra with only their cores dissolved ;
- Pores developed between coalescent rhombohedra and within partially dissolved cement fringes.

This porosity was related to dissolution of dolomite by meteoric waters that should have been infiltrated through the cleavage plane of dolomite crystals.

- **Vuggy porosity**

This is the most common porosity type in the studied dolomites. It consists of irregularly shaped voids with diameters ranging from a few millimeters to about 2 cm. Some vugs are elongated and correspond probably to fractures widened by dissolution. The contours of these vugs are generally rounded, and in some cases rhombohedral to semi-rhombohedral in shape, indicating their origin from dolomite dissolution (**Figure 8** ; C).

- **Fracture porosity**

Fracture-related porosity is as significant as vuggy porosity. However, it can be locally reduced by the precipitation of dolomitic and, occasionally, siliceous cements (**Figure 8** ; D). Fractures and associated microfractures within the Bireno dolomitic unit represent not only a highly developed porous network but also an effective permeability system that facilitated the circulation of dissolving fluids and carbonate- and silica-rich solutions throughout the carbonate body.

Figure 8. Photomicrographs of different porosity seen in the MEB: A- Inter-crystalline porosity, B- Intra-crystalline porosity, C- Vuggy porosity, D- Fracture porosity

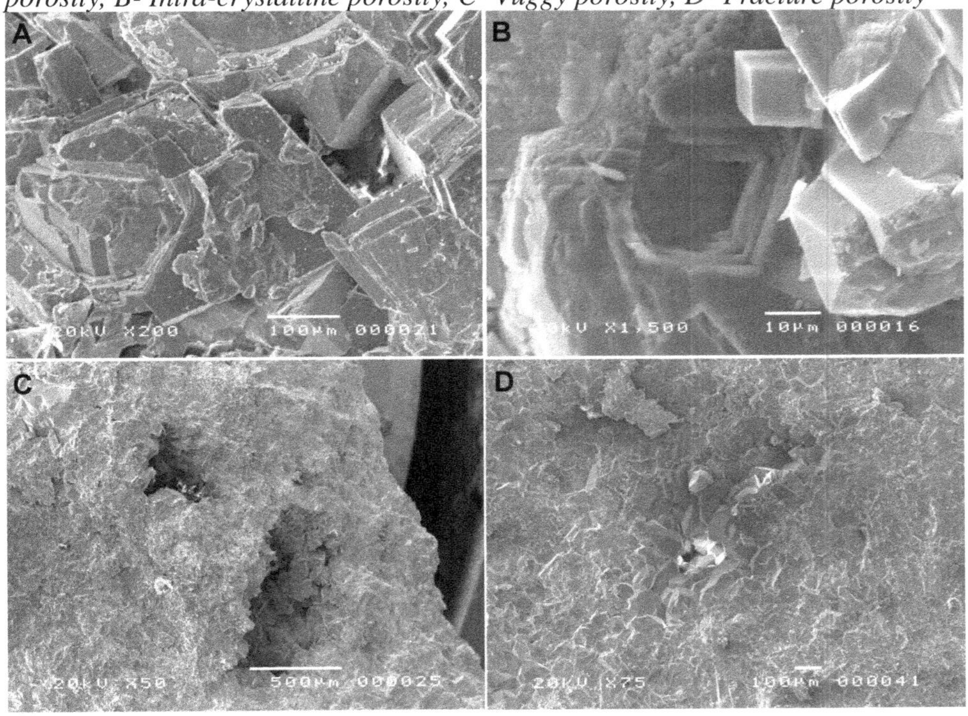

5.4.3- Diagenetic Processes Affecting the Dolomitic Unit of Bireno Member

Through our petrographic and microstructural investigations, we identified a variety of diagenetic processes associated the Bireno dolomites, including in addition to dolomitization, cementation, recrystallization, dissolution, and silicification.

We observed that these processes did not occur in isolation but interacted through successive stages, progressively overprinting the original depositional fabrics. By characterizing these diagenetic transformations, we were able to assess how they controlled the petrophysical properties of the Bireno dolomitic unit, particularly porosity and permeability. Ultimately, our objective was to reconstruct the diagenetic sequence in this unit, with a special emphasis on the role of the fracture-controlled dolomitization as a key factor shaping reservoir quality.

According to Choquette and Pray (1970), the fundamental issue of porosity in carbonate systems lies not in the initial creation of pores but rather in their long-term preservation. Prior to diagenesis, carbonate sediments may exhibit porosity values comparable to those of sandstone or other siliciclastic sediments. However, diagenetic processes exert a profound control on pore evolution, either through the generation of new voids (porogenesis) or through the occlusion and destruction of existing pores (poronecrosis). The diagenetic sequence that we established for the Bireno dolomitic unit clearly illustrates this duality, emphasizing the interplay between cementation, dissolution, dolomitization, recrystallization, and silicification. By integrating petrographic and microstructural evidence, we aimed not only to document these processes but also to reconstruct the diagenetic history of this unit. Ultimately, this approach allows us to evaluate how successive diagenetic transformations have shaped porosity evolution and, by extension, controlled the reservoir potentiality of these carbonates.

5.4.3.1- Diagenetic Factors Preserving or Creating Porosity (Porogenesis)

- *Recrystallization*

Recrystallization, often grouped with chemical inversion under the term neomorphism, is a common diagenetic phenomenon in the carbonates of the Bireno Member at Jebel M'rhila. It refers to a change in the crystalline morphology of a mineral without significant alteration of its overall chemical composition, which may result in the partial or complete obliteration of primary structures and textures (Wolf, 1965 ; Neumann, 1965; Bathurst, 1971; Purdy, 1968). This process, generally aggrading, is typically associated with an increase in crystal size (Folk, 1965) and is controlled by several factors, including precursor texture, fluid circulation, temperature, CO_2 pressure, and the concentration of Mg-rich waters (Neumann, 1965). During burial, dolomicrite (Ø < 5 µm) may be transformed into dolomicrosparite (5 < Ø < 15 µm), and subsequently into dolosparite (Ø > 15 µm). In the studied dolomites, subhedral and anhedral dolosparite appears to result from the recrystallization of an original dolomicrite, as indicated by the occurrence of iron oxides along dolomite cleavage planes, reflecting iron expulsion during the process. By contrast, euhedral, particularly zoned dolosparite, is more likely the product of direct dolomite precipitation (cementation). SEM observations reveal that dolosparite is arranged in a crystalline mosaic, with crystals displaying irregular mutual orientations (**Figure 9**; A). Such an arrangement leaves intercrystalline voids that were absent or only very fine in the original dolomicrite, demonstrating that recrystallization, far from being exclusively destructive, may also act as a porogenetic process by generating secondary porosity.

- *Dissolution*

The dissolution of calcium carbonates becomes effective when they come into contact with waters undersaturated in $CaCO_3$. The presence of dissolved CO_2 further promotes dissolution by enriching the water in carbonic acid (H_2CO_3), which is particularly aggressive toward carbonates. In the Bireno dolomites at Jebel M'rhila, dissolution is observed both at the macroscopic scale, notably along fissures and fractures, and at the microscopic scale, where it affects recrystallized dolosparite and dolomite cements. It is expressed by the development of intracrystalline porosity (**Figure 9** ; B), by rhombohedra with corroded or stepped edges along cleavage planes (staged dissolution, **Figure 9**; C), as well as by broken rhombohedra or crystals with partially dissolved cores, probably corresponding to relic of non-recrystallized dolomicrite, which is more susceptible to dissolution than dolosparite (Touir et al., 2009) (**Figure 9** ; D). In some cases, dissolution affects dolomite cement fringes forming a rim around primary dolosparite crystals, which may become detached, thereby indicating the lower stability of dolomite cement compared to recrystallized dolosparite. This phenomenon is particularly pronounced along fractures, which often appear widened and may even evolve into dissolution vugs and microcaves. Dissolution has therefore enhanced pre-existing porosity by creating vuggy, inter-crystalline, and intracrystalline porosity. However, it has also generated carbonate-rich solutions that subsequently precipitated as cements, locally reducing porosity. Altogether, these observations suggest that meteoric waters enriched in dissolved CO_2 circulated preferentially along fractures before diffusing into the dolomitic body, where they played a major role in porosity evolution.

- *Fracturation*

Fracturing, alongside dissolution, represents one of the main diagenetic processes responsible for porosity development in the Bireno dolomites (**Figure 9** ; E). These fractures, distributed in several families, are interpreted as early fractures related to extensional tectonic activity during the Middle Turonian in central Tunisia (Touir et al., 1987; Chihi and Ben Ayed, 1991), although they may have been reactivated during later tectonic phases. SEM observations reveal that micro-fractures affecting recrystallized dolosparite exhibit walls rounded by dissolution, locally coated with dolomite cement in the form of rhombohedra or half-rhombohedra attached to the walls. These micro-fractures do not cross-cut dolomite cement crystals or dolomitic cements, suggesting that fracturing occurred after the recrystallization of dolomicrite but prior to dissolution and cementation. By acting as preferential pathways for circulating dissolving fluids, fractures not only generated vuggy porosity through widening but also indirectly promoted the formation of other porosity types. Furthermore, the presence of fractures is a key factor enhancing the permeability of these carbonates.

Figure 9. Diagenetic factors preserving porosity (porogenesis) seen in the MEB: A- Recrystallization, B-C and D- Dissolution, E- Fracturation

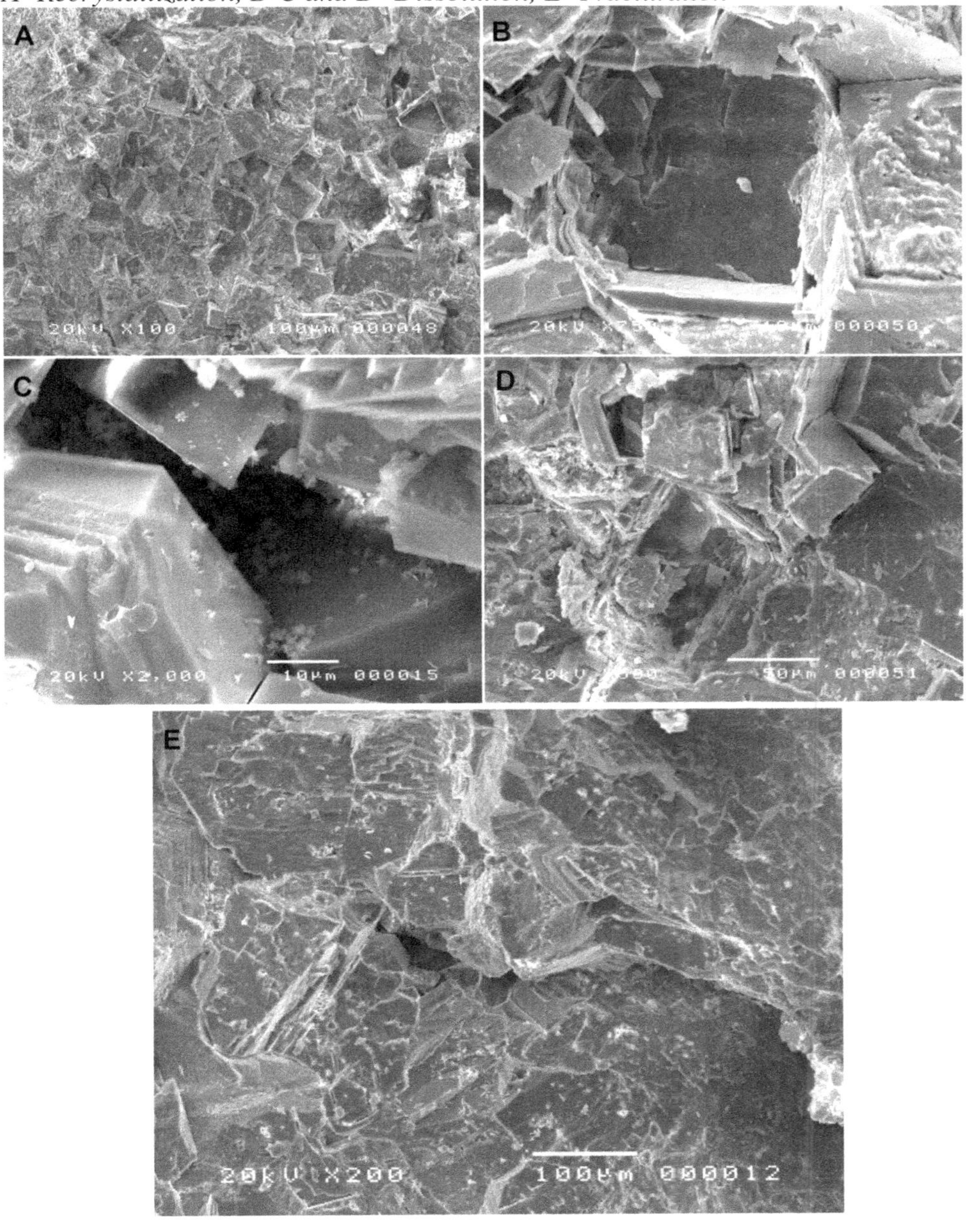

5.4.3.2- Diagenetic Factors Reducing or Destroying Porosity (Poronecrosis)

- Cimentation

Cementation in the Bireno dolomitic unit is represented by several types of cements precipitated from saturated ionic solutions, including dolomitic, calcitic and accessorily siliceous, and aragonitic cements. These cements, generally localized along the walls of vuggy pores and fractures, are distinguished from recrystallization products by their limpidity, their relatively large crystal size (up to millimetric scale), their euhedral texture, and their commonly drusy arrangement. Dolomitic cement occurs in two main forms: (i) thin growth fringes (5–10 µm thick) surrounding dolosparite crystals, or complete rhombohedra with sharp borders (**Figure 10** ; A) ; and (ii) a later cement that fills vuggy pores and fractures previously enlarged by dissolution, thus postdating both fracturation and dissolution (**Figure 10** ; B). This latter type tends to occlude most of the porosity. Silica cement occurs as euhedral hexagonal crystals averaging 150 µm in size (**Figure 10** ; C), whereas aragonite appears as botryoidal cement forming radial aggregates of fibrous crystals (**Figure 10** ; D) (in Chaouachi, 1988 ; Alonso-Zarza and Martin-Perez, 2008 ; GE et al., 2021). Calcite, by contrast, is mainly late to telogenetic in origin (Touir et al., 2009). Overall, while fracturing enhanced the circulation of mineralizing fluids and thus favored cement precipitation, it ultimately led to a significant reduction of porosity, highlighting the ambivalent role of cementation in reservoir quality evolution.

- Compaction

The compaction process represents a major diagenetic phenomenon that can be related both to lithostatic stresses, generated by the overburden pressure of superimposed sediments, and to tectonic stresses induced by regional deformation. In the dolomites of the Bireno unit, compaction generally tends to reduce porosity, particularly in dense and massive rocks where fluid circulation is already limited. The most characteristic expression of this process is the development of stylolites (**Figure 7** ; C), which result from stress-induced dissolution along discontinuity planes. The nature and orientation of stylolites provide valuable indicators of the origin of the stresses: stylolitic seams parallel to bedding are usually indicative of lithostatic compaction, whereas oblique or discordant stylolites reflect a tectonic influence. In the case of the Bireno dolomites, we identified the coexistence of both types of stylolites, suggesting that the reservoir experienced mechanical compaction related to burial, as well as tectonic compaction associated with subsequent phases of regional structural reactivation.

Figure 10. Diagenetic factors reducing porosity (poronecrosis) seen in the MEB: A and B- Cementation with Dolomitic cement, C and D- Cementation with Silica cement

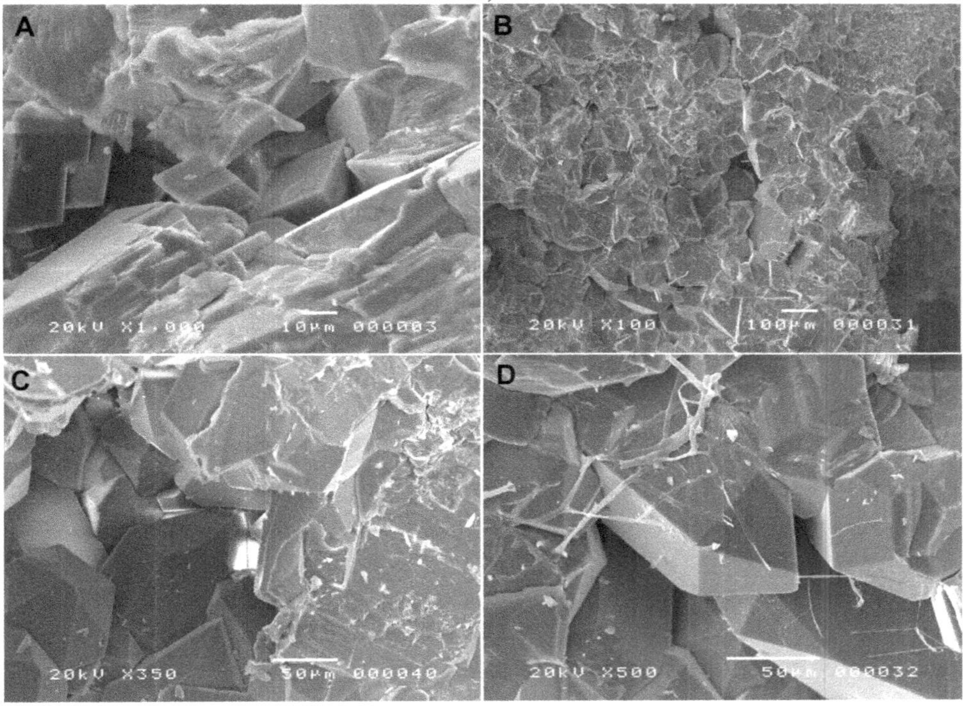

- *Silicification*

Silicification affects the Bireno dolomite unit at Jebel M'rhila, particularly in the Oued Gourbej section. In this locality, the dolomites contain approximately 10% detrital quartz grains and also display clear evidence of silicification. This process is expressed both by the precipitation of silica cements within microfractures, represented by euhedral hexagonal crystals (**Figure** 10 ; C-D) characterized by their differential extinction and gray birefringence color, and by the epigenetic replacement of dolomite by silica (**Figure** 7 ; E).

In the Bireno dolomites, silicification products gradually decrease from the top to the base of the dolomitic body, reflecting the continental origin of silica. Silicifying fluids were most likely meteoric waters enriched in silica through the leaching of nearby sandstone reliefs during the emergence of the Bireno platform during the global sea-level fall SB 90.5 Ma (Haq et al., 1987). These silica-bearing waters plausibly invaded the Bireno Member through the associated fracture networks.

From a petrophysical perspective, silicification represents a porosity-reducing diagenetic process, since the precipitation of silica cements within microfractures and the replacement of dolomite crystals tend to occlude available pore space and diminish reservoir quality.

6- DISCUSSION: DIAGENETIC SEQUENCE RECONSTRUCTION

The Bireno dolomites at Jebel M'rhila record a complex diagenetic history shaped by multiple successive and/or concomitant early and late processes. Petrographic investigations, supported by both polarizing microscopy and Scanning Electron Microscopy (SEM), enabled us to reconstruct the succession of these processes and to assess their impact on porosity and permeability changes (Tucker & Wright, 2009; Flügel and Munnecke, 2010 ; Touir et al., 2009).

In the Bireno platform, Touir et al. (2009) identified three successive dolomitization phases associated with the sea-level fall at sequence boundary SB 90.5 Ma, which resulted in the formation of mixing waters marine and meteoric or evaporitic fluids.

Dolomitization represents the major diagenetic process, responsible for the replacement of bioclastic and rudist-bearing limestones by dolomicrite. This early stage was linked to a shallowing-up followed by a relative sea-level fall. This phase was followed by widespread recrystallization of dolomicrite into dolosparite, a process that generated an initial intercrystalline porosity (Sibley & Gregg, 1987; Machel, 2004). The last dolomitization phase was preceded by fracturing, after which the circulation of various fluids induced dissolution and led to the precipitation of clear dolomite cement within the fractures. Subsequent fracturing, likely related to Turonian extensional tectonics, further enhanced the reservoir potential by facilitating fluid circulation. These fractures, sometimes reactivated, not only acted as conduits for permeability but also provided preferential sites for cement precipitation (Moore, 2001; Warren, 2000).

Dolomitic and siliceous cementation, localized mainly along fractures and vugs, occurred in several stages. An early phase is represented by clear crystalline fringes surrounding rhombohedra and voids. This phase, predating later fracturing, was followed by a pronounced dissolution event that enlarged inter- and intra-crystalline pores and generated abundant secondary vuggy porosity (Choquette and Pray, 1970). However, a subsequent phase of cementation by calcite, dolomite, and silica, interpreted as late to telogenetic, significantly reduced the effective porosity by occluding much of the previously created pore network (Bathurst, 1975; Lucia, 1999). Silicification, in particular, played locally a key role in pore occlusion, either by precipitation of euhedral hexagonal crystals or by replacement (epigenesis) of dolomite (Folk & Pittman, 1971; Land, 1985 ; Touir et al., 2009).

Compaction also contributed to porosity reduction, as evidenced by the occurrence of stylolites. Stylolites parallel to bedding planes reflect lithostatic pressure, whereas oblique ones point to tectonic stresses (Bjørlykke, 1998; Croizé et al., 2010). Although compaction did reduce pore volume, its overall impact appears to be less significant than the cumulative effect of cementation phases.

In summary, the various diagenetic processes that affected the Bireno dolomitic unit at Jebel M'rhila have produced a set of diagenetic features that were identified and described under the polarizing microscope and through Scanning Electron Microscopy (SEM). The diagenetic evolution of the Bireno dolomites illustrates a dynamic balance between porosity-enhancing processes (dolomitization, fracturing, dissolution) and porosity-reducing processes (cementation, silicification, compaction). Careful analysis of the spatial and chronological relationships between these diagenetic features allowed us to establish a comprehensive diagenetic sequence (Machel, 1999; Flügel and Munnecke, 2010).

In Jebel M´rhila, the petrographic relationships analysis between these diagenetic products allowed us to reconstruct the most complete diagenetic sequence encountered in the Bireno Member. In fact, this sequence was established not only on the basis of the spatial arrangement (micro-stratigraphy) of diagenetic products and features, but also considering local tectonic and paleogeographic data. It comprises the following consecutive stages:

1) Dolomitization of the bioclastic and rudist-bearing limestones of the Bireno middle unit, leading to the formation of replacement dolomicrite during marine shallowing-up and following relative sea-level fall ;
2) Partial to complete recrystallization of the dolomicrite into dolosparite during early burial ;
3) Compaction and stylolitization of recrystallized dolomite, reflecting progressive mechanical pressure and chemical compaction during diagenesis
4) Fracturing of the dolomitic bar during the Middle Turonian extensive tectonics, it affects in recrystallized dolosparite ;
5) Early phase of dolomite cementation due to the marine shallowing-up, consisting of clear crystalline fringes around primary rhombohedra, particularly along fractures and vugs. These cements are not cut by fractures and thus must postdate fracturing ;
6) Dissolution by meteoric water during the subaerial exposure of the Bireno platform during SB 90,5 Ma, it affected dolosparite rhombohedra as well as fractures and fissures, generating significant secondary porosity (vuggy, intra- and inter-crystalline);
7) Second phase of cementation linked with the previous resulting solutions rich in Mg and Ca involving the precipitation of calcite, dolomite, and silica, which

largely occluded the remaining voids and significantly reduced porosity in the Bireno dolomites ;

8) Second phase of meteoric dissolution affecting dolomite cement rhombohedra and their fringes, producing rounded crystal edges. Dissolution also created voids within rhombohedral cores and between central rhombohedra and cement fringes. Stepwise dissolution observed in some rhombohedra follows the cleavage planes of dolomite.

7- CONCLUSIONS

- The petrographic and geochemical investigations of the Bireno Member dolomites at Jebel M'rhila in Central Tunisia has enabled the reconstruction of a comprehensive diagenetic sequence involving dolomitization, recrystallization, fracturing, dissolution, cementation, and silicification. Integrated optical and SEM analyses allowed for a detailed identification of diagenetic products and for establishing their spatial and chronological relationships.
- Reservoir quality in the Bireno dolomites results from the interplay between porosity-enhancing processes (dolomitization, fracturing, dissolution) and porosity-reducing factors (cementation, silicification, compaction). This study highlights the importance of sequential diagenetic analysis for understanding the petrophysical evolution of carbonate systems and provides useful insights for predicting reservoir potential in comparable settings along the southern Tethyan margin.
- Dolomitization generated early intercrystalline porosity, and fracturing provided the main pathways for fluid migration, strongly controlling the distribution and enhancement of secondary porosity types, including vacuolar, intercrystalline, and intracrystalline forms. Dissolution processes further contributed to significant porosity development..
- Conversely, porosity was progressively reduced by several destructive mechanisms: stylolitic compaction, multiple phases of carbonate and silica cementation, and epigenetic silicification. These processes considerably decreased both porosity and permeability, leading to pronounced heterogeneity within the dolomite reservoir.

Conflict of Interest Statement

We wish to confirm that there are no known conflicts of interest associated with this publication, and that no significant financial support has been received that could have influenced its outcome.

ACKNOWLEDGEMENTS

I wish to express my heartfelt gratitude to the Doctoral School of Fundamental Sciences, Faculty of Sciences of Sfax, University of Sfax for providing the framework and resources that made this research possible.

I also wish to thank the Laboratory of Spectroscopy and Optical Characterisation of Materials (LaSCOM), Faculty of Sciences of Sfax, University of Sfax, for the opportunity to pursue my postdoctoral research within its team.

REFERENCES

Abdallah, H., Sassi, S., Meister, C., & Souissi, R. (2000). Stratigraphie séquentielle et paléogéographie à la limite Cénomanien-Turonien dans la région de Gafsa-Chotts (Tunisie centrale). *Cretaceous Research*, *2000*(21), 35106.

Alonso-Zarza, A. M., & Martin-Perez, A. (2008). Dolomite in caves: Recent dolomite formation in oxic, non-sulfate environments. Castañar Cave, Spain. *Sedimentary Geology*, *205*(3-4), 160–164.

Bachari, M., Negra, M. H., Geraud, Y., Grosheny, D., & Soltani, A. (2022). Petrophysics and sedimentology of the Turonian-Coniacian Bireno and Douleb reservoir analogues in north-central Tunisia. *Journal of African Earth Sciences*, *196*, 104728.

Badiozamani, K. (1973). The dorag dolomitization model-application to the middle Ordovician of Wisconsin. *Journal of Sedimentary Petrology*, *4*(43), 965–984.

Bathurst, R. G. C. (1971). *Carbonate Sediments and their diagenesis*. Elsevier.

Bathurst, R. G. C. (1975). *Carbonate Sediments and their diagenesis, Developments in gfsedimentology n° 12*. Elsevier.

Ben Ayed N., Ghali A. et Touir J. (1991). Sur les orientations des paléocontraintes tectoniques durant la sédimentation du CénomanienTuronien de Tunisie. Association Tunisienne des Etudes Internationales de Géologie, Tunis, 31 Mai et 1er Juin 1991.

Bjørlykke, K. (1998). Compaction, porosity reduction and fluid flow in sedimentary basins. *Marine and Petroleum Geology*, *15*, 321–340.

Bouaziz, S., Barrier, E., Soussi, M., Turki, M. M., & Zouari, H. (2002). Tectonic evolution of the northern African margin in Tunisia from paleostress data and sedimentary record. *Tectonophysics*, *357*(1-4), 227–253.

Bourrouilh F. (1973). Dolomitisation actuelle dans le monde. une revue-sc. Terre, Nancy, Annales, t. XVIII, N°3, p. 279-298, 4 fig.

Chaouachi, M. C. H. (1988). Etude sédimentologique des séries du Permien supérieur du Jebel Tebaga de Medenine Sud-Est de la Tunisie. Genèse, Diagenèse et Potentiel Réservoir des corps récifaux, Thèse 3ème cycle, Tunis, 285 p.

Chihi L., Dlala M. et Ben Ayed N. (1984). Manifestations tectoniques synsédimentaires et polyphasées d'age Crétacé moyen dans l'Atlas Tunisie centrale (région de Kasserine).C.R.A.S., Paris, t. 298, N°4, p. 141146.

Chihi L. et Ben Ayed N. (1991). Le rôle de la fracturation précoce sur la distribution des structures récentes le long du décrochement de Kasserine. Annales tectoniques, vol. V, N°1, pp. 6473.

Choquette P. W. et Pray L. C. (1970). Geological nomenclature and classification of porosity in sedimentary carbonates. An. Assoc. petrol. Geol. Bull., 54, 2, 207250.

Croize, D., Bjørlykke, K., & Jahren, J. (2010). Experimental mechanical compaction of clay mineral aggregates—Changes in physical properties of mudstones during burial. *Marine and Petroleum Geology*, *27*, 1765–1774.

Flügel, E. (2004). Diagenesis, porosity, and dolomitization. In *Microfacies of carbonate rocks: Analysis, interpretation and application* (pp. 267–338). Springer Berlin Heidelberg.

Flügel, E., & Munnecke, A. (2010). Microfacies of carbonate rocks: analysis, interpretation and application (Vol. 976, p. 2004). Berlin: springer.

Folk, R. L. (1965). Some aspects of recrystallization in ancient limestones, Soc, Econ. Paleont. *Mineral. Spec. Pub.*, *13*, 1448.

Folk, R. L., & Pittman, J. S. (1971). Length-slow chalcedony: A new testament for vanished evaporites. *Journal of Sedimentary Petrology*, *41*, 1045–1058.

Folk, R. L., & Siedelcka, A. (1974). The Bschizohaline^ environment: Its sedi mentary and diagenetic fabrics as exemplified by Late Paleozoic rocks of Bear Island, Svalbard. *Sedimentary Geology*, *11*, 1–15.

Folk, R. L., & Land, L. S. (1975). Mg/Ca Ratio and salinity; two controls over crystallization of dolomite. *Bulletin of the American Association of Petroleum Geologists*, *59*, 6068.

Ge, Y., Della Porta, G., Pederson, C. L., Lokier, S. W., Hoffmann, R., & Immenhauser, A. (2021). Botryoidal and spherulitic aragonite in carbonates associated with microbial mats : Precipitation or diagenetic replacement product? *Frontiers of Earth Science*, *9*, 698952.

Gili, E., Masse, J. P., & Skelton, P. (1995, November). Rudists as gregarious sediment-dwellers, not reef-builders, on Cretaceous carbonate platforms. *Palaeogeography, Palaeoclimatology, Palaeoecology*, *118*(3-4), 245–267.

Gregg J. M. and Sibley D. F. (1984). Epigenetic dolomitization and the origin of xenotopic dolomite texture. Jour. sedim. Petrology, 54, 3, 908931.

Haj Ali, H., Essefi, E., Belghithi, H., Yaich, C., & Touir, J. (2016a). Repercussions of the dolomitization of the fractured Middle Turonian succession of Bireno platform central part of Tunisia. *Arabian Journal of Geosciences*, *9*(3), 213.

Haj Ali, H., Belghithi, H., Ouali, J. A., & Touir, J. (2016b). Relationship between fracturing and porogenesis in a carbonate reservoir: Example from the Middle Turonian Bireno Member in Jebel M'rhila, Central Tunisia. *Journal of African Earth Sciences*, *121*, 342–357.

Hanshaw, B. B., Back, W., & Deike, R. G. (1971). A geochemical hypothesis for dolomitization by groundwater. *Economic Geology and the Bulletin of the Society of Economic Geologists*, *66*, 710–724.

Haq, B. U., Hardenbol, J. A. N., & Vail, P. R. (1987). Chronology of fluctuating sea levels since the Triassic. *Science*, *235*(4793), 1156–1167.

Hood, S. D., Nelson, C. S., & Kamp, P. J. J. (2004). Burial dolomitisation in a nontropical carbonate petroleum reservoir : The Oligocène Tikorangi Formation, Taranaki Basin, New Zealand. *Sedimentary Geology*, *172*, 117138.

HU, Y., Cai, C., Pederson, C. L., Liu, D., Jiang, L., He, X., & Immenhauser, A. (2020). Dolomitization history and porosity evolution of a giant, deeply buried Ediacaran gas field (Sichuan Basin, China). *Precambrian Research*, *338*, 105595.

Land, L. S. (1985). The origin of massive dolomite. *Journal of Geological Education*, *33*(2), 112–125.

Lucia, F. J. (1999). Rock-Fabric, Petrophysical Parameters, and Classification. In *Carbonate Reservoir Characterization* (pp. 23–57). Springer Berlin Heidelberg.

Machel, H. G. (1999). Effects of groundwater flow on mineral diagenesis, with emphasis on carbonate aquifers. *Hydrogeology Journal*, *7*, 94–107.

Machel, H. G. (2004). "Concepts and models of dolomitization: A critical reappraisal." In The Geometry and Petrogenesis of Dolomite Hydrocarbon Reservoirs. *Geological Society of London, Special Publications*, *235*, 7–63.

Merino, E. (2006). Self-accelerating dolomite-for-calcite replacement and displacive zebra veins: Dynamics of burial dolomitization. *Geochimica et Cosmochimica Acta Supplement*, *70*(18), A417–A417.

Moore, C. H. (2001). Diagenetic environments of porosity modification and tools for their recognition in the geologic record. Carbonate reservoirs porosity evolution and diagenesis in a sequence stratigraphic framework. *Developments in Sedimentology*, *55*, 61–88.

M'rabet, A., Ghenima, R., & el Euchi, H. (1996). MidCretaceous source rocks and reservoirs in central Tunisia. The fifth Tunisian petroleum exploration conference, ETAP, N°11, p2945.

Neumann, A. C. (1965). Processes of recent carbonate sedimentation in Harrington sound, Bermuda. *Bulletin of Marine Science*, *15*, 987–1035.

Pervinquiere, L. (1903). Etude géologique de la Tunisie centrale thèse. Sc. Paris, de Rudeval édit., Vol. gr. in. 4, 360 p.

Philip, J. (1972). Les formations calcaires à rudistes du Crétacé supérieur provençal et rhodanien. Thèse Doc. Etat, Marseille, 443 p., VIII pl.

Purdy, E. G. (1968). Carbonate diagenesis : An environmental survey. *Geologica Romana*, *7*, 183–228.

Purser, B. H. (1978). Early diagenesis and the preservation of porosity in Jurassic limestones. Jour. *Petroleum Geology*, *1*(2), 8394.

Reeder, R. J., & Prosky, J. L. (1986). Compositional sector zoning in dolomiteJour. Sed. *Petro.*, *56*, 237–247.

Saidi, F., Ben Ismail, M. H., & M'rabet, A. (1997). Le Turonien de Tunisie centrooccidentale : Faciès, paléogéographie et stratigraphie séquentielle d'une plate forme carbonatée ennoyée. *Cretaceous Research*, *1997*(18), 6385.

Sibley, D. F., & Gregg, J. M. (1987). Classification of dolomite rock textures. *Journal of Sedimentary Petrology*, *57*, 967–975.

Touir, J., M'rabet, A., & Ben Ayed, N. (1987). Sédimentation et tectonique précoce des séquences turoniennes du Jebel M'rhila, Tunisie centrale. 8th IAS Regional Meeting of SedimentologyTunis 87.

Touir, J., Ben Haj Ali, N., Donze, P., Maamouri, A. L., Memmi, L., M'rabet, A., & Zaghbib-Turki, D. (1989). Biostratigraphie et sédimentologie des séquences du Crétacé supérieur du Jebel Mrhila-Tunisie centrale. *Géologie Méditerranéenne*, *16*(1), 55–66.

Touir, J., & Soussi, M. (2003). The Growth and Migration of two Turonian Rudist-Bearing Carbonate Platforms in central TunisiaEustatic and tectonic controls. North African Cretaceous Carbonate platform. *System*, •••, 5381.

Touir, J., Soussi, M., & Troudi, H. (2009). Polyphased dolomitization of a shoal rimmed carbonate platform: Example from the middle Turonian Bireno dolomites of central Tunisia. *Cretaceous Research*, *30*, 785–804.

Troudi, H., Negra, M. H., & M'rabet, A. (2000). Predictive evaluation of lowermiddle Turonian carbonate reservoir in central Tunisia through integration of deposition environment and diagenesis. *Annales des Mines et de la Géologie*, *2000*(40), 81–98.

Troudi, H. (1990). *Les faciès cénomanoturoniens du Jebel Essif (Tunisie centrale). Stratigraphie séquentielle, sédimentologie et aspect réservoir*. D. E. A., Fac. Sc.

Troudi, H. (1997). Les réservoirs et faciès associés du Crétacé supérieur en Tunisie centrale : Sédimentologie, Stratigraphie séquentielle et Diagenèse. Thèse, Sc., Fac. Sc. Tunis, 287 p.

Troudi, H., Saidi, M., Akrout, N., & Kilani, F. (2002). Jurassic and cretaceous petroleum systems in central Tunisia. The eigfth tunisian petroleum exploration and production conference. *ETAP*, (18), 1926.

Tucker, M. E., & Wright, V. P. (2009). *Carbonate sedimentology*. John Wiley & Sons.

Ward, W. C., & Halley, R. B. (1985). Dolomitization in a mixing zone of near seawater composition, Late Pleistocene, northeastern Yucatan peninsula, J. Sedim. *Petrol.*, *55*, 407420.

Warren, J. (2000). Dolomite: Occurrence, evolution and economically important associations. *Earth-Science Reviews*, *52*, 1–81.

Wilson, E. N., Hardie, L. A., & Phillips, O. M. (1990). Dolomitization front geometry, fluid flow patterns, and the origin of massive dolomite: The Triassic Latemar buildup, northern Italy. *American Journal of Science*, *290*(7), 741–796.

Wolf, K. H. (1965). Petrogenesis and paleo-environment of Devonian algal limestones of New South Wales. *Sedimentology*, *4*, 113–178.

Yang, L., Yu, L., Chen, D., Liu, K., Yang, P., & Li, X. (2020). Effects of dolomitization on porosity during various sedimentation-diagenesis processes in carbonate reservoirs. *Minerals (Basel)*, *10*(6), 574.

Zagrarni, M. F. (1999). Sédimentologie, stratigraphie séquentelle et diagenèse des faciès du Crétacé supérieur du Jebel Bireno. Paléogéographie des plates-formes carbonatées du Cénomanien supérieur Coniacien en Tunisie centrale. Unpublished PhD thesis, University of Tunis-El Manar, 220 p.

Chapter 3
Thermal Barrier Coatings: Conventional, New Materials and Methods, Properties, and Its Applications

Kiran Prakasha A.
https://orcid.org/0000-0003-0308-5479
JSSATE Bengaluru, Sahyadri College of Engineering and Management, Mangaluru, India

Vinay B. U.
https://orcid.org/0009-0008-2512-1921
Sahyadri College of Engineering and Management, Mangaluru, India

Shivakumar K. Malladad
Sahyadri College of Engineering and Management, Mangaluru, India

Gagan Shetty
https://orcid.org/0000-0001-5593-9929
Sahyadri College of Engineering and Management, Mangaluru, India

ABSTRACT

There are different methods of coating are used to provide the protection from corrosion, wear, erosion along with the provision for lubrication and thermal insulation for the materials especially having applications in high temperature structural members. Among these methods Thermal Barrier Coating (TBCs) most commonly used coating with complex structure and they are much helpful to use in high temperature component applications in the field of aircraft and gas turbine engines. TBCs can make use of either ceramic or metallic multi layers for the formation of protective coating. The coating made from these will helpful for providing the insulation

DOI: 10.4018/979-8-3373-3506-3.ch003

against hot gas or steam streams which being exposed by turbine blades, casing, combustion engine and its auxiliary equipment. Thus, the efficiency of performance of these systems can be improved along with the durability. In order to improve the properties of the TBCs, it is much essential to understand the changes that occurs with the structure and properties while operating under extreme temperature and pressure conditions.

1. INTRODUCTION

There are different methods of coating are used to provide the protection from corrosion, wear erosion along with the provision for lubrication and thermal insulation for the materials especially having applications in high temperature structural members. Among these methods Thermal Barrier Coating (TBCs) most commonly used coating with complex structure and they are much helpful to use in high temperature component applications in the field of aircraft and gas turbine engines. TBCs can make use of either ceramic or metallic multi layers for the formation of protective coating. The coating made from these will helpful for providing the insulation against hot gas or steam streams which being exposed by turbine blades, casing, combustion engine and its auxiliary equipment. Thus, the efficiency of performance of these systems can be improved along with the durability. In order to improve the properties of the TBCs, it is much essential to understand the changes that occurs with the structure and properties while operating under extreme temperature and pressure conditions.

Thermal Barrier Coatings (TBCs) represent advanced material technologies that are essential in contemporary high-temperature engineering applications. Their primary role is to shield underlying metallic components—such as those in gas turbine engines, jet engines, and power generation turbines—from excessive thermal loads and harsh environmental conditions. The increasing demand for higher operating temperatures in gas turbines—aimed at enhancing thermal efficiency, reducing emissions, and augmenting power output—has significantly propelled the development and optimization of TBC systems. Modern turbine engines are required to function at temperatures surpassing the melting points of their metallic components, often exceeding 1400°C. Without thermal protection, the performance and service life of engine components such as turbine blades, vanes, and combustion liners would be compromised by thermal fatigue, oxidation, and corrosion. TBCs enable components to operate reliably under these severe conditions by providing a thermal gradient, which reduces the effective temperature experienced by the underlying metal substrates (Vaßen & Stöver, 2001).

1.1 Role in Turbine Blades and Hot Gas Path Components

In turbine engines, the first stage of the turbine—directly downstream from the combustor—is exposed to the hottest gases, making it one of the most demanding environments for any material. TBCs applied to turbine blades in this region allow the engine to operate at temperatures above the melting point of the superalloy substrate. This is achieved by providing thermal insulation, typically resulting in a temperature drop of up to 150–200°C across the coating. Similarly, hot gas path components—such as vanes, combustor liners, shrouds, and transition pieces—benefit from TBC systems that reduce thermal fatigue and corrosion, thereby extending maintenance intervals and improving engine efficiency. TBCs also permit the use of less cooling air, which can be redirected for combustion, thereby increasing thrust or output. This leads to enhanced fuel efficiency and reduced environmental impact in both aviation and power generation sectors.

1.2 Material Selection for High-Temperature Durability

The most commonly used ceramic material for TBCs is 8YSZ, due to its relatively low thermal conductivity (~2 W/m·K), high melting point (~2700°C), and excellent thermal shock resistance. However, as turbine inlet temperatures continue to rise, YSZ begins to reach its temperature stability limit, especially above 1200°C, where phase transformations and sintering can reduce its insulating capability and structural integrity.

To address these challenges, researchers are exploring alternative ceramic materials, including: Gadolinium Zirconate ($Gd_2Zr_2O_7$): Offers lower thermal conductivity and higher phase stability.

Lanthanum Zirconate ($La_2Zr_2O_7$): Known for high temperature capability and better CMAS (calcium-magnesium-alumino-silicate) resistance.

Rare-Earth Pyrochlores: These exhibit ultra-low thermal conductivity and are stable at elevated temperatures.

High-Entropy Oxides: A new class of materials combining multiple cations for better thermal and mechanical performance. These emerging materials provide potential routes to TBC systems capable of withstanding next-generation turbine operating conditions.

1.3 Deposition Techniques and Microstructural Engineering

The performance of thermal barrier coatings (TBCs) is profoundly affected by the deposition technique employed, which dictates microstructural characteristics such as porosity, columnar structure, crack density, and layer adhesion. The two most prevalent industrial methods are:

Air Plasma Spray (APS): This technique produces a lamellar structure with inter-splat porosity, which can be advantageous for thermal insulation and strain tolerance. It is relatively cost-effective and widely utilized for land-based gas turbines.

Electron Beam Physical Vapor Deposition (EB-PVD): This method yields columnar, vertically oriented grains that accommodate thermal strain more effectively, rendering it suitable for rotating engine components in aircraft turbines. It offers superior erosion resistance and cyclic durability compared to APS coatings.

Other innovative techniques include Solution Precursor Plasma Spray (SPPS), Suspension Plasma Spray (SPS), and High-Velocity Oxy-Fuel (HVOF) spraying, each providing unique microstructural control that can enhance specific aspects of TBC performance, such as thermal conductivity, adherence, or erosion resistance.

1.4 Challenges in TBC Performance

Despite their advantages, TBCs encounter several degradation mechanisms under operational conditions:

TGO Growth and Stress Accumulation: The growth of the thermally grown oxide during thermal cycling leads to stress accumulation and potential spallation of the ceramic layer.

Phase Instability: In yttria-stabilized zirconia (YSZ), the transformation from the tetragonal to monoclinic phase can induce volumetric changes, resulting in cracking.

Sintering: At elevated temperatures, the ceramic top coat tends to densify, reducing porosity and thereby increasing thermal conductivity, which compromises insulation effectiveness.

CMAS Attack: Ingested sand and dust particles, particularly in aircraft engines, can melt and react with the ceramic layer, leading to premature failure.

These issues are the focus of ongoing mechanical and materials research aimed at enhancing the thermal cycling resistance, phase stability, and chemical durability of TBCs.

1.5 Mechanical Research on TBCs

- Thermo-mechanical fatigue modeling to predict failure mechanisms.
- Finite element simulations to study stress evolution across the coating layers.
- Experimental testing under thermal shock, erosion, and high-cycle fatigue to assess durability.
- Microstructural optimization through controlled deposition parameters and novel materials.

Emerging directions include multilayer and graded coatings, self-healing ceramics, smart TBCs with embedded sensors, and machine-learning-assisted design for accelerated material discovery. These innovations hold promise for making TBCs even more integral to the future of sustainable and high-performance turbine technology.

The thermochemical properties of both metal and ceramics are different which makes them to prevent from the fabrication and application of the components having composite structure which is being subjected to large variation in temperature, load and pressure values over certain duration of time. In case of components of gas turbines which are used for aircraft and marine propulsion, the hot gas stream paths are being provided with TBCs made from ceramics which is having lower thermal conductivity will be used as thermal insulation layer to these metallic components. TBCs can be used over the super alloy components will be helpful to reduce the surface temperature of the super alloys, if proper internal cooling is provided such that the thickness of coating may varies from 100 to 500mm. This makes the modern gas turbines to operate at elevated temperature about 1300°C so that it improves the engine performance and efficiency. At the same time, during low temperature applications TCBs will be helpful to reduce the metal temperature so that durability and life of the engines can be increased. In few cases, diesel engine components also find the applications of TBCs, which is improving fuel economy and emission characteristics while operating at high temperature.

The TBCs are having two layers of ceramic and two layer of metallic materials such that each layer having different mechanical, thermal and physical properties so as to provide the complex structure than the individual metallic or ceramic components. Whatever the TBCs produced over the components of high temperature gas turbine engine must withstand the extreme elevated temperature, pressure, oxidation, temperature cycling, stress conditions. It is expected to perform for large number of landing and take-offs for the case of aircraft applications and several hours upto 30000 need to be operated in case of industrial applications.

Even though TBCs found dramatic increased application in past decade, there were lot of incidents were premature failure of coatings took place during their service over certain components. Due to failure of coated layer, there will be direct exposure of bare metals with the hot gas stream causing unpleasant condition for operations. The intensity at which TBCs fails to perform may vary depending on the substrate materials, temperature of exposure, duration, temperature cycling. etc., There are certain factors which are responsible for the failure of these coated layers. Few of the most important factors are i) Oxidation of the substrate metals ii) Thermal expansion iii) Continuous changes in properties of TBCs, its structure, microstructure of layers. With these premature failure reasons TBCs are found limited applications. Although whichever the component is provided with TBCs does not realised the effective advantages of high temperature withstanding capacity for longer duration of operating time, oxidation resistance. It is much essential to learn and understand the principle behind the TBC materials and their failure to realise their effective utilisation over certain substrate over certain life time under extreme temperature operating conditions. This is why it found a focused topic for research activities. The complex varying structure making TBCs more challenging task under elevated operated conditions. One who deals with this type of coating must be understanding the principle and phenomenon of diffusion, oxidation, phase transformation, elastic deformation, plastic deformation, creep deformation, thermal expansion, thermal conduction, radiation, fracture, fatigue, and sintering etc., TBCs allow components to operate at much higher temperatures than would be possible with the base metal alone. This leads to improved thermal efficiency, especially in turbines and engines (Vaßen & Stöver, 2001), (Zhu & Miller, 2004) (Sampath et al., 2012).

2. STRUCTURE OF TBCS

Let us consider the four layers of TBCs which are made from two metals and two ceramic layers with different functions and properties as shown in figure. The four layers are substrate, bind coat, thermally grown oxide and uppermost ceramic coat.

The Ni-Co based super alloy is used as substrate material. In order to provide the temperature gradient across the wall of the component it is preferred to provide either air cooling or internal hollow channels. Additional elements can be added to enhance the properties such as oxidation resistant, corrosion resistant, high temperature strength, castability and ductility.

The second layer is **bond coat** which is made from metal having thickness of 75 to 150mm having better oxidation resistant properties. It can be made from NiCrAlY or NiCoCrAlY alloy which can be deposited with aid of electron beam vapour deposition or plasma spray methods.

The third layer **Thermally Grown Oxide (TGO)** is formed when the operating temperature of the bond coat of the gas turbine engine crosses 700°C may lead to oxidation of bond-coat and formation of TGO occurs. It is formed between the bond coat and ceramic to coat having thickness of 1 to 10mm. (Kulkarni et al., 2003).

Figure 1 explains Cross-sectional scanning electron micrograph (SEM) of an electron- beam physical-vapor deposited (EB-PVD) TBC.

Figure 1. Cross-sectional scanning electron micrograph (SEM) of an electron-beam physical-vapor deposited (EB-PVD) TBC, superimposed onto a schematic diagram showing the temperature reduction provided by the TBC. The turbine blade contains internal hollow channels for air-cooling, whereas the outside hot-section surface is thermal barrier coated, setting up a temperature gradient across the TBC.

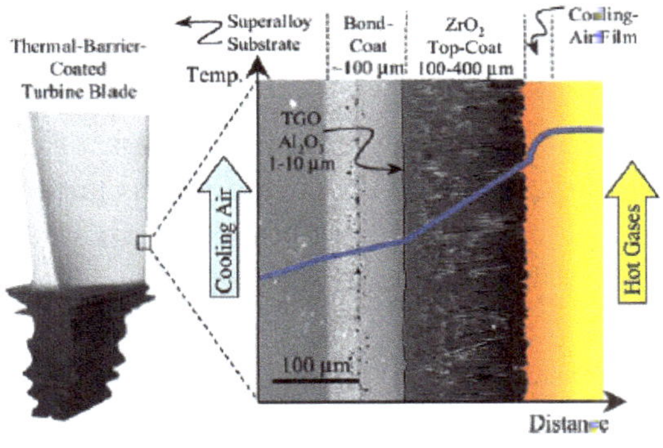

Figure 2. Cross-sectional SEM of an air-plasma sprayed (APS) TBC that has been subjected to 120 thermal cycles. Each 60-min cycle constitutes 50 min at 1120¡C followed by 10 min of forced-air cooling to ambient temperature.

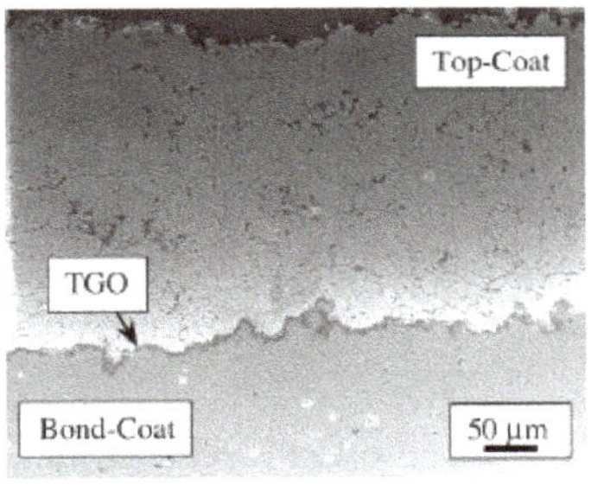

The oxidation in the bond-coat takes place as a result of presence of interconnected porosity on the ceramic top-coat through which oxygen is trying to penetrate from the outer environment of engine towards the bond-coat surface. Most of the time it can be prevented by using more denser top-coat layer, even though there will be the effect of diffusion of ionic oxygen present in the ZrO_2-based ceramic top-coat which renders it "oxygen transparent" phenomenon.

The last layer which is made from ceramics is known as **ceramic top-coat**. This is made from

Y_2O_3-stabilized ZrO2 (YSZ) which is providing better thermal insulation. Certain desirable and attractive properties of YSZ material makes it suitable for a proper material for the top coat surface. These ceramics are much denser and having lower thermal conductivity at the temperature of 1000°C, because of the presence of larger concentration of defects they are scatters the phonons which are responsible for heat conduction. These YSZ also possess greater coefficient of thermal expansion which will help to compensate the stresses arising due to mismatched thermal expansion between the top coat and metal underlying (Vaßen et al., 2009).

Thermal Barrier Coatings (TBCs) represent advanced material systems extensively utilized in high-temperature applications, including gas turbines, aircraft engines, and power generation turbines. These multilayer coatings serve to protect metallic components from extreme heat and oxidation, thereby enhancing their durability and efficiency. The typical structure of TBCs comprises four principal layers: the substrate, bond coat, thermally grown oxide (TGO), and ceramic top coat. The substrate is typically a superalloy, such as nickel-based alloys, which provides structural strength. A bond coat, often composed of MCrAlY alloy (where M = Ni or Co), is applied to improve adhesion and act as a barrier to oxygen diffusion. During operation, the bond coat reacts with oxygen to form a TGO layer, primarily consisting of alumina (Al_2O_3), which offers a protective oxide interface. The outermost layer is the ceramic top coat, commonly made of yttria-stabilized zirconia (YSZ), which provides thermal insulation due to its low thermal conductivity.

Mechanical research in TBCs is focused on optimizing microstructure, adhesion, crack resistance, and thermal cycling behaviour. Techniques such as finite element modelling (FEM), nano indentation, and thermal shock testing are employed to investigate stress distribution, creep behaviour, and delamination mechanisms. Research also explores novel materials, such as rare-earth zirconates, and improved deposition techniques, including Electron Beam Physical Vapour Deposition (EB-PVD) and Air Plasma Spray (APS).

The advantages of TBCs include enhanced thermal efficiency, extended component life, and improved fuel economy by enabling engines to operate at higher temperatures. They also provide oxidation and corrosion resistance and reduce cooling air requirements. However, limitations exist. The TGO layer can grow excessively, leading to spallation due to thermal mismatch stresses. Ceramic layers may crack or delaminate under cyclic loading, and the bond coat can degrade over time due to oxidation or phase transformation. Additionally, manufacturing complexity and high costs can pose barriers to widespread implementation (Kulkarni et al., 2003) (Sampath et al., 2012).

2.1 Processing of TBCs- YSZ Material

2.1.1 EB-PVD Process- YSZ Material

In this process the core made from YSZ material is heated by using electron beam. It is heated above the melting point temperature which results in the evaporation of material. These evaporated atoms will undergo condensation over substrate and surface diffusion of nuclei takes place and growth initiates after the formation of layer over the substrate. Later the Grain growth starts along certain favourable orientations. Following diagram shows the microstructure of a PVD YSZ coating. A columnar structure is observed for this substrate coating. There are certain factors such as substrate surface orientation, temperature of substrate, speed of rotation of substrate which are the few of the processing parameters affecting the orientation of column structure. In order to obtain more uniformed denser coatings, it is preferred to have lower roughness value for surface of the bond coat (Gell et al., 2018).

Figure 3. Micrograph of a PVD coating produced by the DLR Koln, Germany, on a nickel-based superalloy (IN738) with a plasma sprayed NiCoCrAIY bond coat

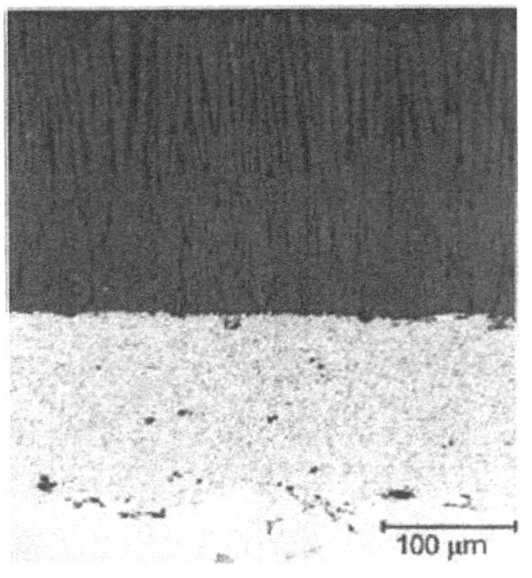

(produced by FZ Jtilich).

Because of better strain tolerance characteristics, PVD Coating methods are effectively used in gas turbine applications, especially in high pressure turbine air foils of aircraft engines. Comparatively this PVD process is more expensive than APS Process while dealing with the larger component formation, but have the tendency in order to close the tiny holes which are present in the aerofoils.

Electron Beam Physical Vapor Deposition (EB-PVD) is a high-vacuum coating technique extensively employed for the application of ceramic thermal barrier coatings (TBCs), particularly Yttria-Stabilized Zirconia (YSZ), onto turbine engine components such as blades and vanes. This method offers superior microstructural characteristics compared to conventional thermal spray processes, rendering it particularly suitable for high-performance, high-temperature applications in the aerospace and power generation sectors (Stern, n.d.).

2.1.2 Process Parameters

In the EB-PVD process, a focused electron beam is utilized to heat and evaporate a YSZ ceramic ingot situated within a vacuum chamber. The evaporated material subsequently condenses onto the heated substrate, forming a uniform ceramic coating. Key process parameters include:

Vacuum Pressure: Typically in the range of 10^{-4} to 10^{-5} Torr, ensuring the free travel of vapor species without collisions.

Substrate Temperature: Maintained between 800–1000°C to promote columnar grain growth and strong coating adhesion.

Electron Beam Power: Regulated (20–60 kW) to achieve sufficient vaporization of the YSZ source.

Deposition Rate: Usually 1–5 μm/min, depending on power and geometry.

Rotation and Manipulation of the Substrate: Ensures uniform coverage and controlled coating thickness.

These parameters are critical for tailoring the coating's thickness, density, porosity, and columnar structure.

2.1.3 Coating Characteristics

YSZ coatings produced by EB-PVD are characterized by their columnar microstructure, which enhances strain tolerance and thermal cycling durability. Key characteristics include:

Low Thermal Conductivity: Typically 1.5–2.5 W/m·K, due to inherent porosity and microstructural gaps.

High Strain Tolerance: The columnar grains accommodate thermal expansion mismatches between the coating and the substrate.

Good Adhesion: Strong bonding to the bond coat due to high substrate temperature during deposition.

Smooth Surface Finish: Suitable for aerodynamic surfaces in turbine engines.

This structure minimizes delamination and spallation under rapid thermal cycles, a key requirement in jet engine environments.

EB-PVD-deposited YSZ coatings are primarily utilized in:

Aero-engine Turbine Blades and Vanes: Where temperature resistance, oxidation protection, and thermal shock resistance are critical.

Land-based Gas Turbines: In power generation systems operating at extreme temperatures.

Advanced Military Engines: Due to their ability to operate in extreme flight conditions.

These coatings enable operation at higher turbine inlet temperatures, improving engine efficiency and reducing fuel consumption (Padture et al., 2002).

2.1.4 Limitations

Despite its advantages, the EB-PVD process has several limitations:

High Equipment and Operational Cost: Requires expensive vacuum chambers and high-power electron beam sources.

Line-of-Sight Deposition: Limits coating on complex geometries or internal surfaces.

Lower Deposition Rates: Compared to plasma spraying, making it less economical for thick coatings or large components.

Limited Material Choices: Mostly suitable for materials that can be vaporized without decomposition.

2.2 APS Thermal Barrier Coatings- YSZ Material

In this process, the particles to be deposited is sent through a carrier gas into plasma torch. The plasma flame from the torch is melting the particles and plasma gases will accelerate the melted particles over the substrate surface. These melted particles along with the flame will be in molten state, upon cooling it, a layer will be deposited over the substrate material. The entire process will be forming a lamellar structure which contains large number of pores having diameters above 1 micrometre. Certain number of micro cracks can be observed on the coated materials due to thermal stress which are induced as a result of rapid cooling during the deposition process by spraying. Due to these micro cracks, the thermal conductivity is severely reducing as half of that of PVD Process. Even though, its value can be improved by introducing the sintering effect so as to find the applications in high temperature applications. Let us consider the diagram below, which is having substrate as NI, NiCoCrAlY as bond coat with YSZ-TBC.

Figure 4. Microstructure of plasma-sprayed TBC system

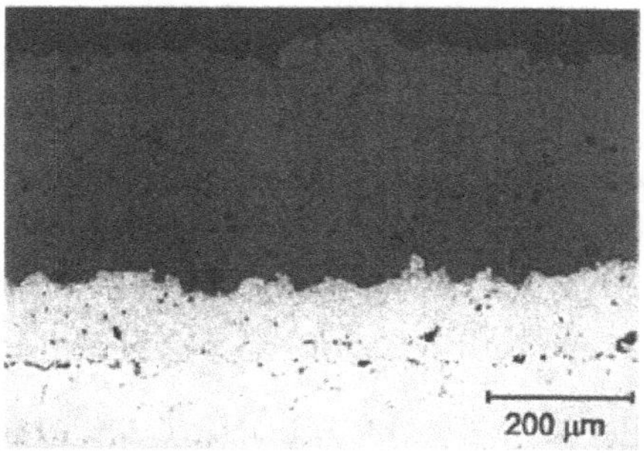

The thickness of APS TBCs are greater than that of PVD Coating and find applications in the combustion chamber of diesel engines.

Atmospheric Plasma Spraying (APS) is a prevalent technique for the application of Thermal Barrier Coatings (TBCs), particularly Yttria-Stabilized Zirconia (YSZ), onto metallic substrates. This thermal spray process employs a high-energy plasma arc to melt and propel ceramic powder particles, which subsequently flatten and solidify on the surface of a component, thereby forming a protective coating. APS is extensively utilized in the aerospace, automotive, and power generation sectors for components such as turbine blades, combustor liners, and hot section parts (Jonnalagadda, 2019) (Kulkarni et al., 2003) (Clarke et al., 2012)..

2.2.1 Process Parameters

The APS process is conducted in open air and involves several critical parameters that influence the final coating's structure and performance:

- Plasma Gas Composition: Argon serves as the primary gas, often combined with hydrogen or helium to enhance energy and improve melting. The gas flow rates regulate the plasma temperature and velocity.
- Arc Power: Typically ranging from 30 to 80 kW, it determines the energy available to melt the YSZ powder.
- Powder Feed Rate: Ranging from 20 to 60 g/min, it must be optimized to ensure uniform melting and deposition.
- Stand-Off Distance: The distance between the spray gun and the substrate, typically 80–150 mm, affects coating thickness, density, and splat morphology.
- Substrate Preheating and Surface Preparation: The substrate is generally grit-blasted and preheated to promote adhesion and reduce thermal stress.

These parameters are meticulously controlled to produce a consistent and high-quality TBC.

2.2.2 Coating Characteristics

YSZ coatings deposited via APS typically exhibit the following features:

- Lamellar Microstructure: Formed by overlapping splats of molten ceramic, resulting in a layered structure.
- Porosity: Generally between 10–20%, which aids in reducing thermal conductivity and enhancing thermal insulation.
- Thermal Conductivity: Low, typically around 1.5–2.0 W/m·K, making it suitable for insulating metallic substrates.
- Thickness: Can range from 100 to 500 μm, depending on the application and design requirements.
- Mechanical Properties: Good bond strength with the assistance of a metallic bond coat (often NiCrAlY), though inferior strain tolerance compared to Electron Beam Physical Vapor Deposition (EB-PVD) coatings.

The microstructure's porosity and microcracks facilitate strain absorption and reduce thermal conductivity, but also contribute to potential failure under thermal cycling.

2.2.3 Applications

APS YSZ coatings are extensively employed in:

- Aero-engine turbines, including blades, nozzles, and vanes.
- Land-based gas turbines for power generation, enhancing service life and efficiency.
- Automotive engines, particularly in diesel pistons and exhaust systems.
- Industrial equipment in high-temperature environments such as kilns and reactors.

The versatility of APS renders it suitable for both small-scale and large-area components.

2.2.4 Limitations

Despite its advantages, APS coatings possess several limitations:

- Lower Thermal Cycling Resistance: Compared to EB-PVD, due to their layered structure and weaker splat bonding.
- Higher Surface Roughness: May necessitate post-processing for components with stringent aerodynamic or dimensional requirements.
- Sintering Over Time: Reduces porosity, increases thermal conductivity, and diminishes thermal protection efficiency.
- Line-of-Sight Limitation: Cannot coat internal or highly complex geometries without reorientation.

2.3 Failure Mechanism of TBCs

Even though TBCs finds large applications in different high temperature components, they are subjected to several failure during the performance. Some of the factors which are responsible for the failure of YSZ TBCs are bond coat oxidation, corrosion and erosion, degradation due to the limited high temperature capability of YSZ.

2.3.1. Bond Coat Oxidation

One of the methods to find the failure of TBCs is the formation of oxide scales at the interface of bond -coat and TBC. These oxide scales are a-alumina which is responsible for concentration of higher level of stress to the components over the TBCs. It is observed that TGO growth will be leading to stress conversion in the bond-coat region on few locations, leading to the promotion of crack growth within and around TBC hence failure takes place.

Bond coat oxidation constitutes a significant failure mechanism in Thermal Barrier Coatings (TBCs) applied via both Electron Beam Physical Vapor Deposition (EB-PVD) and Atmospheric Plasma Spray (APS) utilizing Yttria-Stabilized Zirconia (YSZ). During high-temperature operations, the metallic bond coat, typically composed of MCrAlY or aluminide, undergoes a reaction with oxygen, resulting in the formation of a thermally grown oxide (TGO) layer, predominantly consisting of α-Al_2O_3. While initially protective, the continuous growth of this oxide layer induces stress accumulation at the TGO/ceramic interface due to thermal expansion mismatches and volumetric changes during cyclic heating and cooling. In APS coatings, the splat-like lamellar structure, characterized by higher porosity, can exacerbate stress concentration at the TGO interface, thereby accelerating delamination or spallation. Conversely, EB-PVD coatings, with their columnar microstructure, are more adept at accommodating thermal strains, thus delaying failure; however, they remain vulnerable to TGO-induced spallation after extended exposure. Furthermore, bond coat oxidation depletes aluminum, diminishing the capacity to form a stable alumina scale, thereby exacerbating oxidation and hastening coating degradation. In summary, bond coat oxidation persists as a predominant degradation mode in both EB-PVD and APS YSZ coatings, particularly under conditions of thermal cycling and oxidizing environments(Jonnalagadda, 2019).

2.3.2 Corrosion and Erosion

As we know TBCs are built on hot gas path components of engines. There might be certain corrosive environments where these TBCs are operates. In such condition also TBCs will operate upto certain extent. But only restriction is that, these corrosive environments may contain certain impurities such as phosphorous, sulphur, vanadium etc., which may react with the TBCs may lead to severe corrosion by forming oxide scales. The particle erosion mechanism in TBCs surface may cause the loss of material thus wear of the TBC as well as substrate material of the various components of the gas turbine engines.

Thermal Barrier Coatings (TBCs) composed of Yttria-Stabilized Zirconia (YSZ) and applied through Electron Beam Physical Vapor Deposition (EB-PVD) or Plasma Spray (APS) are essential for safeguarding high-temperature components, such as turbine blades, against heat and oxidation. Nevertheless, corrosion and erosion represent significant degradation mechanisms that can substantially curtail the service life and performance of these coatings, particularly in harsh environments like gas turbines, jet engines, and industrial combustion systems (Jonnalagadda, 2019) .

2.3.2.1 Corrosion-Induced Failure

A primary corrosive threat to TBCs is the infiltration of CMAS (Calcium-Magnesium-Alumino-Silicate), a type of molten deposit ingested from sand, dust, or volcanic ash during engine operation. At elevated temperatures (typically exceeding 1200 °C), CMAS melts and penetrates the porous YSZ coating. In APS coatings, characterized by a lamellar structure and relatively high porosity, molten CMAS can readily infiltrate, react with the zirconia, and destabilize its phase structure. This chemical attack diminishes the coating's insulating capability and can result in brittle or glassy phases that promote spallation. In EB-PVD coatings, the columnar structure offers somewhat enhanced resistance to CMAS infiltration due to narrow inter-column gaps. However, once CMAS penetrates, it tends to travel deeper along the grain boundaries, disrupting the ceramic's thermal properties. Moreover, prolonged exposure to corrosive species such as vanadium, sulfur, or alkali metals from low-grade fuels can chemically degrade the TBC and oxidize the underlying bond coat. Corrosion also accelerates the formation of an unstable or non-protective thermally grown oxide (TGO), particularly if the aluminum in the bond coat is depleted due to corrosive reactions. This weakens the bond between the ceramic top coat and the metallic substrate, eventually leading to delamination.

2.3.2.2 Erosion-Induced Failure

Erosion is a mechanical degradation process caused by high-velocity impacts of solid particles (such as sand or soot) on the coating surface. APS YSZ coatings, due to their micro-cracked and porous lamellar structure, generally exhibit greater resistance to erosion because the splat interfaces and pores absorb impact energy and reduce crack propagation. However, this structure may also allow microcracks to connect over time, leading to surface roughening and material removal. In contrast, EB-PVD YSZ coatings, with their smooth surface and columnar grains, are less resistant to erosion. Although the columnar structure offers excellent strain tolerance, it has limited capacity to absorb impact energy, rendering it more susceptible to surface damage and chipping under erosive conditions. This can reduce the coating thickness and eventually expose the bond coat or substrate to high-temperature gases, leading to rapid oxidation and mechanical failure (Jonnalagadda, 2019).

2.3.3 Degradation Due to the Limited High Temperature Capability of YSZ

Degradation of YSZ top coat take place at higher elevated operating temperature especially above 1200°C.This is due to the phase transformation. It happens when t-phase is transforming its phase into tetragonal and cubic phase which will be in equilibrium condition in the temperature above 1200°C.Whenever the phase transformation takes place from tetragonal to m-phase, the volume increases by 4%. This increase in the volume causes the disintegration of the TBCs.

Yttria-Stabilized Zirconia (YSZ) is the predominant ceramic material employed in Thermal Barrier Coatings (TBCs) due to its superior thermal insulation, relatively high toughness, and phase stability up to approximately 1200 °C. Nevertheless, beyond this temperature threshold, YSZ undergoes significant degradation, impacting both Electron Beam Physical Vapor Deposition (EB-PVD) and Plasma Spray (APS) coatings. As gas turbine inlet temperatures increase to enhance efficiency, the limited high-temperature capability of YSZ emerges as a critical failure mechanism for TBCs (Jonnalagadda, 2019).

2.3.3.1 Phase Instability at Elevated Temperatures

YSZ typically exists in the metastable tetragonal (t') phase, which imparts toughness through stress-induced transformation toughening. However, at sustained temperatures exceeding 1200–1250 °C, YSZ undergoes a phase transformation to the monoclinic phase, particularly if the stabilizing yttria content is insufficient or altered by diffusion or reactions with infiltrants. This transformation is accompanied by a volume expansion (~3–5%), resulting in microcracking, increased porosity, and eventual spallation of the ceramic layer. This phenomenon affects both EB-PVD and APS coatings but is particularly problematic in APS, where the lamellar structure is more susceptible to cracking due to internal stresses.

2.3.3.2 Sintering and Loss of Porosity

High temperatures also accelerate the sintering of the YSZ ceramic, leading to densification of the microstructure. Both EB-PVD and APS coatings rely on controlled porosity to achieve low thermal conductivity. In EB-PVD, the columnar grains become less distinct and more tightly packed; in APS, the inter-splat porosity and microcracks close up. Consequently, the thermal conductivity increases, reducing the coating's effectiveness as an insulator. Additionally, sintering renders the coating more brittle, diminishing its strain tolerance during thermal cycling, which increases the risk of delamination.

2.3.3.3 Degradation of Thermal Insulation Properties

YSZ's low thermal conductivity (~1.5–2.0 W/m·K) is a primary reason for its use in TBCs. However, as the coating degrades due to phase transformation and sintering, this value can increase significantly, diminishing its protective capabilities. The inability to maintain a thermal gradient leads to elevated temperatures at the bond coat and substrate, which accelerates oxidation, TGO growth, and other failure mechanisms.

In EB-PVD coatings, the columnar microstructure provides better strain tolerance and slightly higher sintering resistance due to less intergranular contact area. However, EB-PVD coatings also exhibit reduced erosion resistance and can experience rapid conductivity increases under long-term high-temperature exposure. APS coatings, while possessing better erosion resistance and higher initial porosity, are more prone to sintering-related degradation due to their splat-based, porous microstructure. At high temperatures, splat interfaces bond more tightly, eliminating the beneficial microcracks and pores, making the coating denser and more brittle.

3. TBCS- WITH NEW MATERIALS

3.1 Modified Chemical Compositions

To enhance the YSZ coating performance it is found from the literature that, the sintering stage must be reduced on the YSZ coatings. The rate of sintering on coatings can be reduced by having lower quantity of impurity phases. But there is certain restriction in order to reduce the sintering effect so as to reduce the impurity quantities.

In recent developments, Oxides of calcium or magnesium are being used as stabilizers for zirconia. By comparing this composition with yttria doped zirconia which shows the performance at lower level. To enhance the phase stability, it is recommended to prefer scandia in place of yttria while considering the previous case. $NaV0_3$ find the application to improve stability against thermal and hot corrosion.

There are certain cases where ceria is doped to YSZ which can be used as TBC material which helps to enhance and provide the better thermal cycling behaviour.

There is certain other type of stabilizers such as Ta_20s and Nb_20 can be doped to zirconia to increase the thermal stability at the elevated temperature range of 1400°C(Sampath et al., 2012) (Padture et al., 2002) (Clarke et al., 2012) (Vaßen & Stöver, 2001) .

3.2 New Chemical Compositions

There are so many new chemical compositions are being invented and many are under research work. Few of the materials are mentioned below. Hafnia can be used as a substitute material for zirconia whose chemical nature is almost similar and can be doped on an Y_2O_3 which has got better performance but will not able find the improvement over YSZ coatings.

There are materials which is having composition $CaO\text{-}Si0_2\text{-}Zr0_2$ system also found. They have better oxidation resistant even up to 1100°C which is much better than YSZ. But they lack in perform under corrosive surroundings.

Figure 5. Comparison of thermal conductivity of new materials with standard YSZ.

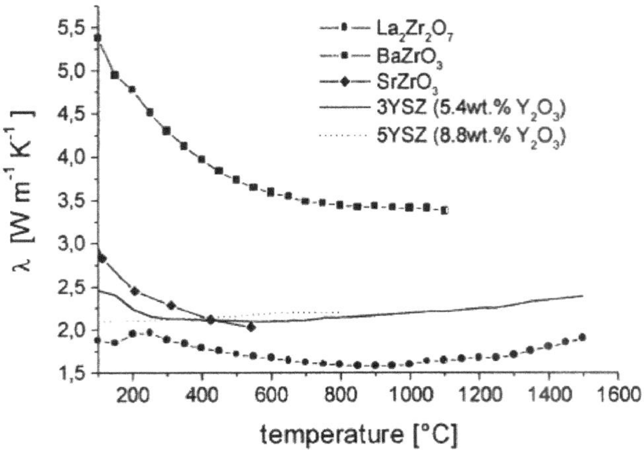

The attractive and high meting point having greater than 2000°C capability of a perovskite structure also find an application for the coating materials.

Figure 6. Comparison of Coefficient of thermal expansion of new materials with standard YSZ.

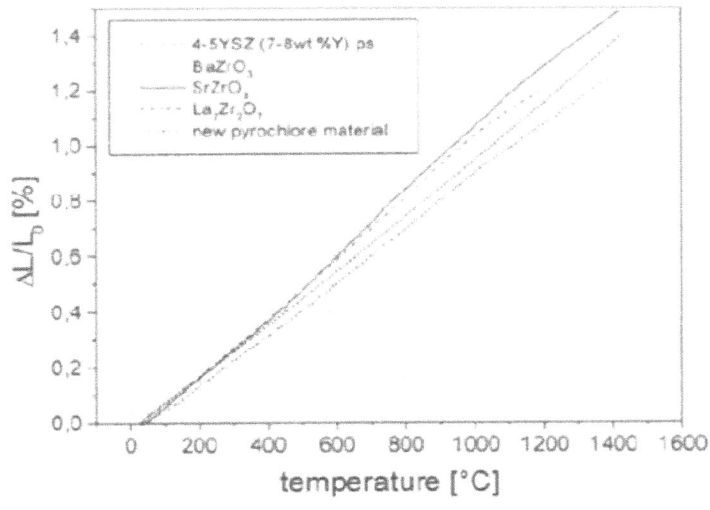

$La_2Zr_2O_7$, lanthanum zirconate, used as material which is having lower thermal conductivity

Garnets also utilised as TBC material having higher thermal conductivity at elevated temperature.

Due to improper cyclic behaviours of thicker mullite coating does not find continuous application in gas turbine engines as compared to YSZ Coatings.

For increased modulus of elasticity, density and better thermal conductivity with lower coefficient of thermal expansion, Lanthane Aluminate may be used having magneto plumbite structure in place of YSZ.

It also investigated the better thermal conductivity and coefficient of thermal expansion for $SrZr0_3$ but with poor thermal stability(Stern, n.d., 1996) (Gell et al., 2018) (Vaßen & Stöver, 2001).

3.3 Multilayers and Graded Coatings

It is possible to reduce the stress by introducing an intermediate graded material in between metallic and ceramic material coating layers, i.e., in between top coat and bond coat this graded layer should be introduced. Even though oxidation of the bond coat takes place due to larger surface are of exposure for the graded coating layer. These multilayer or graded coats will be used within the top coat of ceramic layer. It is preferred to use two layers of top coat with upper surface is being provided with the ceramic layer which is exposed to the flow of high temperature gas path. So that this surface will be much more resistant to both erosion and corrosion resistance. It is also recommended to choose the newly developed materials under ceramics which will have enhanced properties under high temperature conditions but due to poor mechanical strength, worse thermal expansion coefficient these components may lead to undergo thermal failure. By selecting a standard YSZ along with greater thermal expansion coefficient and better mechanical properties will be helpful to avoid the crack growth in bond coat region.

In the realm of advanced gas turbines and aero-engines, traditional single-layer Thermal Barrier Coatings (TBCs) frequently fall short in addressing the escalating demands for durability, efficiency, and high-temperature performance. To address these challenges, multi-layer and compositionally graded TBCs have been developed. These coatings optimize thermal, mechanical, and chemical performance by integrating multiple materials or gradually varying the composition throughout the coating's thickness. Such systems are particularly adept at managing thermal expansion mismatches, stress distribution, and material compatibility between the ceramic top coat and the metallic substrate (Gell et al., 2018) (Clarke et al., 2012) (Vaßen & Stöver, 2001).

3.3.1 Process Parameters

Multi-layer and graded TBCs can be fabricated using several deposition techniques, including:

- Atomic Plasma Spray (APS)
- Electron Beam Physical Vapor Deposition (EB-PVD)
- Suspension Plasma Spray (SPS)
- Solution Precursor Plasma Spray (SPPS)
- Key process parameters for these coatings include:
- Feedstock Material Composition: Utilization of multiple powders or suspensions with varying compositions (e.g., differing yttria or rare-earth content).
- Deposition Sequence: In multi-layer coatings, materials are deposited in distinct layers, each possessing specific properties.
- Gradation Rate: In graded coatings, the composition is gradually varied from the bond coat to the surface.
- Spray Parameters: These include torch power, gas flow, and stand-off distance, which must be dynamically adjusted to accommodate changing material characteristics.
- Substrate Temperature: This must be meticulously controlled to prevent cracking and ensure robust adhesion between layers.

The complexity of these coatings necessitates precise control of process variables to ensure uniform transitions between layers and to avoid interfacial defects.

3.3.2 Coating Characteristics

Multi-layer and graded coatings offer several advantages over traditional single-layer systems:

- Improved Thermal Shock Resistance: Gradual compositional changes reduce stress concentrations.
- Enhanced Adhesion: Minimizing abrupt differences in thermal expansion enhances bond strength.
- Tailored Functionality: Different layers can provide specific functions such as oxidation resistance, thermal insulation, or CMAS resistance.
- Controlled Thermal Conductivity: By adjusting porosity and material composition, heat transfer through the coating can be optimized.
- Damage Tolerance: Layered structures can inhibit crack propagation through deflection or energy absorption.

For instance, a graded coating might transition from a NiCrAlY bond coat to YSZ and then to a $Gd_2Zr_2O_7$ or $La_2Zr_2O_7$ outer layer to enhance performance in aggressive environments.

3.3.3 Applications

Multi-layer and graded TBCs are increasingly employed in:

- Aero-engine turbine blades and vanes for enhanced thermal protection and extended life cycles.
- Power generation gas turbines, where prolonged exposure to high temperatures and corrosive gases is prevalent.
- Advanced combustion systems, including hypersonic vehicle engines and space propulsion.
- Diesel engine components and thermal insulation in industrial furnaces.

Their design flexibility renders them ideal for next-generation thermal protection systems.

3.3.4 Limitations

Despite their benefits, multi-layer and graded coatings encounter certain challenges:

- Complex Manufacturing: Requires precise control and sequencing during deposition, increasing production time and cost.
- Interfacial Defects: Improper bonding between layers can lead to delamination under thermal cycling.
- Material Compatibility: Differences in sintering behaviour or chemical reactivity between layers can compromise long-term stability.
- Limited Industrial Adoption: Higher complexity and cost restrict their widespread use to critical or high-performance applications.

3.4 Nanostructured Materials

Nanostructured materials find application in formation of high temperature TBCs. Because of phonon scattering phenomenon at the grain boundaries, it is possible to reduce the thermal conductivity characteristics of the materials.

It has been observed that the thermal conductivity is drops down in SiC and up to 30% in case of nanocrystalline in theoretical concepts. The presence of the oxygen vacancies in the YSZ coatings may be the reason for reduction in thermal conductivity.

The grain growth is occurring for YSZ/$Al0_3$ composite structure. Nanostructured ceramics with multilayer can also be utilised to form the TBCs. Several experiments were done on YSZ/Al03 which showed that there is no significant resistant at the interface of this multilayer systems.

Before selecting a nanostructured material for the TBCs, one should be able to understand the nanophase structure and its effect on thermal properties as well as the materials are capable of operating at elevated temperature for longer life time in gas turbine engine components.

Quasi-crystalline materials can be utilised to manufacture TBCs which is having almost same characteristic as that of YSZ with better thermal conductivities. They have better oxidation resistant and corrosion resistant capabilities.

Nanostructured materials have emerged as a promising category of thermal barrier coatings (TBCs) for application in gas turbines, jet engines, and other high-temperature components. In contrast to conventional coatings, which typically rely on microscale lamellar or columnar structures, nanostructured TBCs are characterized by grains, pores, or features within the nanoscale range (1–100 nm). These distinctive microstructures enhance the material's resistance to thermal conductivity, sintering, and cracking. Nanostructured coatings are frequently based on Yttria-Stabilized Zirconia (YSZ) or advanced ceramics such as $Gd_2Zr_2O_7$, offering improved performance under severe thermal and mechanical loads.

Producing nanostructured TBCs necessitates specialized deposition techniques and meticulous control of spray parameters.

Common methods include: Atmospheric Plasma Spray (APS) utilizing nano powders or agglomerated nanoscale particles. Suspension Plasma Spray (SPS), which employs a liquid suspension containing nanoscale ceramic particles. Solution Precursor Plasma Spray (SPPS), where a liquid precursor is injected into the plasma jet, forming nanoparticles during deposition (Vaßen & Stöver, 2001).

3.4.1 Key Parameters Influencing Coating Quality Include:

- Powder Feedstock Characteristics: Particle size, morphology, and agglomeration affect melting and deposition.
- Suspension Viscosity and Solids Loading: In SPS, the suspension must be optimized for atomization and spray ability.
- Spray Distance and Velocity: These control the cooling rate and splat formation on the substrate.
- Substrate Temperature: Needs to be sufficiently high to allow good splat bonding but low enough to prevent premature sintering.

Fine-tuning these parameters is critical for forming a stable, crack-resistant, and porous nanostructured layer.

Nanostructured TBCs exhibit several distinctive and advantageous characteristics:

- Lower Thermal Conductivity: Nanoscale grain boundaries scatter phonons more effectively, reducing heat transfer.
- Improved Sintering Resistance: Nanoporous structures are more stable at high temperatures, maintaining low conductivity longer.
- Higher Toughness: Enhanced strain tolerance and crack deflection from nanoscale features improve resistance to thermal cycling.
- Tailored Porosity: Nanostructured coatings can have finely distributed porosity that contributes to thermal insulation and shock resistance.
- Refined Microstructure: Homogeneous and dense grain distributions help delay the onset of degradation mechanisms such as CMAS infiltration.

These benefits render nanostructured coatings particularly valuable for advanced, high-efficiency engine components.

Nanostructured thermal barrier coatings are being applied in:

- Aerospace turbine blades, nozzles, and vanes exposed to extreme heat and cyclic stress.
- Land-based power generation turbines, where durability and long service life are essential.
- Automotive turbocharger parts and diesel engine components, improving heat resistance and fuel efficiency.
- Aerospace hypersonic systems, where rapid temperature changes demand high thermal shock tolerance.

Their enhanced performance characteristics make them well-suited for next-generation propulsion and energy systems.

Despite their advantages, nanostructured TBCs face several challenges:

- Complex Fabrication: Requires specialized feedstocks, suspension formulations, and precise spray control.
- Higher Cost: Nanopowder production and advanced spraying methods increase overall cost.
- Scalability Issues: Large-scale, uniform deposition over complex geometries can be difficult.
- Long-Term Stability: Some nanostructures may still coarsen or sinter over prolonged high-temperature exposure.

REFERENCES

Clarke, D. R., Oechsner, M., & Padture, N. P. (2012). Thermal-Barrier Coatings for More Efficient Gas-Turbine Engines. *MRS Bulletin*, *37*(10), 891–898. Advance online publication. DOI: 10.1557/mrs.2012.232

Gell, M., Wang, J., Kumar, R., Roth, J., Jiang, C., & Jordan, E. H. (2018). Higher Temperature Thermal Barrier Coatings with the Combined Use of Yttrium Aluminum Garnet and the Solution Precursor Plasma Spray Process. *Journal of Thermal Spray Technology*, *27*(4), 543–555. DOI: 10.1007/s11666-018-0701-7

Jonnalagadda, P. (2019). *Thermal Barrier Coatings : Failure Mechanisms and Life Prediction*. DOI: 10.3384/diss.diva-154777

Kulkarni, A., Wang, Z., Nakamura, T., Sampath, S., Goland, A., Herman, H., Allen, J., Ilavsky, J., Long, G., Frahm, J., & Steinbrech, R. W. (2003). Comprehensive microstructural characterization and predictive property modeling of plasma-sprayed zirconia coatings. *Acta Materialia*, *51*(9), 2457–2475. https://doi.org/https://doi.org/10.1016/S1359-6454(03)00030-2. DOI: 10.1016/S1359-6454(03)00030-2

Padture, N., Gell, M., & Jordan, E. (2002). TBCs for Gas-Turbine Engine Applications. *Science*, *296*, 280–284. DOI: 10.1126/science.1068609 PMID: 11951028

Sampath, S., Schulz, U., Jarligo, M., & Kuroda, S. (2012). Processing Science of Advanced Thermal-Barrier Systems. *MRS Bulletin*, *37*(10), 903–910. Advance online publication. DOI: 10.1557/mrs.2012.233

Stern, K. H. (n.d.). *Coatings Ceramic Protective*.

Vaßen, R., Giesen, S., & Stöver, D. (2009). Lifetime of Plasma-Sprayed Thermal Barrier Coatings: Comparison of Numerical and Experimental Results. *Journal of Thermal Spray Technology*, *18*(5-6), 835–845. DOI: 10.1007/s11666-009-9389-z

Vaßen, R., & Stöver, D. (2001). *Conventional and New Materials for Thermal Barrier Coatings.*, DOI: 10.1007/978-94-010-0702-3_21

Zhu, D., & Miller, R. A. (2004). *Development of Advanced Low Conductivity Thermal Barrier Coatings. July.*

Chapter 4
Evaluation of PM10 Air Quality in Akure, Nigeria

Francis Olawale Abulude
https://orcid.org/0000-0003-0795-3966
Science and Education Development Institute, Akure, Nigeria

ABSTRACT

Determining the elemental components of PM10 and sources were the goals of this study. Three sites were chosen: Oba-Ile, National Museum and Monuments, and the Federal University of Technology, Akure (FUTA). With Nano sampler, samples were taken eight hours daily for ten months. Metals were identified with Energy Dispersive X-ray Fluorescence (EDXRF). PM10 content was estimated gravimetrically. Effects of meteorological parameters on PM10 was determined. Museum had the highest average PM10 results (125.20 µg/m3), which might be because of human activity. Findings showed that Akure's PM10 quality was unhealthy compared to International Air Quality Index and significantly higher than WHO recommended limit. According to Principal Component Analysis (PCA), natural and anthropogenic activities were the sources of PM10. In summary, there is reason for worry regarding the environmental quality of the studied locations because PM10 and elements values were high. The findings of this investigation may be utilized to raise awareness of air pollution in Nigeria.

INTRODUCTION

The health effects of air pollution on people and property make it a significant worldwide problem (Zhang et al., 2016). Environmental quality has recently deteriorated globally as a result of industrial development and urbanization (Wang et al., 2019). Globally, PM pollution is estimated to be the cause of about 2.4 million

DOI: 10.4018/979-8-3373-3506-3.ch004

deaths annually (WHO, 2005). Ndamitso et al. (2016) state that when PM is imbibed, it travels through the lungs and has a negative effect on them, which leads to health issues like cancer, asthma, cardiovascular disease, and respiratory disorders. Rich et al. (2019) found that PM, namely $PM_{2.5}$, was the cause of increased sickness and admission rates.

In the past few decades, Akure, Nigeria, has seen an upsurge in heavy traffic, agricultural, industrial activity, and several road developments. Dust, metals, and gaseous emissions are undoubtedly carried in such areas by wind or discharge water bodies. Contamination results from the high quantities of these particles that are deposited on soils, rivers, road shoulders, and other materials. In certain regions, samples of food, soil, water, and air are typically enriched with respect to a number of metals (trace) (Alpdogan et al., 2019). These metals frequently have detrimental impacts on both people and structures. Determining the amounts and locations of these metals as well as environmental deposit in the surrounding —especially in the air and water—is therefore essential. This study determines the sources of particulate matter (PM_{10}) in Akure, Ondo State, Nigeria, and evaluates its elemental composition. The particular goals are:

i. Determining PM_{10} the samples mass levels.
ii. Examine PM's interaction with meteorological variables
iii. Determine the metal concentrations of PM_{10} collected using EDXRF technique.
iv. Identify air contaminants sources using Principal Component Analysis (PCA) and area metals assessment (Enrichment and Contamination factors).

LITERATURE REVIEW

According to Amodio et al. (2014), the atmosphere serves as a medium for the elimination of chemicals that are produced by both natural and man-made processes and wind up in soil and water bodies. Dry or wet depositions are the method used to filter out airborne suspended components. Deposition rates are known to be impacted by meteorological conditions.

In their study, Ndamitso et al. (2015) found that whereas Na+, total nitrogen, NO_3^{2-}, and SO_4^{2-} are higher in wet depositions, PM, total phosphorus, Ca^{2+}, Mg^{2+}, and K^+ are higher in dry depositions. Wet deposition and pollution levels under high precipitation have been shown to be correlated. PM and other aerosols, as well as organic materials, remain in the environment until they are removed by dry or wet deposition. Numerous studies have been conducted as a result of the significance placed on air pollution because of the health issues it causes. The studies on environmental issues from various nations are shown in Table 1. According to previous studies (Alahmr et al., 2012; Kumar et al., 2014) the number of sampling periods used ranged from a week to a month. Following sample collection, the samples are kept in airtight containers or low humidity until the analyses are completed.

Table 1. Studies on environmental issues in different countries

References	Country	Samples	Measurement	Instrumentation
Manousakas *et al.*, 2013	Greece	PM10	10 elements	ED-XRF
Ahmadlsiyaka *et al.*, 2014	Malaysia	Data on air pollution (2007-2011)	Data	Models
Farao *et al.*, 2014	Italy	$PM_{10,}$ $PM_{2.5}$	Metals Ionic species	EPD-XRF ICP-OES ICP-MS
Xiao *et al.*, 2014	China	PM_{10}& $PM_{2.5}$ (Roadside Rural & Urban)	Heavy metals	ICP-MS
Kulshrestha *et al.*, 2014	India	$PM_{2.5}$, $PM_{10.5}$	Heavy metals	AAS
Co *et al.*, 2014	Vietnam	Ambient PM $PM_{2.5}$& $PM_{10.5}$	13 element ions	XRF Ion Chromatography
Kchih *et al.*, 2015	Tunisia	PM_{10}, $PM_{2.5}$	18 Elements anions	ED – XRF IC

Source: Ndamitso *et al.*, 2015

It is necessary to use a large number of stations to sample for PM deposition studies in the majority of industrialized zones and areas with irregular topography. The security, physical access and electricity, population exposure area, meteorological data, methodology, and equipment are other factors that must be carefully considered. Nevertheless, under specific field situations, these rules and regulations may change or even be completely disregarded.

For a variety of reasons and objectives, different samplers and equipment are used in PM pollution research. For instance, a minivol portable air sampler is used to gather PM samples. Total Atmospheric Deposition (TAD) uses a simple sampler that consists of a funnel and container placed inside an enclosure (Onwudiegwu et al., 2016). This is due to the fact that these particles have different physical and chemical properties. After analysis, source makers are employed in PM pollution studies to determine the origins of the constituents in the PM samples. The source makers linked to emission sources are displayed in Table 2.

Table 2. Source Makers Associated with Emission Sources

Sources	Markers (Chemical Species)
Soil	Ca, Al, Sm, Si, Ti, Sc, Fe
Road dust	Al, Sc, Ca, Sm, Ti, Fe, Si Na, Na^+, Cl, Cl^-, Br, I, Mg, Mg^{2+}
Sea salt	Mn, V, Fe, Ni, Cr, S, Fe, As, SO_4^{2-} Al, Sc, Se, Co, As, Ti, Th, S
Oil burning	Mn, Cr, Fe, Zn, W, Rb Zn, Cu, As, Sb, Pb, Al
Coal burning	Sb, As, Pb, Ca
Iron and Steel industries	K, Zn, Pb, Sb K, C_{ele}, C_{org}, Br, Zn, Guaicols & Syringols
Non-ferrous metal industries (smelters) Glass industry Cement industry Refuse incineration Biomass burning	
Automobile gasoline (exhaust)	C_{ele}, Br, Ce, La, Pt, SO_4^{2-}, NO_3^-, PAHs
Automobile diesel (exhaust)	C_{org}, C_{ele}, S, SO_4^{2-}, NO_3^-
Secondary aerosols	SO_4^{2-}, NO_3^-, NH_4^+

Sources: (Onabowale and Owoade, 2015).

Airborne pollutants can be either naturally occurring inorganic, organic, or biological compounds, or they can be foreign substances or energy. Because of their detrimental effects on the environment, they are given attention and awareness. This research is concerned with inorganic pollutants that have a mineral origin and contain metals, salts, and minerals. A variety of anthropogenic activities release inorganic contaminants into the environment (Figure 1). Because many companies are releasing metal-containing discharges into their environments that are not being treated appropriately, the environment is becoming more concerned about the growing amounts of these metals.

Figure 1. Heavy metal origins and the environmental process that produces these metals

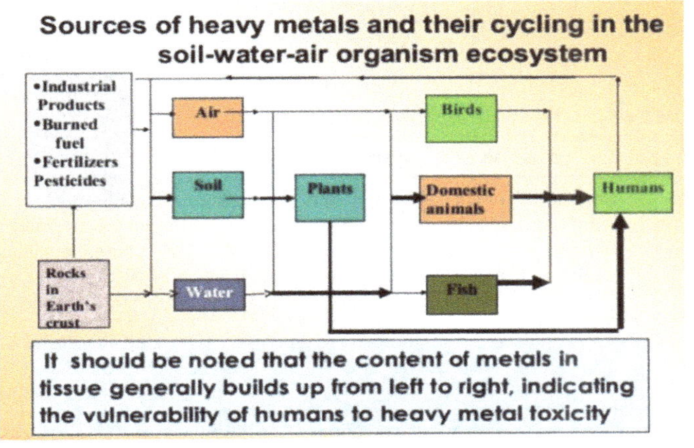

Trace elements were investigated among Jordanian jewelry artisans by Alkhatib et al. (2014). It was noted that the study's components were within the suggested ranges. According to the metals found in the workers' jewelry and urine samples, these elements appear to be stable. When jewelry is processed, no cadmium is released into the air that could harm the workers or the jewelry.

It has been established that traffic-related activities contribute to urban particles. According to Adebiyi et al. (2015), road dust and urban soils contain trace elements contamination from human activity. According to reports, excessive concentrations of heavy elements have a significant impact on materials near the roadside, human health, and air quality. According to Ezeh et al. (2012), the PM collected in Lagos, Nigeria, has high amounts of lead, ranging from 7 to 53 $\mu g/m^3$. The processes of atmospheric deposition, sedimentation, impaction, and interception result in the concentration of heavy metals in PM.

Pongpiachan and Iijima (2016) used observations and multivariate statistical techniques to identify 18 distinct elements in PM_{10} collected in Bangkok, Thailand. As a result of their introduction into the ambient air, they concluded that anthropogenic activities were the cause. It was discovered from the data that the mean values of Al, V, Cr, Mn, and Fe were higher than those observed in other developed cities. According to their findings, industrial and high-density road traffic activities were the main sources of the heavy metals' entry into the atmosphere. A variety of smelting and industrial processes, including the generation of electricity, can leak lead, arsenic, cadmium, and mercury into the environment.

Particulate matter is the metal carrier to the lung structure, and its composition dictates its potential health hazard, according to a review paper by Fortoul et al.

(2015). Asthma, rhino conjunctivitis, and dermatitis are associated with occupational exposure to Fe, Zn, and Ni. The sources of some of these particle contaminants and their impacts on people are compiled in Table 3.

Table 3. Sources of particulate pollutants and their health effects

POLLUTANTS	MAJOR SOURCES	EFFECTS
Particulate Matter	Combustion of coal and petroleum products, burning of refuse, domestic burning of fuels and powerhouses, condensing vapour, natural gaseous precursors.	Increased breathing rate and feeling of air starvation, suffocation and aggravation of asthma and chronic bronchitis, respiratory irritation, irritation of the eyes and throat, premature death.
Copper	Industrial plants. Hazardous waste sites, from containers.	Dizziness, headache, bronchitis, stomachaches, diarrhea, vomiting, loss of appetite, kidney and liver damage, death.
Dusts	Asbestos factories, mining activities, road traffic activities, road construction, metallurgical industries, cement industries etc.	Respiratory diseases, toxicity from metallic dusts, pulmonary fibriosis, lung cancer etc.
Nickel	It can be contacted through smoking of tobacco, contacts with shower water, soil, metals containing Ni and breathing air	Blisters, redness, itching, rash and dry patches of the skin
Chromium	Mining sites, stainless steel welding,	Respiratory tract irritation, sinus cancer, cough, skin rashes, shortness of breath, asthma
Manganese	Auto exhaust(from gasoline),paints, storage batteries, pipes	Pain, symptoms of Parkinson disease, blood clotting, birth defects, skin problems, glucose intolerance, neurological symptoms.
Zinc	Coal, zinc mining, incineration of plastic containers, refining of metal tobacco smoke	Cardiovascular diseases and hypertension, interference with and copper metabolism, kidney damage etc.
Titanium	Combustion of coal, diesel and residual oils, tobacco smoke, chemicals and catalysts, steel non-ferrous alloys manufacture	Pain in the chest, Respiratory symptoms, lungs cancer, skin burns, eye irritant
Vanadium	Industrial sources –oil refineries, volcanic emission, continental dust, marine emission.	Sore throat, lung irritation, chest pain, asthma, cough.
Iron	Smelting industries, attrition of automobile tires, burning of automobile oils, phosphate fertilizers and from sewage wastes, Industrial wastes.	Loss of biodiversity, erosion, contamination of surface water, ground water, and soil. conjunctivitis, choroiditis, and retinitis, Anaemia, Lung cancer.

continued on following page

Moses and Orok (2015)

According to Adebiyi et al. (2015), during the study period, Nigerian towns were characterized by elevated air PM concentrations and elements. The concentrations

Table 3. Continued

might have a significant impact on the health of urban people who are exposed to the surrounding areas over time. Additionally, Hsu et al. (2016) found a correlation between the incidence of respiratory issues and PM pollution. According to a health-related study, an increase in PM_{10} of roughly 50μg/m^3 is linked to between 2% and 8% of all human mortality (USEPA, 2002).

Figure 2. The various methods for source apportionment

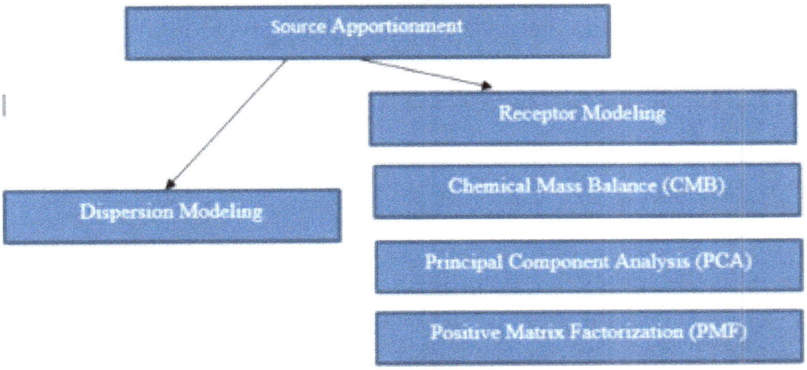

Use of Receptor Models

Figure 3. Source Apportionment Procedures

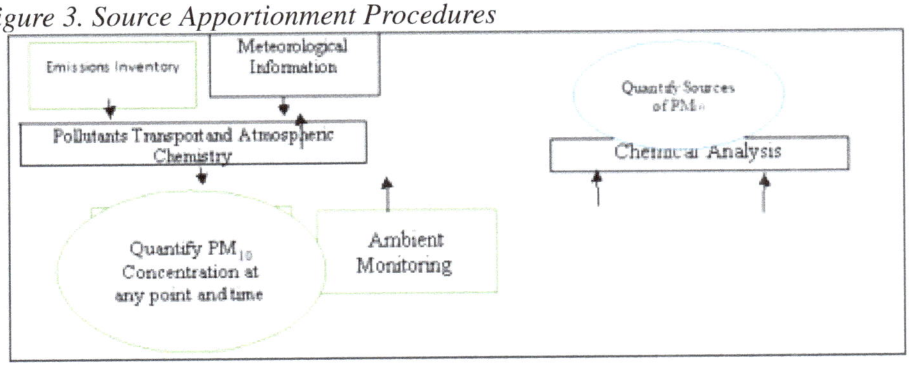

Figure 3 lists the techniques used to estimate pollution sources, and this section discusses the other models that fall under these techniques. The most used model in SA is principal component analysis. PCA makes it possible to better comprehend and assess the relationships between the information sets. The source of PM can be found using PCA with varimax rotation. This multivariate mathematical method is used to factorize data concentrations under the assumption that relationships exist between information concentrations and levels of species mass concentration. Numerous basic information grouping processes are included in the model. The data will first be transformed into a common format, which is represented by

$$X_{it} = \frac{C_{it} - C_i}{d_i} \quad (1)$$

Ci and di are the arithmetic mean and standard deviation of the variable i for each sample used in the analysis, while Cit is the concentration of the variable i in sample t. The expression for the PCA model is

$$X_{it} = \sum_{j=1}^{N} L_{ij} S_{jt} + E_{it} \quad (2)$$

where Lij is the variable i's factor loading in source j with N sources, Sjt is the source j's factor scores for sample t, and Eit is the variable i's residual in sample t that cannot be explained by the j sources or factors (Kothai et al., 2011). After that, eigenvector decomposition is used to solve the equation. The loading factors of each compound evaluated in relation to each rotated principal component are maximized (or minimized) by using Varimax normalized rotation, which gives the factors the best interpretable structure.

EF is used to determine the elements' origins in PM samples. It contrasts the relative abundance of an element in crustal material with its degree of enrichment. Iron (Fe) was used as the reference element (Elemental composition of crustal rock) in this study to determine crustal EFs. Since Ti is a crustal element with minimal to no anthropogenic influence, it can also be utilized as a normalizing agent. Wedephol, 1968; Manson, 1966; and Taylor, 1964 provide the crustal elements.

XRF analysis is a technique used to identify the elements contained in any kind of material, regardless of its form. XRF is fast, precise, easy to use, and accurate. Energy dispersive X-ray fluorescence (EDXRF) and wavelength dispersive X-ray fluorescence (WDXRF) are two of the many types of XRF. In industrial settings, this method is helpful (Bruker, 2019).

MATERIALS AND METHODS

Study Area

The Sunshine State, or Ondo State (Figure 4), was established in February 1976. The National Population Commission (NPC, 2010) reported that the state's total population was 1,745 males, 057 women, and 1,715,820 women in the 2006 census. There are 18 Local Government Areas and 15,195.177 km^2 of area in the state. Numerous mineral resources are abundant in the state. The state contains a lot of potential tourist attractions.

Figure 4. Map of Ondo State depicting the locations

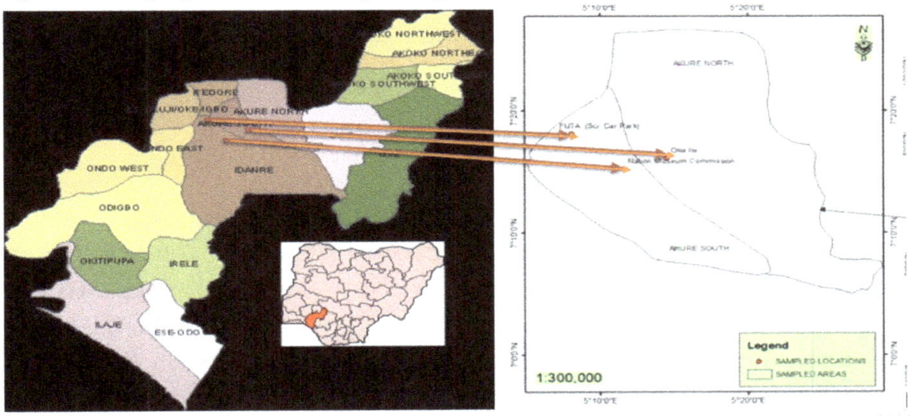

Akure serves as the state's headquarters and is thought to be its major city. With an elevation of 353 meters and a coordinate of 5° 12' 0" East, 7° 15' 0" North, the city is home to 421,100 people (GeoNames Geographical Database, 2012). Akure is in the Africa/Lagos time zone. The rapidly expanding city of Akure serves as a hub for the trade of cash crops, including citrus fruits, cocoa, cola nuts, palm oil and kernels, and bitter cola.

Sampling Area

Three distinct sites inside the town—the Federal University of Technology - FUTA, Oba-Ile, and the National Museum and Monuments—were chosen for the sampling (Table 4 and Figure 4). These sites, respectively, depict various business, home, and educational environments. For ten (10) months (January to October 2018), samplings were conducted every month for eight hours (8am–4pm). Thirty

(30) samples were taken at the conclusion of the sampling period, ten from each sampling region. Finding the sampling locations was done using GPS Map 76CSX (Garmin Ltd., Taiwan).

Table 4. The Sampling Locations

Location no	Location	Description of Sites	No. of samples
1	Oba Ile (SEDInst) (005 14 29.1 E), (07 16 04.4 N)	Residential, traffic, low dense forest area	10
2	Federal University of Technology, FUTA (Science car park),(005 08 06.5 E), (07 18 07.6 N)	Educational setting low traffic area	10
3	National Museum, (005 11 40.2 E), (07 15 11.6 N)	Commercial, high traffic area	10

Sampling of Particulate Matter (PM)

The sampler was connected to an air tube and a vacuum pump in order to capture the PM. To prepare for the sample, sets of 55 mm Teflon filters (one, two, three, four stages, backup filters, and an inertial filter) were set up on the stages. To start the sampling process, the pump was opened and the pressure was set to -0.02 MPa. The flow meter was set to 40 L/min. Following eight hours of operation, the pressure and flow rate were measured and recorded. According to Furuuchi et al. (2017), the filters were taken out of the stages, folded into a semicircle, wrapped in aluminum foil, placed in a zip-lock bag, and stored in the refrigerator before analysis.

Determination of PM Concentration (Mass)

The filters were stabilized by putting them in a chamber with 20 °C and 35% relative humidity (RH) for 48 hours. The PM mass concentration was calculated by weighing the filters both before and after exposure using an analytical balance (Model Sartorius Microbalance ME 5 Balance) to estimate the particle mass gravimetrically. These factors were used to compute the PM mass concentration:

$$\text{Average flow rate} = \text{averaging flow rate sat the beginning and end of 8 h sampling}. \quad (3)$$

$$\text{Average PM mass concentration } (\mu g/m^3) = \frac{\text{Mass}}{\text{total}} \text{volume} \quad (4)$$

(Furuuchi *et al.*, 2017).

Investigation of Meteorological Parameters

The Observatory Unit at Federal University of Technology, Akure, Ondo State, Nigeria, recorded wind direction and speed, rain, and relative humidity (RH). The study displays the readings at 10-minute intervals.

Elemental Analysis

This was accomplished by employing the use of EDXRF instrument (model PX 2CR). The samples were analyzed by exposing 300 mg of the samples in the XRF system's sample holder, which was exposed to X-rays for 1200 counts or 20 minutes in an external chamber setup. The solid-state Si-pin detector system detected the characteristic X-ray of sample, and spectrum acquisition was done using an Amptek model multi-channel analyzer (model: AMPTEK – Oxford Eclipse II). International Atomic Energy Agency (IAEA) thin target mode was used for elemental analysis. Prior to the analysis, certified micrometer® thin films (Table 5) were irradiated for calibration purposes (Ezeh *et al.*, 2017).

Table 5. Calibration results of the equipment before analysis

Element	Certified Values (μg cm^{-2})	Experimental Values (μg cm^{-2})
Ca	47.4	46.5
Fe	46.3	46.1
Pb	52.8	51.3

Calculation

$$\text{Concentration } (\mu\text{ gm}^{-3}) = \frac{\text{Instrument Reading (cm}^2) \times 1000}{\text{Filter Area (cm}^2) \times \text{Volume of air sucked (m}^3)} \quad (5)$$

Principal Component Analysis (PCA)

The number of components was obtained using methods described by Jolliffe et al. (2016) and De La Cruz et al. (2019). The pollutants (PM, K, Ca, Ti, V, Cr, Mn, Fe, Ni, Cu, Zn, and Br) were identified using the method described by James et al. (2015). PCA with VARIMAX rotation was used to identify possible sources of emissions in the study area.

Enrichment Factor (EF)

The ratio that exists between elements of interest's concentration to a reference value is known as its EF. An element with a high environmental value that lacks specific characteristics, such antagonistic or synergistic effects, toward the element being studied can serve as a reference element (Pandey et al., 2015). Fe, which is most frequently employed for normalization, serves as the reference element in this investigation (Gowd et al. 2010).

$$EF = \frac{(C_i/C_{ref})\text{ sample}}{B_i/B_{ref})\text{ Background}} \quad (6)$$

where Ci represents trace element concentration in sample, Cref represents reference element concentration in sample, Bi represents element of interest's background value, and Bref represents the reference element's background value in the study region. Using the criteria provided by Buat-Menard and Chesselet (1979), the EF values acquired for elements of interest were utilized to categorize their origin.

Contamination Factor (CF)

CFs of the elements of interest were used to measure overall contamination of soil by elements, as described by Bhuiyan et al. (2010). These values were computed by dividing concentration of element of interest in sample by its background concentration (Hakanson 1980).

$$CF = C_i / B_i \quad (7)$$

Bi is element's geochemical background value, and Ci is concentration of element under study. In ascending order of pollution, the numbers are as follows: 0 for no, 1 for medium, 2 for moderate, 3 for strong, 4 for strongly polluted, 5 for strong to very strong, and 6 for very strong (Varol, 2011).

Statistical Analysis

The Duncan Multiple Range test was used to examine the data using one-way analysis of variance (ANOVA) at 95% confidence or $p < 0.05$. The correlation coefficients were computed based on the findings. Minitab 16 Statistical Software was used to compute PCA and basic descriptions. Microsoft Excel was used to calculate the wind rise, enrichment factor, and contamination factor.

RESULTS AND DISCUSSION

The statistical results of the PM gathered and examined for this investigation were summarized in Table 6. The mean PM values were 100.5, 125.2, and 112.6 μg/m^3, respectively, and ranged from 39 to 133.2 (FUTA), 49.70 to 196.7 (Museum), and 34.50 to 161.30 (Oba-Ile). The range of the standard deviation and standard error was as follows: For FUTA, Museum, and Oba-Ile, the corresponding values were 10.9, 34.30; 13.10, 41.3, and 10.4, 33.0; the corresponding values for Skewness and Kurtosis were -0.98, -0.01, -1.29, and -0.65, 0.68, and 3.80. Bar charts (Figures 5 and 6) show that the museum had the highest mass, followed by Oba-Ile and FUTA. The same pattern was seen for the months of April, June, August, and October. For the months of March and September, the mass concentration range was Oba-Ile > Museum > FUTA. Lastly, Figure 6 shows that for the month of February, Oba-Ile had the highest concentration of 108.8 μg/m^3.

Table 6. An overview of the particle matter (μg/m^3) measurements statistical results

	FUTA	**Museum**	**Oba Ile**
Mean	100.50	125.20	112.80
Standard Error	10.90	13.10	10.40
Standard Deviation	34.30	41.30	33.00
Variance	1179.40	1709.10	1089.00
Coefficient of Variation (%)	34.17	33.02	29.25
Minimum	39.10	49.70	34.50
Maximum	133.20	196.70	161.30
Q1	64.30	107.00	107.50
Q3	127.50	148.30	131.70
Skewness	-0.98	-0.01	-1.29
Kurtosis	-0.65	0.68	3.80

The Mass Concentrations of the PM_{10} in Akure, Ondo State, Nigeria

Figure 5. Average mass concentrations ($\mu g/m^3$) of PM_{10} in the locations

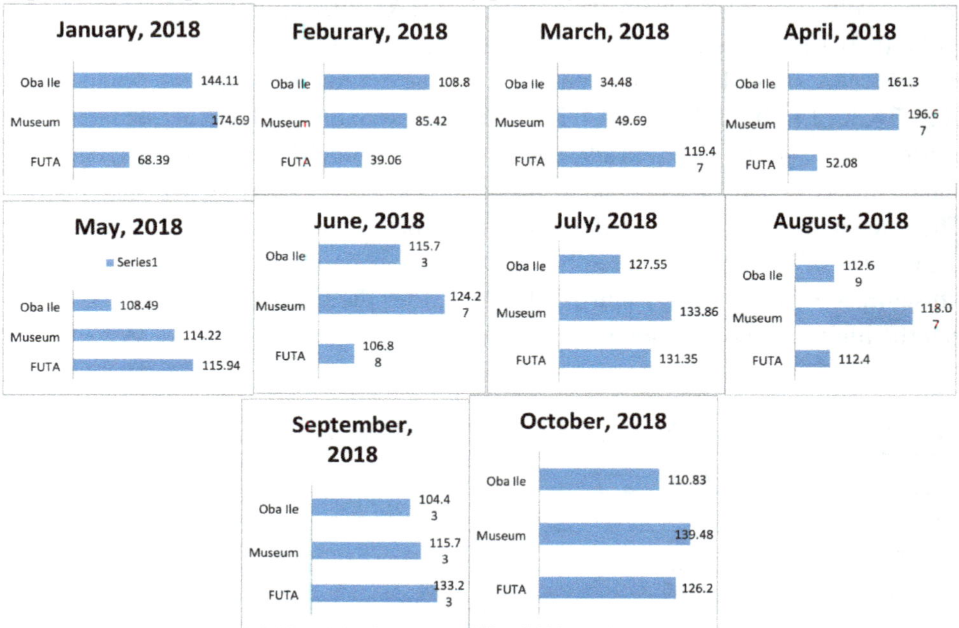

The summer months had the highest levels of PM. According to Wang et al. (2018), April and October had the highest levels of pollution, which may have been brought on by the severe weather and high secondary transformations. The former may be brought on by releases from vehicles and agricultural fires during the April clearing of land for crop cultivation, whilst the later may be brought on by October harvest season (Zhang et al., 2016). The deposition of re-entrained and re-suspended dust from low, medium, and high vehicle emissions as well as activities on unpaved terrain may have contributed to the high PM values found in the Museum area. Seasonal fluctuations in pollution levels are linked to changes in emissions, weather patterns, long-distance transportation types, and secondary production (Abulude et al., 2018a).

All of the locations had values that were higher than the 100 $\mu g/m^3$ that the Central Pollution Control Board of India (CPCB, 2008), the NAAQS (Purohit et al., 2019), more than three times the European limit (EAA, 2014), and the World Health Organization's limit of 50 $\mu g/m^3$ (WHO, 2015) had recommended. Khillare and Sarkar (2012) reported the same findings in an urban setting in Delhi (166.5 μg/

m^3), whereas Cardoso et al. (2018) reported 33-117 μg/m^3, and Guan et al. (2017) reported 115 μg/m^3. The results of this study, however, are higher than those reported by Al-Thani et al. (2018) in recent years (37 μg/m^3) and by Hsu et al. (2016) in Doha, Qatar, and the Western Coastal areas of Taiwan (33.2 and 76.4 μg/m^3, respectively). However, the findings shown here are less than the 285.7±26.3 μg/m^3 found in the industrial area of Delhi by Mandal et al. (2014) and the 584.8 μg/m^3 found in Delhi, India, by Jain et al. (2018). The climate in the various regions, the methods employed, and the anthropogenic and natural activities of the various sites may all contribute to the discrepancies between these and other results.

The Contributions of Each Location to PM in Akure

Figure 7 shows the distributions of PM (%) in the three locations. The location that generated the most PM during the months of January, April, May, June, and December was the Museum Area. In contrast to the Museum's 36% and FUTA's 17% contributions to the PM in February, Oba-Ile, a residential area, contributed 47%. The accident that happened at the Akure-Owo road expressway during the sampling time may have caused traffic to be diverted to this sampling location, which could account for Oba-Ile's large contribution. The rise in PM_{10} deposits in the area during the specified period (February) may have been greatly influenced by the rise in anthropogenic activity (ongoing renovations to a lecture theater).

The Air Quality of the Vicinities of the Sampling Locations

The air quality index (AQI) of the chosen sampling sites for this study is displayed in Table 7. As a basis for comparison, the Mandal et al. (2014) model AQI was employed. When compared to the guidelines, the results show that the air quality in the sampling regions was classified as "unhealthy" because of the particle matter in the surrounding areas. Mandal et al. (2014) state that there are five categories for the PM air quality index: good, moderate, unhealthy, very healthy, and extremely unhealthy. This index was used to compare the results, and those with values over the 50 AQI range were categorized as "Moderate" or "Unhealthy," while those with values below this range were categorized as "Good" (Table 7). PM sizes, especially PM_{10}, are associated with health issues (Liu et al., 2019).

Table 7. Air Quality Index of the Selected Sampling Locations

Air Quality Index				
Sampling Locations	0-50 (Good)	101-200 (Unhealthy)	201-300 (Very) (Unhealthy)	> 301 (Extremely) (Unhealthy)
FUTA	-	100.50±34.00	-	-
Museum	-	125.20±41.30	-	-
Oba-Ile	-	112.80±33.00	-	-

Figure 6. Distribution of PM10 in the locations

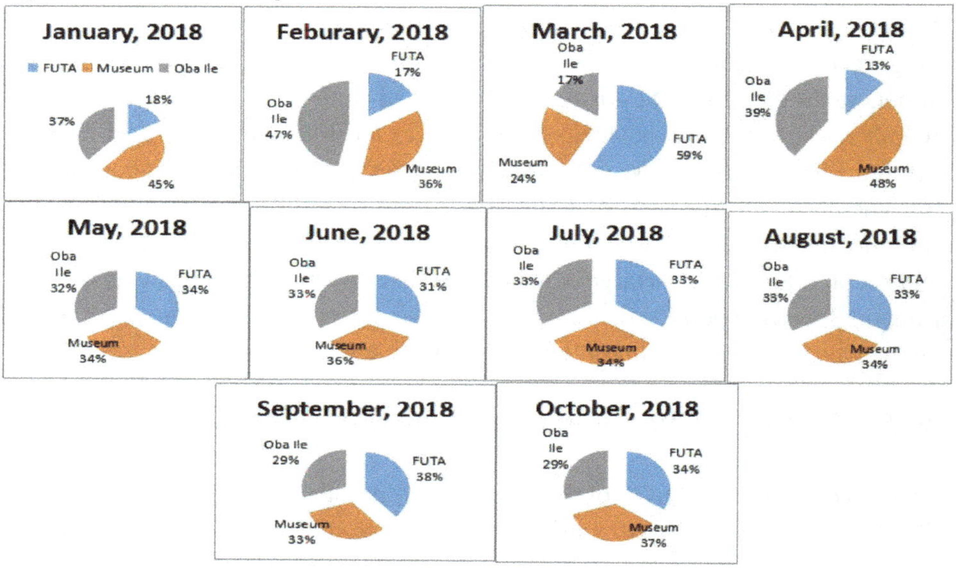

Comparisons of the Results of This Study With Previous Studies and International Standards

The study's comparison of the findings with published air quality guidelines is summed up in Table 8. The range of the values was 34.50±7.0 to 196.70±41.30 μg/m^3. These values fall within the same range as those found by Cardoso et al. (2018) for urban areas in Santiago, Cabo Verde; Al-Thani et al. (2018) in Doha, Qatar; Guan et al. (2017) in Lanzhou, China; Hsu et al. (2016) in Western Coastal, Taiwan; Police et al. (2016) in Jogannapalem, India; and Khillare and Sarkar (2012) in Delhi, India. However, they were different from the findings of Contini et al. (2010) for urban cities in Lecce, Italy; Huang and Wang (2014) in Harbin, China. In general, the outcomes surpass the WHO (2005), India (Purohit et al., 2019), and European standards for air quality (EEA, 2014).

Table 8. Comparison of this study with the results of different countries and standards

Location	Type of Environment	PM_{10} Annual average ($\mu g/m^3$)	References
This Study	Urban	34.50±7.00-196.70±41.50	
Countries			
Port Harcourt, Nigeria	Urban	4-796	Ezeh *et al.*, 2017
Lecce, Italy	Urban	63-754	Contini *et al.*, 2010
Delhi, India	Urban	166.5 ± 54.7	Khillare and Sarkar, 2012
Harbin, China	Urban	250	Huang, and Wang, 2014
Accra, Ghana	Urban	22-655.4	Huges, 2014
North Spain	Urban	19-220	Almeida *et al.*, 2015
Jogannapalem, India	Rural	68 ± 15	Police *et al.*, 2016
Tung Chung, Hong Kong	Sub-urban	23-80	Gao *et al.*, 2016
Western Coastal, Taiwan	-	33.2-76.4	Hsu *et al.*, 2016
Doha, Qatar	Urban	37	Al-Thani *et al.*, 2018
Santiago, Cabo Verde	Urban (Island)	33-117	Cardoso *et al.*, 2018
Air Quality Standard			
EEA	-	40	EEA, 2014
NAAQS	-	100	Purohit *et al.*, 2019
WHO	-	20	WHO, 2005

NAAQS - The U.S. National Ambient Air Quality Standards,
EEA - European Environment Agency
WHO – World Health Organization

Figure 7. Wind Direction (January – October, 2018)

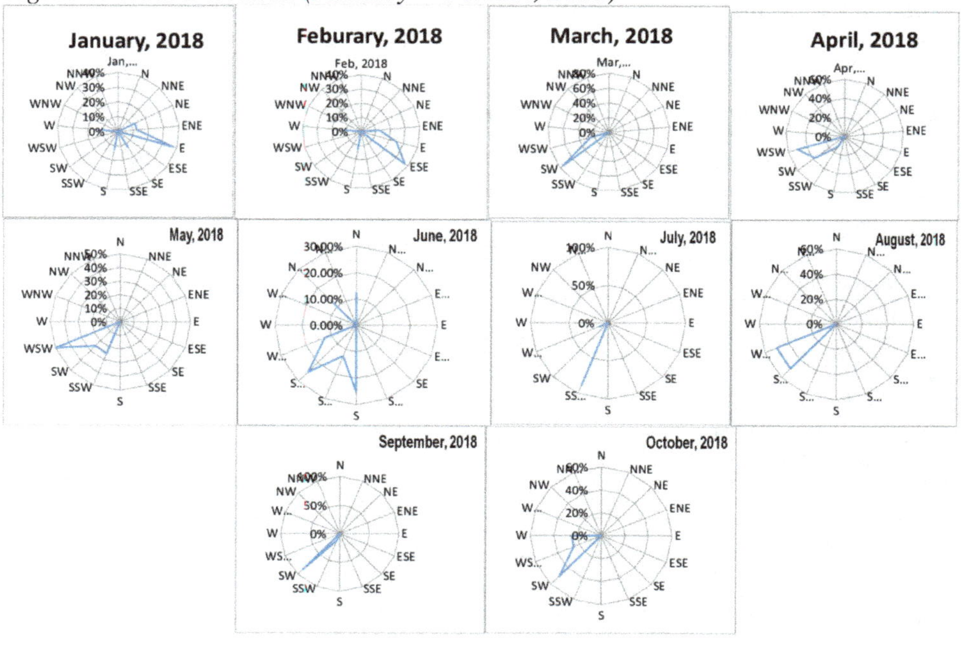

Meteorological Parameters

The weather parameter charts were shown in Figure 8. Rainfall totals (mm) were typically low during the research periods. July saw the most rainfall, with March and October recording roughly 3mm and January showing little to no rainfall. Low rainfall created the highest temperature (°C). January, February, April, and June saw the highest temperatures. This implied that more PM would be released into the sky during dry spells, and that temperatures would be high during these times, with the exception of times when there was a lot of rainfall. May saw the highest wind speed, while June saw the lowest, at 4 m/s. The months of July and September had the highest humidity levels (95%). Table 9 provides a summary of weather parameters. The mean humidity, wind speed, temperature, and precipitation were 77.34%, 5.84 m/s, 26.10 °C, and 3.71 mm, respectively.

The direction of wind summaries, which show percentages and directions as wind climbed, are shown in Figure 8. Every month, wind directions are denoted by the letters "N" through "NNW." Each wind direction's percentage is proportion of each direction's appearances throughout study period. Wind blew from east in January, and from east-south-west (ESE) in February, SW in March, WSW in April,

WSW in May, S in June, SSW in July, SW in August, SW in September, and SW in October. Findings suggest that wind blew primarily from SW and WSW directions during sample process.

Table 9. Summary of Weather Data in the Periods of Study (January – October 2018)

	Rain (mm)	Temperature (°C)	Wind Speed (m/s)	Humidity (%)
Mean				
SE mean	2.62	1.27	0.35	6.81
Std Dev	1.27	4.01	1.12	11.54
Coef Var	223.02	15.36	19.13	27.85
Minimum	0.00	19.75	4.00	21.25
Maximum	27.10	31.23	7.63	94.38
Skewness	3.08	-0.25	-o.22	-2.28
Kurtosis	9.63	-1.50	-0.44	5.93

SE – Standard Error, Std . Dev. – Standard deviation, Coef. Var. – Coefficient of Variation

According to Abu-Allaban and Abu-Qudais (2011), during dry season, wind is a significant factor in the movement of PM. Wind speed (WS) is divided into three categories: light (less than 1.5 m/s), moderate (1.5–3.0 m/s), and strong (greater than 3.0 m/s). Wind speed values recorded in this study fall between these categories, indicating the significance of WS for the dispersion of particles from various locations in this study. The wind rose distributions for the wind direction in this study could be divided into Easterly flow (100 and 120), Southerly (240 and 180), and Westerly (240 and 270). The findings of this study are comparable to meteorological data gathered from Jordan (Abu-Allaban and Abu-Qudais 2011) and Lanzhou, North China by Xiao et al. (2013). Lanzhou's average summertime temperatures ranged from 27 to 30.6°C, with average wind speeds of 0.7 to 1 m/s and relative humidity levels between 42.4 and 60.8%.

Figure 8. Meteorological Data (January – October, 2018)

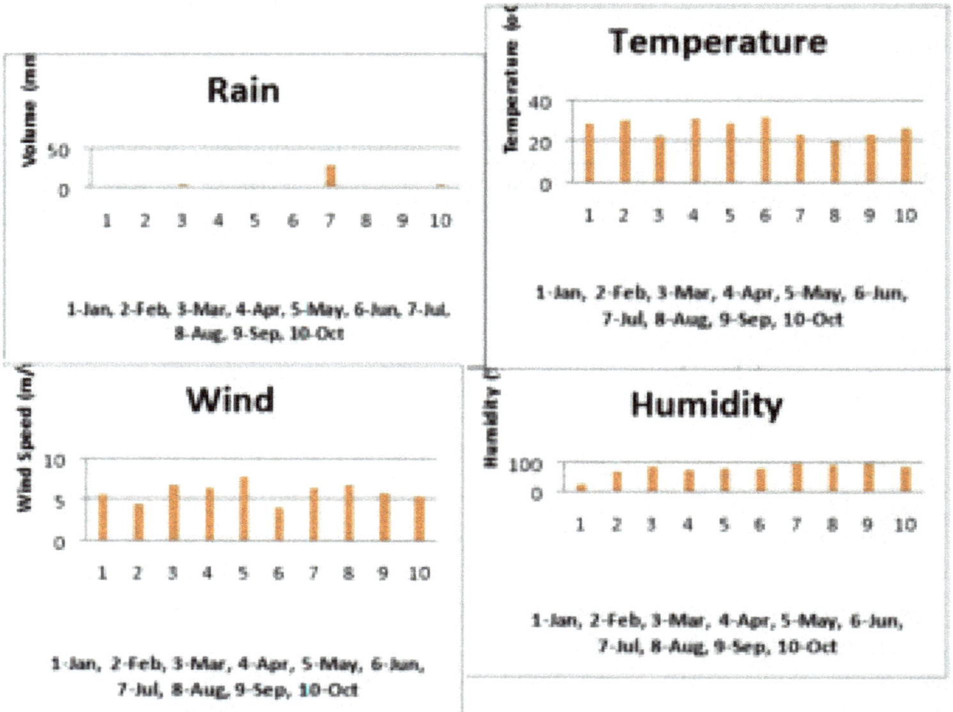

Relationship of PM and Meteorological Data

The correlations between PM and meteorological parameters are displayed in Table 10. Relationships were classified as strong, moderate, or low based on the results. There was a moderate positive connection between temperature and PM_{10} mass concentration. This indicates that when the temperature rises or falls, PM_{10} rises or falls as well. For the Museum, FUTA, and Oba-Ile sampling sites, respectively, wind speed (WS) and PM_{10} exhibit a link (r=0.87, r=0.61, and r = 0.35), with the Museum exhibiting the largest correlation. Additionally, there was a good correlation between the atmospheric temperature and PM_{10} values of the sampling sites, with r values of 0.59, 0.62, and 0.71 for Museum, FUTA, and Oba-Ile, respectively. There was a weak correlation between PM_{10} mass concentration and relative humidity (RH), with r values of 0.05 (FUTA), 0.28 (Museum), and 0.31 (Oba-Ile). There were correlations between the climatic factors, according to results as well. Wind speed and rain had a correlation (r=0.67). Furthermore, there was a positive correlation (r=0.50) between wind and humidity.

Table 10. The PM and Meteorological Data Matrix for Study Areas

	FUTA	Museum	Oba-Ile	Rain	Temperature	Wind	Humidity
FUTA	1						
Museum	0.38	1					
Oba-Ile	0.20	0.00	1				
Rain	0.30	0.89	0.78	1			
Temperature	0.59	0.62	0.71	0.45	1		
Wind	0.35	0.87	0.61	0.67	0.171	1	
Humidity	0.05	0.28	0.31	0.37	0.17	0.50	1

However, there is little correlation between the FUTA PM_{10} and the weather data. Strong correlations found in this study suggest that PM deposition is positively impacted by both high relative humidity (RH) and low wind speed. Li et al. (2015) state that weather conditions have an impact on the amount of aerosol particles, primarily when wind speed is low and relative humidity is high. This study also reveals further relationships between precipitation, humidity, and wind. In general, the correlation coefficients across the various study locations exhibit slight variations, indicating the features of location pollution. Additionally, in FUTA, there is a weak association between precipitation and PM. The study's finding that PM and temperature have a positive link suggests that photochemical reactions are likely significant throughout the sample period (Abulude et al., 2018b). Low precipitation, which raises humidity, may potentially be the cause of the study's weak relationships with relative humidity. The correlation between wind speed and PM_{10} suggests that local sources predominate.

The Levels of Elements in PM_{10} Collected from the Sampling Locations of Akure

The samples' elemental components are shown in Tables 11. Ti levels measured in FUTA (193.65 µg/m^3), Museum (48.12 µg/m^3), and Oba-Ile (204.86 µg/m^3) are displayed in Table 11. The most prevalent metals in the two tables were Ca, Ti, K, and V, whilst Fe, Ni, Cu, and Zn were found in trace levels. The samples contain several components in either low or high amounts, while some were undetectable (ND). The findings of other researchers (Jain et al., 2018; Ballová et al., 2019; De La Cruz et al., 2019) are consistent with this finding. When the elemental results in this study were compared to other PM results by researchers, it was found that most results were low to those reported by Viao et al. (2018) in Acerra, Italy, but higher than those reported in Table 11 by Wang et al., 2012 (China); Tositti et al.,

2013 (Italy); Co et al., 2014 (Vietnam); Onabowale and Owoade, 2015 (Ibadan, Nigeria); Basahi et al., 2017 (Saudi Arabia); and Okuo et al., 2017 (Lagos, Nigeria).

Table 11. Elemental Constituents of PM_{10} Sample (mean levels)

S/N	Elements (μgm^{-3})	FUTA	Museum	Oba-Ile
1	K	61.52	15.20	ND
2	Ca	262.80	66.50	72.12
3	Ti	193.65	48.12	204.86
4	V	43.95	11.94	ND
5	Cr	ND	ND	3.86
6	Mn	ND	ND	3.46
7	Fe	2.13	0.52	ND
8	Ni	0.12	0.09	0.10
9	Cu	0.05	0.06	3.62
10	Zn	1.82	0.61	ND
11	Br	ND	ND	ND

ND – Not detectable

Correlation Coefficients (CC)

To demonstrate their association, correlation coefficients between PM mass and elements were calculated for each sample. The CC results for January are shown in Table 12.

When the CC exceeds 0.90, it is considered very high; when it exceeds 0.70, it is considered high. For PM and Cr, Mn, Ni, and Cu, the r values are above 0.60, indicating that these elements have a strong correlations with the PM of the sampling sites, suggesting that their sources or origins are likely the same. For PM_{10} and Ti, the r values are 0.76 for samples from FUTA and 0.98 for samples from Oba-Ile. Additionally, there were correlations between the elements; for instance, there was a substantial correlation between Ni and Ca (r=0.24), Fe and Ca (r=0.06), Cr and K, (r=0.14). There was also a minor correlation between Ni and Zn (r = 0.29) and Cu and Ni (r = 0.24). Additionally, the elements Fe, Ca, Ti, and Cr; all show weak positive connections.

Table 12. Relationship of PM_{10} with the elements

	FUTA	Museum	Oba Ile	K	Ca	Ti	V	Cr	Mn	Fe	Ni	Cu	Zn	Br
FUTA	1													
Museum	**0.67**	1												
Oba Ile	-	**0.67**	1											
K	**0.86**	0.47	**0.86**	1										
Ca	**0.9**	0.33	**0.9**	0.14	1									
Ti	**0.99**	0.33	**0.98**	0.14	-	1								
V	-	-	-	-	-	-	1							
Cr	**0.97**	0.33	**0.97**	0.14	-	-	-	1						
Mn	**0.97**	0.34	**92**	0.14	-	-	-	-	1					
Fe	**0.94**	0.39	**0.94**	0.08	0.06	0.07	-	0.06	0.06	1				
Ni	**0.76**	0.1	**0.76**	0.38	0.24	0.24	-	0.25	0.24	0.3	1			
Cu	**9.1**	0.32	**9.3**	0.14	-	-	-	-	-	0.06	0.24	1		
Zn	**0.95**	0.34	**0.98**	0.18	-	-	-	-	-	0.01	0.29	-	1	
Br	-	-	-	-	-	-	-	-	-	-	-	-	-	1

Positive correlations (r) between PM_{10} and metals are noted in January; these relationships are strong and low. Ca, Ti, V, Ni, Zn, Fe, and Mn. Pb, Zn, and Ba are traffic markers, according to Fujiwara et al. (2011). Janta and Chantara (2017) stated that Cr, Pb, Cu, and Zn are only obtained from vehicles. Sb, Cu, Zn, Ba, and Fe are indicators of re-suspension brought on by traffic and brake or other vehicle wear, according to a claim made by Querol et al. (2008). According to Enamorado-Báez et al. (2015), Ba, V, Ni, Zn, Cu, and Sb were also linked to traffic emissions. Additionally, V and Ni were claimed to be additional producers of oil combustion processes from diesel engines (Giampaoli et al., 2016). Therefore, it is likely that the elements found in this study, particularly the heavy metals, came from one or more of the sources mentioned by these authors as well as any others mentioned in this work.

Principal Component Analysis (PCA)

Rotated factor loadings with eigenvalues >1 and more than 62% of the total variance are shown in Table 13 results. With strong loadings for PM_{10} from the Museum and Oba-Ile, PC 1 shows the variance at 96.6%. In these sampling areas, Ca and Ti also show substantial loading in the PM_{10}. Ti displays over 31.86%, which could be related to Road/Soil Dust. While its association with numerous other elements suggests some alternative sources, including TVOC, its bulk matrix may be soil.

One interpretation of this element might be that it represents re-suspension of road dust. With the PC2 displaying Ca (-0.60), the PC 4 values of Fe and Cu, which are displayed in table, show a variance of almost 10% RSD may be the cause of these metals' existence. This could be explained by the fact that the majority of the matrix may have come from RSD, and VE.

Table 13. Principal Component Analysis of PM_{10} in the three locations

	PC1 RSD	PC 2	PC 3	PC 4 TVOC
FUTA	0.01	0.03	-0.01	0.05
Museum	-0.01	0.01	0.01	0.01
Oba Ile	0.01	0.01	-0.01	-0.01
K	0.07	0.14	0.08	0.2
Ca	**0.46**	**-0.6**	-0.19	0.03
Ti	**0.85**	0.1	0.15	0.05
V	ND	ND	ND	ND
Cr	0.03	-0.01	-0.06	-0.02
Mn	0.18	0.04	-0.01	0.02
Fe	0.05	0.16	-0.09	**0.49**
Ni	0.01	0.01	-0.01	0.01
Cu	-0.01	0.21	-0.04	**0.66**
Zn	0.01	0.06	-0.02	0.4
Br	ND	ND	ND	ND
Eigenvalue	9	3	1.2	0.91
Variance (%)	96.6	31.82	60	50.8

FUTA: Federal university of Technology, RSD – Road/Soil Dust, VE – Vehicle Emission/wear, BoB–Burning of biomass, TVOC – Traffic/Oil combustion

To identify sources of air pollutants in environment, principal component analysis is used globally (Chen et al., 2015; Liu et al., 2019). Without sacrificing the original data, it converts a large number of indications into meaningful representative indicators using projection dimension reduction techniques (Querol et al., 2008). There were numerous contributing sources of metals in PM, according to earlier research (Onabowale and Owoade, 2015; De La Cruz et al., 2019;). Table 2 provides a summary of the elements' sources and markers. PCA with eigenvalues over 1 could be used to determine the elements linked to the sources in order to objectively identify the sources contributing to the items in PM.

Enrichment Factor (EF)

Based on the typical metal concentrations of the earth crust, the EF for the elements in the PM samples were as follows, with iron serving as the reference element (Aromolo et al., 2015). All elements' enrichment factors in the PM_{10} from the FUTA locations were higher than those in other places, according to the results (Table 14). The high elements loading to the ambient atmosphere is explained by the heavy traffic, BoB, and other associated activities in FUTA and the Museum. In every area, the K, Cr, Mn, Ti, Ni, Cu, and Zn, enrichment factors were exceptionally high. Lower EF were observed for Ti and Mn.

Table 14. Enrichment Factor of PM_{10} in the three locations

	FUTA	Museum	Oba Ile
K	288.73	24.75	156.37
Ca	41.83	12.29	27.06
Ti	129.53	181.40	100.76
V	-	-	1006.92
Cr	364.71	109.06	-
Mn	95.16	22.26	-
Fe	1	1	1
Ni	1147.06	-	211.07
Cu	305.88	44.59	252.60
Zn	882.36	78.81	1058.82
Br	-	40.96	-

In particulate matter research, the EF of elements is an indicator used to verify the existence and severity of anthropogenic contamination depositions. According to the methodology outlined by Ezeh et al. (2015), reference element utilized was Fe. The elements' likely anthropogenic origin is confirmed by the high EF found specifically for zinc. According to Ezeh et al. (2015), the EF of Mn, Fe, Ti, and K might have come from soil and road dusts, but EF of Cu and Zn might have come from burning trash and biomass. According to Bahloul et al. (2015), and Pongpiachan and Iijima (2016), the EF results in this study can be divided into four groups: minimally enriched ($EF < 10$), enriched $10<EF<100$, moderately enriched ($100 < EF < 1000$) and highly enriched; $EF>1000$). K, V, Ni, Ti, Ni, Cu, Zn, and Cr were therefore enriched (100<EF<K1000), but Ca and Mn were little enriched $10<EF<100$. Finding is consistent with findings published by Khademi et al. (2019). Natural sources are the source of an element's EF of less than 10 (Aromolo et al.,

2015). Therefore, it's possible that the elements K, Ca, Ti, Ni, Cu, and Cr originated mostly from human activity.

Contamination Factors (CF)

Table 15 shows the average contamination factors for the samples obtained from January to October. Ti is the most elevated pollutants in all of the PM samples, followed by Cr and Cu, according to the computed CFs. Samples from FUTA and the Museum areas were found to be particularly polluted with the elements Cu, Ti, Zn. Similarly, because of their high CF values in the samples. Ti has the greatest CF value (> 5) in this study's locations collections, suggesting an environment that is substantially enriched. However, the enrichment of the Ti may come from released waste.

Table 15. Contamination Factor of PM_{10} in the three locations

	FUTA	Museum	Oba Ile
K	0.002	0.003	0.002
Ca	0.05	0.07	0.04
Ti	28.55		
V	-	-	-
Cr	1.13	1.22	0.86
Mn	-	0.25	0.21
Fe			
Ni	0.003	0.004	0.003
Cu	0.077	0.111	0.128
Zn	0.014	0.027	0.014
Br	-	-	-

Based on the idea that the degree of element enrichment in the PM is correlated with a particular source, CF determines the possible sources that contribute to the elements in the PM at the locations. Six categories of contamination, all anthropogenic, were then used to identify the sources: < 1 background concentration, 1-2 depletion to minimal enrichment, 2-5 moderate enrichment, 5-20 significant enrichment, 20-40 very high enrichment, and > 40 extremely high enrichment (Sutherland 2000). Based on research by Hakanson (1980) and Shariati et al. (2019), the results are thus divided into four groups: moderate: $3 \leq CF < 6$, Considerable, CF>6, extremely high, and minimally enriched: CF<1. K, Cr, Ni, Cu, Mn, Zn, and Br are lowly contaminating elements (CF<1) in this instance. Ti is extremely contaminating (CF>6).

CONCLUSION

The results of this urban air PM monitoring research showed Akure, Ondo State's PM_{10} concentrations. FUTA, Museum, and Oba-Ile had respective findings ($\mu g/m^3$) of 39.17-133.2, 49.70-196.7, and 34.50-161.30. The WHO and other standard limitations were broken by the monthly mean results. The PM in the chosen areas was classified as "Good" or "Unhealthy" by the AQI. High traffic and other human activities were cited as the reason why the Museum area was deemed "unhealthy." The components in the samples were quantified using the XRF method. Some of the elements were seen to beyond the conventional limitations. Some of the elements had a high EF. Particularly during the dry season, when there was little precipitation and high temperatures, the meteorological data demonstrated correlations with the elements and PM. The high correlations (r=>0.5) could be attributed to chemical similarity or a shared source. High traffic and vehicle emissions, the burning of biomass and garbage, dust from roads and soil, and oil combustion (anthropogenic activities) were the sources identified by PCA. These sources may also be natural. The outcomes of this study will help with climate change forecasting and mitigation.

REFERENCES

Abu-Allaban, M., & Abu-Qudais, H. (2011). Impact assessment by ambient air quality by Cement Industry: A case study in Jordan. *Aerosol and Air Quality Research, 11*(7), 802–810. DOI: 10.4209/aaqr.2011.07.0090

Abulude, F. O., Ndamitso, M. M., Iyaka, Y. A., & Abdulkadir, A. (2018a). Study of particulate matter of Akure, Nigeria using a sharp-cut inertial filter combined with an impactor-a preliminary study. *Journal of Atmospheric and Solar-Terrestrial Physics, 179*, 396–404. DOI: 10.1016/j.jastp.2018.09.005

Abulude, F. O., Ndamitso, M. M., Iyaka, Y. A., & Abdulkadir, A. (2018b). An assessment of meteorological conditions in Akure, Ondo State, Nigeria. *Bartin University International Journal of Natural and Applied Sciences, 1*(2), 73–78.

Adebiyi, F. M., Ezeh, G. C., Adeyemi, F. A., & Obioh, I. B. (2015). Trace elements in particulate matter of ambient air at petroleum filling stations. *Toxicological and Environmental Chemistry, 97*(7), 847–856. DOI: 10.1080/02772248.2015.1071377

Al-Thani, H., Koc, M., & Isaifan, R. J. (2018). Investigations on Deposited Dust Fallout in Urban Doha: Characterization, Source Apportionment and Mitigation. *Environment and Ecology Research, 6*(5), 493–506. DOI: 10.13189/eer.2018.060510

Alahmr, F. O. M., Othman, M., Abd Wahid, N. B., Abdul Halim, A. A., & Latif, M. T. (2012). Composition of dust falls around semi-urban areas in Malaysia. *Aerosol and Air Quality Research, 12*(4), 629–642. DOI: 10.4209/aaqr.2012.02.0027

Alkhatib, A. J., Boran, A. M., Al-Hourani, Z., Al-Balwi, A., Muhammad, A., Bello, A. A. M., Qasem, A., Alkhatatbeh, M. A., & Adamu, A. (2014). Occupational Exposure To Nickel, Cadmium And Copper Among Workers In Jewelry Manufacturing. *European Scientific Journal, 10*(15), 159–169.

Alpdogan, G., San, N., & Zor, S. D. (2019). Analysis of Some Trace Metals in Fish Species after Preconcentration with Congo Red on Amberlite XAD-7 Resin by Flame Atomic Absorption Spectrometry. *Journal of Chemistry*. Advance online publication. DOI: 10.1155/2016/1365171

Amodio, M., Catino, S., Dambruoso, P. R., deGennaro, G., Di Gilio, A., Guingato, P., Laiola, E., Marzocca, A., Mazzone, A., Sardaro, A., & Tutino, M. (2014). Atmospheric deposition: Sampling procedures, analytical methods, and main recent findings from the scientific literature. *Advances in Meteorology, 2014*, 1–28. DOI: 10.1155/2014/161730

Aromolo, R., Savi, F., Salvati, L., Ilardi, F., Moretti, V., Fares, S. (2015). Particulate matter and meteorological conditions in Castelporziano forest: a brief commentary. 26: 269-273.

Bahloul, M., Chabbi, I., Dammak, R., Amdouni, R., Medhioub, K., & Azri, C. (2015). Geochemical behavior of PM_{10} aerosol constituents under the influence of succeeding anticyclonic/cyclonic situations: Case of Sfax City, Southern Tunisia. *Environmental Monitoring and Assessment*, *187*(12), 757. DOI: 10.1007/s10661-015-4980-x PMID: 26577217

Ballová, Z., Janiga, M., & Hančinský, R. (2019). Comparison of Element Concentrations (Ba, Mn, Pb, Sr, Zn) in the Bones and Teeth of Wild Ruminants from the West Carpathians and the Tian-Shan Mountains as Indicators of Air Pollution. *Atmosphere (Basel)*, *10*(2), 64. DOI: 10.3390/atmos10020064

Basahi, J. M., Ismail, I. M., Hassan, I. A., Almeelbi, T., Haiba, N. S., & Hammam, E. (2017). Total suspended particulate matter (TSP) and its associated heavy metals in atmosphere on the Western Coast of Saudi Arabia. *Polish Journal of Environmental Studies*, *26*(5), 2419–2424. DOI: 10.15244/pjoes/69102

Bhuiyan, M. A. H., Parvez, L., Islam, M. A., Dampare, S. B., & Suzuki, S. (2010). Heavy metal pollution of coal mine-affected agricultural soils in the northern part of Bangladesh. *Journal of Hazardous Materials*, *173*(1-3), 384–392. DOI: 10.1016/j.jhazmat.2009.08.085 PMID: 19744789

Bruker (2019). X-ray Fluorescence (XRF). Elemental Analysis, Process and Quality Control. www.bruker.com

Buat-Menard, P., & Chesselet, R. (1979). Variable influence of the atmospheric flux on the trace metal chemistry of oceanic suspended matter. *Earth and Planetary Science Letters*, *42*(3), 399–411. DOI: 10.1016/0012-821X(79)90049-9

Cardoso, J., Almeida, S. M., Nunes, T., Almida-Silva, M., Cerqueira, M., Alves, C., Rocha, F., Chaves, P., Reis, M., Salvador, P., Artinano, B., & Pio, C. (2018). Source apportionment of atmospheric aerosol in a marine dusty environment by ionic/composition mass balance (IMB). *Atmospheric Chemistry and Physics*, *18*(17), 13215–13230. DOI: 10.5194/acp-18-13215-2018

Chen, P., Bi, X., Zhang, J., Wu, J., & Feng, Y. (2015). Assessment of heavy metal pollution characteristics and human health risk of exposure to ambient $PM_{2.5}$ in Tianjin, China. *Particuology*, *20*, 104–109. DOI: 10.1016/j.partic.2014.04.020

Co, H. X., Dung, N. T., Oanh, N. T. K., Hang, N. T., Phuc, N. H., & Lee, H. A. (2014). Level, and composition of ambient particulate matter at a mountainous rural site in Northern Vietnam. *Aerosol and Air Quality Research, 14*(7), 1917–1928. DOI: 10.4209/aaqr.2013.09.0300

Contini, D., Genga, A., Cesari, D., Siciliano, M., Donateo, A., Bove, M. C., & Cuascito, M. R. (2010). Characterization and source apportionment of PM_{10} in an urban background site in Lecce. *Atmospheric Research, 95*(1), 40–54. DOI: 10.1016/j.atmosres.2009.07.010

CPCB. (2008). Stationary Sources Emission Profiles. Central Pollution Control Board, New Delhi, India. https://cpcb.nic.in/Stationary_Sources_Emission_Profiles.xls, Last Access: 09.10.2014.

De La Cruz, A. H., Roca, Y. B., Suarez-Salas, L., Pomalaya, J., Tolentino, D. A., & Gioda, A. (2019). Chemical Characterization of $PM_{2.5}$ at Rural and Urban Sites around the Metropolitan Area of Huancayo (Central Andes of Peru). *Atmosphere (Basel), 10*(1), 21. DOI: 10.3390/atmos10010021

EEA (European Environment Agency). (2014). Air quality in Europe — 2014 report. Report No. 5/2014, Copenhagen (84 pages).

Enamorado-Báez, S. M., Gómez-Guzmán, J. M., Chamizo, E., & Abril, J. M. (2015). Levels of 25 trace elements in high-volume air filter samples from Seville (2001–2002): Sources, enrichment factors and temporal variations. *Atmospheric Research, 155*, 118–129. DOI: 10.1016/j.atmosres.2014.12.005

Ezeh, G. C., Abiye, O. E., & Obioh, I. B. (2017). Elemental analyses and source apportionment of $PM_{2.5}$ and $PM_{2.5-10}$ aerosols from Nigerian urban cities. *Cogent Environmental Science, 3*(1), 1323376. DOI: 10.1080/23311843.2017.1323376

Ezeh, G. C., Obioh, I. B., Asubiojo, O. I., & Abiye, O. E. (2012). PIXE characterization of PM_{10} and $PM_{2.5}$ particulates sizes collected in Ikoyi, Lagos, Nigeria. *Toxicological and Environmental Chemistry, 94*(5), 884–894. DOI: 10.1080/02772248.2012.674133

Fortoul, T. I., Rodriguez-Lara, V., Gonzalez-Villalva, A., Rojas-Lemus, M., Colin-Barenque, L., Bizarro-Nevares, P., García-Peláez, I., Ustarroz-Cano, M., López-Zepeda, S., Cervantes-Yépez, S., López-Valdez, N., Meléndez-García, N., Espinosa-Zurutuza, M., Cano-Gutierrez, G., & Cano-Rodríguez, M. C. (2015). Health Effects of Metals in Particulate Matter. *Current Air Quality Issues. Ch, 25*, 572–607. DOI: 10.5772/59749

Fujiwara, F. G., Gómez, D. R., Dawidowski, L., Perelman, P., & Faggi, A. (2011). Metals associated with airborne particulate matter in road dust and tree bark collected in a megacity (Buenos Aires, Argentina). *Ecological Indicators*, *11*(2), 240–247. DOI: 10.1016/j.ecolind.2010.04.007

Furuuchi, M. M., Hata, M., Nuttapon, K., & Hongtieab, S. (2017). Nano sampler (NS) Standard Operate Procedure. Kanazawa University, 1-6.

GeoNames Geographical Database. **(2012). Population of Akure, Nigeria.** https://population.mongabay.com/population/nigeria/2350841/akure (Retrieved 9th January 2016).

Giampaoli, P., Wannaz, E. D., Tavares, A. R., & Domingos, M. (2016). Suitability of *Tillandsia usneoides* and *Aechmea fasciata* for biomonitoring toxic elements under tropical seasonal climate. *Chemosphere*, *149*, 14–23. DOI: 10.1016/j.chemosphere.2016.01.080 PMID: 26844661

Gowd, S. S., Reddy, M. R., & Govil, P. K. (2010). Assessment of heavy metal contamination in soils at Jajmau (Kanpur) and Unnao industrial areas of the Ganga Plain, Uttar Pradesh, India. *Journal of Hazardous Materials*, *174*(1-3), 113–121. DOI: 10.1016/j.jhazmat.2009.09.024 PMID: 19837511

Guan, Q., Cai, A., Wang, F., Yang, L., Xu, C., & Liu, Z. (2017). Spatio-temporal variability of particulate matter in the key part of Gansu Province, Western China. *Environmental Pollution*, *230*, 189–198. DOI: 10.1016/j.envpol.2017.06.045 PMID: 28651090

Hakanson, L. (1980). An ecological risk index for aquatic pollution control. A sediment logical approach. *Water Research*, *14*(8), 975–1001. DOI: 10.1016/0043-1354(80)90143-8

Hsu, C. Y., Chiang, H. C., Lin, S. L., Chen, M. J., Lin, T. Y., & Chen, Y. C. (2016). Elemental characterization and source apportionment of PM_{10} and $PM_{2.5}$ in the western coastal area of central Taiwan. *The Science of the Total Environment*, *15*(541), 1139–1150. DOI: 10.1016/j.scitotenv.2015.09.122 PMID: 26473714

Huang, L., & Wang, G. (2014). Chemical characteristics and source apportionment of atmospheric particles during the heating period in Harbin, China. *Journal of Environmental Sciences (China)*, *26*(12), 2475–2483. DOI: 10.1016/j.jes.2014.04.008 PMID: 25499495

Jain, S., Sharma, S. K., Mandal, T. K., & Saxena, M. (2018). Source apportionment of PM_{10} in Delhi, India using PCA/APCS, UNMIX, and PMF. *Particuology*, *37*, 107–118. DOI: 10.1016/j.partic.2017.05.009

James, G., Witten, D., Hastie, T., & Tibshirani, R. (2015). *An Introduction to Statistical Learning with Applications in R* (Olkin, G. C. S. F. I., Ed.). Vol. 6). Springer.

Janta, R., & Chantara, S. (2017). Tree bark as a bio indicator of a metal accumulation from road traffic and air quality map: A case study of Chiang Mai, Thailand. *Atmospheric Pollution Research*, *8*(5), 956–967. DOI: 10.1016/j.apr.2017.03.010

Jolliffe, I. T., Cadima, J., & Cadima, J. (2016). Principal component analysis: A review and recent developments. *Philosophical Transactions. Series A, Mathematical, Physical, and Engineering Sciences*, *A374*(2065), 20150202. DOI: 10.1098/rsta.2015.0202 PMID: 26953178

Kchih, H., Perrino, C., & Cherif, S. (2015). Investigation of desert dust contribution to source apportionment of PM_{10} and $PM_{2.5}$ from Southern Mediterranean Coast. *Aerosol and Air Quality Research*, *15*(2), 454–464. DOI: 10.4209/aaqr.2014.10.0255

Khademi, H., Gabarron, M., Abbaspour, A., Martínez-Martínez, S., Faz, A., & Acosta, J. A. (2019). Environmental impact assessment of industrial activities on heavy metals distribution in street dust and soil. *Chemosphere*, *217*, 695–705. DOI: 10.1016/j.chemosphere.2018.11.045 PMID: 30448749

Khillare, P. S., & Sarkar, S. (2012). Airborne inhalable metals in residential areas of Delhi, India: Distribution, source apportionment, and health risks. *Atmospheric Pollution Research*, *3*(1), 46–54. DOI: 10.5094/APR.2012.004

Kothai, P., Saradhi, I. V., Pandit, G. G., Markwitz, A., & Puranik, V. D. (2011). **Chemical** characterization and source identification of particulate matter at an urban site of Navi Mumbai, India. *Aerosol and Air Quality Research*, *11*(5), 560–569. DOI: 10.4209/aaqr.2011.02.0017

Kumar, B., Verma, K., & Kulshrestha, U. (2014). Deposition and mineral characteristics of atmospheric dust in relation to land use and land cover change in Delhi (India). *The Geographical Journal*, *2014*, 1–11. Advance online publication. DOI: 10.1155/2014/325612

Li, Y. J., Lee, B. P., Su, L., Fung, J. C. H., & Chan, C. K. (2015). Seasonal characteristics of fine particulate matter (PM) based on high-resolution time-of-flight aerosol mass spectrometric (HR-ToF-AMS) measurements at the HKUST Supersite in Hong Kong. *Atmospheric Chemistry and Physics*, *15*(1), 37–53. DOI: 10.5194/acp-15-37-2015

Liu, H., Scneider, P., Haugen, R., & Vogt, M. (2019). Performance Assessment of a Low-Cost $PM_{2.5}$ Sensor for a near Four-Month Period in Oslo, Norway. *Atmosphere (Basel)*, *10*(2), 41. DOI: 10.3390/atmos10020041

Mandal, P., Saud, T., Sarkar, R., Mandal, A., Sharma, S. K., Mandal, T. K., & Bassin, J. K. (2014). High seasonal variation of atmospheric C and particle concentrations in Delhi, India. *Environmental Chemistry Letters*, *12*(1), 225–230. DOI: 10.1007/s10311-013-0438-y

Moses, E. A., & Orok, U. B. (2015). Contamination and health risk assessment of suspended particulate matter (SPM) in Uyo, Nigeria Delta, Nigeria. *Journal of Scientific Research and Reports*, *6*(4), 276–286. DOI: 10.9734/JSRR/2015/16296

National Population Commission (NPC). (2010). *Federal Republic of Nigeria. 2006 Population and Housing Census. Priority Table* (Vol. III). Population Distribution – By Sex, State, LGA and Senatorial District.

Ndamitso, M.M., Abdulkadir, A., Abulude, F.O. (2015). Total atmospheric deposit source apportionment: A review. *Environmental Skeptics and Critics,* 2016, 5(4): 63-78.

Ndamitso, M. M., Abdulkadir, A., & Abulude, F. O. (2016). Source Apportionment: Case Studies of Selected African Countries. *International Journal of Advancements in Research & Technology*, *5*(3), 1–25.

Okuo, J. M., Chiedu, I. E., Anegbe, B., Oyibo, F. O., & Ojo, W. (2017). Elemental Characterization and Source Identification of Fine Particulate Matter (PM2.5) in an Industrial Area of Lagos State, Nigeria. *Physical Science International Journal*, *16*(2), 1–11. DOI: 10.9734/PSIJ/2017/36683

Onabowale, M. K., & Owoade, O. K. (2015). Assessment of residential indoor-outdoor airborne particulate matter in Ibadan, Southwestern Nigeria. *Donnish Journal of Physical Sciences*, *1*(1), 1–7.

Onwudiegwu, C. A., Ezeh, G. C., & Obioh, I. B. (2016). Trace Metals in Total Atmospheric Depositions (TAD) of a Nigerian Island. *Journal of Atmospheric Pollution*, *4*(1), 15–22. DOI: 10.12691/jap-4-1-2

Pandey, B., Agrawal, M., & Singh, S. (2015). Ecological risk assessment of soil contamination by trace elements around the coal mining area. *Journal of Soils and Sediments*. Advance online publication. DOI: 10.1007/s11368-015-1173-8

Police, S., Sahu, S. K., & Pandit, G. G. (2016). Chemical characterization of atmospheric particulate matter and their source apportionment at an emerging industrial coastal city, Visakhapatnam, India. *Atmospheric Pollution Research*, *7*(4), 725–733. DOI: 10.1016/j.apr.2016.03.007

Pongpiachan, S., & Iijima, A. (2016). Assessment of selected metals in the ambient air PM_{10} in urban sites of Bangkok (Thailand). *Environmental Science and Pollution Research International, 23*(3), 2948–2961. DOI: 10.1007/s11356-015-5877-5 PMID: 26631022

Purohit, P., Amann, M., Kiesewtter, G., Chaturvedi, V., Rafj, P., Dholakia, H. H., Koti, P. N., Klimont, Z., Borken-Kleefeld, J., Gomez-Sanabria, A. G., Schopp, W., & Sander, R. (2019). *Pathways to Achieve National Ambient Air Quality Standards (NAAQS) in India. International Institute for Applied System Analysis (IIASA). The Council on Energy, Environment and Water*. CEEW.

Querol, X., Alastuey, A., Moreno, T., Viana, M. M., Castillo, S., Pey, J., Rodríguez, S., Artiñano, B., Salvador, P., Sánchez, M., Garcia Dos Santos, S., Herce Garraleta, M. D., Fernandez-Patier, R., Moreno-Grau, S., Negral, L., Minguillón, M. C., Monfort, E., Sanz, M. J., Palomo-Marín, R., & Sánchez de la Campa, A. (2008). Spatial and temporal variations in airborne particulate matter (PM_{10} and $PM_{2.5}$) across Spain 1999–2005. *Atmospheric Environment, 42*(17), 3964–3979. DOI: 10.1016/j.atmosenv.2006.10.071

Rich, D. Q., Zhang, W., Lin, S., Squizzato, S., Thurston, S. W., van Wijngaarden, E., Croft, D., Masiol, M., & Hopke, P. K. (2019). Triggering of cardiovascular hospital admissions by source-specific fine particle concentrations in urban centers of New York State. *Environment International, 126*, 387–394. DOI: 10.1016/j.envint.2019.02.018 PMID: 30826617

Shariati, S., Pourbabaee, A. A., Alikhani, H. A., & Rezaei, K. A. (2019). Investigation of Heavy Metal Contamination in the Surface Sediments of Anzali Wetland in North of Iran. *Pollution, 5*(1), 211–224. DOI: 10.22059/poll.2018.257276.438

Sutherland, R. A. (2000). Bed sediment-associated trace metals in an urban stream, Oahu, Hawaii. *Environmental Geology (Berlin), 39*(6), 611–637. DOI: 10.1007/s002540050473

Taylor, S. R. (1964). Abundance of chemical elements in the continental crust: A new table. *Geochimica et Cosmochimica Acta, 28*(8), 1273–1285. DOI: 10.1016/0016-7037(64)90129-2

Tositti, L., Brattich, E., Maisiol, M., Baldacci, D., Ceccato, D., Parmeggiani, S., Stracquadanio, M., & Zappoli, S. (2013). Source apportionment of particulate matter in a large city of southeastern Po Valley (Bologna, Italy). *Environmental Science and Pollution Research International, 21*(2), 872–890. DOI: 10.1007/s11356-013-1911-7 PMID: 23828727

USEPA (2002). Methods for Measuring the Acute Toxicity of Effluents and Receiving Waters to Freshwater and Marine Organisms. EPA-821-R-02-012.

Varol, M. (2011). Assessment of heavy metal contamination in sediments of the Tigris River (Turkey) using pollution indices and multivariate statistical techniques. *Journal of Hazardous Materials, 195*, 355–364. DOI: 10.1016/j.jhazmat.2011.08.051 PMID: 21890271

Viao, P., Magli, E., Caliendo, G., Corvino, A., Fiorino, F., Frecentese, F., Saccone, I., Santagada, V., Severino, B., Onorati, G., Freda, G., Manzo, C., & Perissutti, E. (2018). Heavy Metals Size Distribution in PM_{10} and Environmental-Sanitary Risk Analysis in Acerra (Italy). *Atmosphere (Basel), 9*(2), 5. DOI: 10.3390/atmos9020058

Wang, L., Li, W., Sun, Y., Tao, M., Xin, J., Song, T., Li, X., Zhang, N., Ying, K., & Wang, Y. (2018). $PM_{2.5}$ Characteristics and Regional Transport Contribution in Five Cities in Southern North China Plain, During 2013–2015. *Atmosphere (Basel), 9*(4), 157. DOI: 10.3390/atmos9040157

Wang, Q., Kwan, M., Zhou, K., Fan, J., Wang, Y., & Zhang, D. (2019). The impacts of urbanization on fine particulate matter ($PM_{2.5}$) concentrations: Empirical evidence from 135 countries worldwide. *Environmental Pollution, 247*, 989–998. DOI: 10.1016/j.envpol.2019.01.086 PMID: 30823354

Wang, Z., Wu, T., Shi, G., Fu, X., Tian, Y., Feng, Y., Wu, X., Wu, G., Bai, Z., & Zhang, W. (2012). Potential source analysis for PM_{10} and $PM_{2.5}$ in autumn in a Northern city in China. *Aerosol and Air Quality Research, 12*(1), 39–48. DOI: 10.4209/aaqr.2011.04.0045

Wedephol, K. H. (1968). In Ahren, L. H. (Ed.), *Origin and Distribution of the Elements* (p. 99). Pergimon Press.

WHO. (World Health Organization). (2015). World health statistics 2015.World Health Organization.

World Health Organization (WHO). (2005). *Air Quality Guidelines for Europe. WHO Regional Publications, European Series No. 91*. WHO Regional Office for Europe.

Xiao, Z., Shao, L., Zhang, N., Wang, J., & Wang, J. (2013). Heavy metal compositions and bio-reactivity of airborne PM_{10} in a valley-shaped city in Northwestern China. *Aerosol and Air Quality Research, 13*(3), 116–1125. DOI: 10.4209/aaqr.2012.10.0287

Zhang, J., Cheng, M., Ji, D., Liu, Z., Hu, B., Sun, Y., & Wang, Y. S. (2016). Characterization of submicron particles during biomass burning and coal combustion periods in Beijing, China. *The Science of the Total Environment, 562*, 812–821. DOI: 10.1016/j.scitotenv.2016.04.015 PMID: 27110992

Chapter 5

Emerging Trends in Self-Healing Materials for Smart and Sustainable Applications

G. Boopathy
https://orcid.org/0000-0003-2515-7277
Vel Tech Rangarajan Dr. Sagunthala R&D Institute of Science and Technology, India

Chandra Bose G.
https://orcid.org/0009-0008-2455-6907
Vel Tech Rangarajan Dr. Sagunthala R&D Institute of Science and Technology, India

V. Srinivasan
https://orcid.org/0009-0006-2988-6476
Vel Tech Rangarajan Dr. Sagunthala R&D Institute of Science and Technology, India

C. Suresh
https://orcid.org/0000-0001-7549-3506
Global Academy of Technology, India

ABSTRACT

The chapter gives a detailed review of self-healing materials (SHMs) in terms of their mechanisms, classifications, and smart applications in various fields. It also emphasizes on the intrinsic and extrinsic healing systems which are activated by stimuli like heat, light, pH and electrical or mechanical forces. Important design

DOI: 10.4018/979-8-3373-3506-3.ch005

concepts such as dynamic bonding, environment responsiveness and biomimicry such as skin, bone and spider silk are discussed. Some of the latest developments in the multifunctional SHMs, Nano composites, environmentally friendly materials, and self-governing structures are also analyzed in the chapter. Applications are described in construction, electronics, biomedical, energy storage and environmental remediation. The problems of scalability, material fatigue, and regulatory problems are covered, as well as the future direction of the use of AI-driven design, Industry 4.0 integration, and circular economy strategies. SHMs find themselves at the centre of sustainable and intelligent material systems.

INTRODUCTION

Self-healing materials (SHMs) are termed as innovative materials which can heal themselves and the imposed damages in it without any external assistance or intervention. This also involves the ability to restore their initial properties after physical damage caused by mechanical loads, the environment and usage through time. SHMs have significance because they can extend the lifetime as well as safety of numerous products and structures considerably reducing the yearly required maintenance and replacement costs. This feature is particularly beneficial in the areas of construction, electronics and biomedical applications, where material failure can cause enormous financial loss and be hazardous to life. Even if the crack or fatigue is microscopic, catastrophic failure can be initialized and it has been especially well known in aerospace, biomedical implants and infrastructure. SHMs can resolve this problem since they offset damage and initiate the in-situ repair mechanisms to increase the safety and to decrease the maintenance cost.

Self-healing technologies are motivated by the need for more sustainable materials that will reduce the environmental impact of manufacture and disposal. However, traditional materials do not have the ability to heal themselves, which contributes towards waste, and environmental degradation. Including self-healing capabilities in materials allows us to increase their lifespans, coherent with circular economy practices targets for resource efficiency and sustainability. For instance, conventional materials can be replaced by self-healing polymers, which not only provide functional performance but also extend the composite usable life. Figure 1 illustrates a schematic diagram of the self-healing mechanism.

Figure 1. Schematic diagram of self-healing mechanism

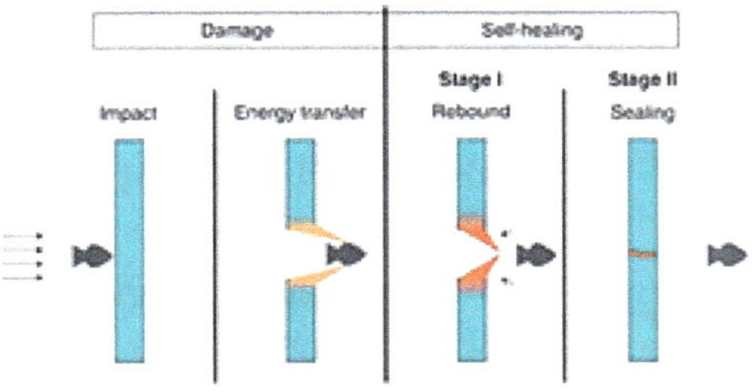

(Source: Srivastava et al., (2024) from "Self-Healing Materials: Mechanisms, Characterization, and Applications: A detailed Review (https://www.researchgate.net/publication/379330262_Self-Healing Materials: Mechanisms, Characterization, and Applications: A detailed Review)" licensed under CC BY 4.0)

In construction, and infrastructure, for example, there are self-healing concrete techniques that deal with the common problem of cracking by environmental factors such as shrinkage or freeze thaw cycling. Such technologies try to decrease maintenance costs and increase longevity of structures by letting the materials repair themselves without much time to lose after damage (Oh & Kim, 2025, Ihenketu, 2024). Smart materials with self-repair potential reduce the need for frequent intervention with resultant large savings in labour and material costs. Furthermore, self-healing concrete leads to decreased carbon footprints from traditional methods of repair so as to catch up with the world sustainable construction practices (Barros et al., 2023).

SHMs emerge as promising in electronics to improve reliability of devices, especially in the case of wearable technology and soft robotics. These materials have the capacity to recover from mechanical damage which aids product durability and performance, in line with the growing need for long lasting electronic components. For instance, soft but tough SHMs have been developed and applied in capacitive sensors for which contact failure is inevitable (Sun et al., 2023). Because of its inherent ability to recover functions lost due to damage, it helps innovation, and matches with consumer expectations for durability. In addition, self-healing capabilities are being incorporated in a variety of applications contributing to progress in smart material systems for improving performance in fields such as biomedical engineering. With self-healing polymers, big breakthroughs in tissue engineering or wearable healthcare devices can be made with a material which has to be mechanically durable, but also very comfortable, lightweight and also still highly performing (Morang et al., 2024).

This chapter aims at explaining the definition and importance of SHMs, their significance towards sustainability and their potential to propel smart material systems. Our goal is to establish a complete basis for understanding mechanistic design principles in self-healing processes as well as the possible use of SHMs in a variety of fields. This overview will set the ground for more discussions related to the fundamentals, challenges and directions of SHMs research.

FUNDAMENTALS OF SELF-HEALING MECHANISMS

To develop efficient SHMs from a fundamental point of view, one needs to understand the underlying mechanism. This section discusses the major factors that lead to self-healing behaviour such as how these materials are classified, what are the external stimuli that can trigger healing, what are the key design factors that affect healing performance, as well as the biological factors that motivated their development. This set of foundations enables development of smart and sustainable self-healing technologies in multiple industries.

Classification of Self-Healing Materials

Two main categories of SHMs can be considered, i.e., extrinsic self-healing systems and intrinsic self-healing systems with different mechanisms to heal damage.

Extrinsic Self-Healing Systems

The extrinsic SHMs utilize the healing agents which are externally introduced into the polymer matrix. Normally these agents are encapsulated within microcapsules or embedded in the materials via vascular networks. However, when the damage occurs (for instance cracking or puncturing), the capsules rupture and these healing agents are released into the affected area, and hence facilitate a chemical reaction to restore integrity of the material (Du et al., 2020).

For instance, research demonstrated the use of self-healing mechanism using hydrogen bonding and electrostatic forces in polyaniline based self-healing conductive materials which utilized such extrinsic methods of repair (Wang et al., 2023). Likewise, self-healing agents can be encapsulated and activated by capillary action in a fast response to damage (Perin et al., 2023). However, the recurrent healing cycles and poor performance of the healing agent with time in the majority of the extrinsic systems can affect negatively the overall performance of the material (Terryn et al., 2021).

Intrinsic Self-Healing Systems

In contrast, intrinsic self-healing systems utilize the material intrinsic properties for self-repair in absence of any external healing agent. Designed mainly from the materials that are based on dynamic covalent bonding or reversible non-covalent interactions (hydrogen bonds) within a polymer network. In the case of damage, these dynamic bonds are able to reform and the material can return to its original structure and properties (Zhou et al. 2022), (Khan and Rabnawaz 2021).

One advantage of the intrinsic self-healing materials is that they can heal their damage repeatedly in the same site, which cannot be done by the extrinsic systems. Researchers like Paolillo et al emphasized the possibility of autonomous and rapid repair mechanisms in intrinsic self-healing epoxies for possible aerospace applications (Paolillo et al., 2021). These materials, which are equipped with self-repair functionality activated by external stimuli, e.g., temperature or humidity, can make the self-healing process more robust, and in potentially more demanding environments (Zhou et al., 2022). Table 1 gives the comparative analysis of Intrinsic and Extrinsic self-healing systems

Table 1. Comparison of Intrinsic and Extrinsic self-healing systems

Aspect	Intrinsic Self-Healing Systems	Extrinsic Self-Healing Systems
Healing Efficiency	High repeatability and consistency due to reversible molecular interactions; limited to small-scale damage.	One-time healing using embedded agents; effective for larger cracks but less repeatable.
Cost and Complexity	Simpler fabrication; requires precise molecular design and tuning of chemical interactions.	More complex manufacturing involving microcapsules or vascular networks; higher production cost.
Application Suitability	Ideal for repeated healing applications such as soft robotics and wearable electronics.	Suitable for high-impact, structural applications like construction and aerospace where immediate healing is needed.

Heating Stimuli

SHMs rely on different stimuli to move the healing process forward automatically to correct damage. The effectiveness of self-healing mechanisms is often influenced by the type of external stimulus applied. Here we discuss five primary categories of healing stimuli: thermal, light, pH, electrical, and mechanical activation.

Thermal Activation

Thermal activation is one of the most common stimuli used to initiate self-healing in polymers. Many self-healing materials are designed to respond to changes in temperature, where increased thermal energy enhances molecular mobility, allowing for the reformation of damaged bonds (Jin et al., 2023). For instance, Huang et al. demonstrated that increasing the healing temperature of a polymer elastomer to 80°C significantly improved its healing efficiency, achieving as high as 93.6% in just one hour (Huang et al., 2024). The facilitated movement of polymer chains at elevated temperatures allows for quicker and more efficient healing processes.

In addition to external heating, certain self-healing systems operate effectively at room temperature by leveraging dynamic bonds that can reorganize quickly even at lower thermal conditions. Previous research has shown that there are range of temperatures above the glass transition temperature (Tg) of the material where enhanced chain mobility allows for effective self-healing (Zhou et al., 2024).

Light Activation

A new effective method of initiating self-healing is light activation. An example is in the polymer which can be healed by specific wavelengths of light which initiate the photochemical reaction in the material. As an example, in a study on nanocomposite materials, it was identified that the light triggered self-healing mechanism showed substantial recovery in relatively less time compared to traditional heating techniques (Motora et al., 2021). This method is compatible and can be applied in areas where heating is not permitted. The spectral properties of the materials are optimised, designed systems can only react to a specific type of light, which can increase the robustness of self-healing mechanisms (Peng et al., 2023).

pH Activation

Organic pH –responsive self-healing materials are engineered as to respond to changes in acidity or alkalinity. These materials may contain pH sensitive crosslinkers to allow it to be in a chemical environment that aids healing (Ahn et al., 2022). Such systems can also include healing agents which become active only when the pH of the surrounding environment falls in a certain range and thus provide self-healing effectively (Liu et al., 2024). Such a healing mechanism is particularly beneficial for biomedical purposes because the pH changes due to physiological conditions enables timely repair and thereby maintenance of device functionality in the body (Datta et al., 2025).

Electrical Activation

Electrical activation of self-healing materials is very important when conductive polymers and devices are employed in the research. Reformation of polymer structures can be induced by the application of an electrical field and a healing can result through ionic movement or Joule heating (Sima et al., 2021). Microcapsules are embedded in materials in electrical breakdown scenarios such as stress propagation to the area of damage through using healing agents, taking advantage of the electrical stress to drive the healing agents to the area of damage (Sima et al., 2021). Most importantly, smart devices may have interactive functionality through electrical stimuli, and fast responses to incipient damage are very important to the durability and reliability of electronics (Utrera-Barrios et al., 2020).

Mechanical Activation

Mechanical activation allows healing to be stimulated by means of applied mechanical stress. It harnesses physical deformation or changes in stress as it promotes the movement and rearrangement of polymer chains at the damage site (Liu et al. 2021). Recent developments indicate that the employed mechanical fields can activate critical molecular interactions required for healing, thus enabling self-healing in performance driven applications (Wen et al., 2023). Fracturing of microcapsules containing healing agents upon mechanical activation are capable of allowing localized repair in the presence of mechanical disruptions and maintaining functionality at the same time in certain systems (Gao et al., 2020). As a result, this presents a dynamic healing process that can be incorporated to overall material performance and, in general, load bearing applications.

Key Design Principles for Self-Healing Efficiency

Specific design principles are selected for SHMs to make them effective in recovering after damage. The efficiency of these materials for achieving healing is determined by several important factors, and these can determine the applicability and value of these materials for practical, long-term use.

Dynamic Bonding Mechanisms

Dynamic bonding mechanisms are incorporated as one of the fundamental principles of designing SHMs. Among these are reversible covalent bonds, hydrogen bonds and ionic interactions, allowing the material to be able to recover from damage. Such dynamic covalent bonds can rearrange and reattach in response to an external stimulus, thereby the material can heal autonomously (Sun et al., 2023). For instance, materials relying on Diels-Alder reaction have efficacious self-healing properties upon compression, or thermal stimuli that enables swift recovery of mechanical properties (Motora et al., 2021).

Material Architecture

SHMs should have the right structural design and composition in order to facilitate effective healing. Efficient delivery of healing agents when damage happens can be achieved in architectures including microcapsules or vascular networks. It has been shown in research that using vascular networks in soft materials may enable higher self-healing efficiency, i.e., the high recovery of mechanical properties (Sun et al., 2023). It was notable that these capsules are spatially arranged and distributed with a different size which has a great impact on its healing performance, that is, how well it can respond to the damage (Liu et al., 2024).

Healing Efficiency Metrics

The quantification of healing efficiency is considered as a key for evaluating the performance of these SHMs. The healing efficiency (η) is defined as the ratio of recovery of mechanical properties (e.g., stress, strain, etc) to their original state (Ahmed et al., 2023). Researchers can therefore better understand the capabilities of the material and compare different self-healing systems quantitatively based on this metric. However, healing efficiencies above 98% have been demonstrated in tensile tests in multiple studies in which innovative designs are applied (Sun et al., 2023).

Environmental Responsiveness

SHMs are very efficient because they have the ability to react to the environmental stimuli. Improving healing rates and making them more robust in varying situations, materials can be engineered to respond to changes in temperature, pH, moisture or light. For instance, moisture was found to promote chain mobility in some hydrophilic self-healing polymers and aid healing process (Davydovich & Urban, 2020). Drawing on the response to physical conditions, designers can generate the systems that adaptively preserve their integrity in the presence of loads during the course of their service life.

Integration of Multi-Functionality

The integration of self-healing capabilities with other functional properties improves the versatility of materials. For applications in flexible electronics, soft robotics, and biomedical devices, smart materials can be engineered to simultaneously show self-healing abilities, conductivity, and good mechanical robustness. For example, SHMs combined together with electroconductive properties ensure autonomous repair with preservation of operational characteristics required for wearable sensors (Terryn et al., 2021).

Biomimetic Design Inspiration

A powerful approach to the design of SHMs is inspired by biological systems. Nature has provided us with self-repairing materials, and these give us hints as to how best to achieve restoration of capability following damage. One example is emulating complex dynamic bonding behaviours of biological tissues such as skin or muscle and, thus, creating a foundation to develop new SHMs (Wang et al., 2023). Through biomimetic approach, the designers can employ natural repair mechanisms that have evolved to ensure optimality in terms of healing.

Biomimetic Inspirations in Self-Healing Materials

Biomimicry is the process of seeking inspiration in nature to address human issues has also played its part in the development of SHM. New materials are being used as attempts to emulate the efficiency and effectiveness of the natural healing processes including skin healing and bone regeneration. This section discusses those biomimetic inspirations which are central in driving the design and functionality of self-healing systems.

Skin Healing

The human skin is a complicated system and provides a great model for SHMs because it is able to repair itself after it is hurt. For example, from the inspiration of skin regenerative processes, dynamic imine linkages strained chitosan-based hydrogels with self-healing property has been developed. The potential for chitosan hydrogels in biomedical applications, particularly on wound healing, which create an environment suitable for cell recover and exhibits antimicrobial property is emphasized by (Liu et al., 2022). Hydrogels exhibiting natural processes of dermal healing and efficient mechanical properties for the tissue regeneration are presented. The skin healing inspired self-healing system comes from that of dynamic bonds that are used for quick reestablishment of the material integrity after cuts or abrasions. With the elasticity and ability to recover functionality, properties of elasticity such as human skin, the applications in soft robotics and wearable technology (Sayadi et al., 2024).

Bone Regeneration

The regeneration in bone tissue is another major biomimetic inspiration. Fractures in the bones are healed naturally over time as some cellular pathways that are activated and the remodelling of bone tissues takes place. Some materials which are designed with intrinsic self-healing capabilities are inspired by these biological mechanisms and utilise dynamic physical interactions to emulate the regenerative responses observed in skeletal systems. Figure 2 highlights the role of biomolecules in the effective bone regeneration system.

Figure 2. Illustration depicting the biomolecules used to support bone tissue regeneration and strategies for immobilizing bioactive compounds on biomaterials, with a focus on the coupling of bioactive molecules to scaffold materials for controlled release and effective bone regeneration.

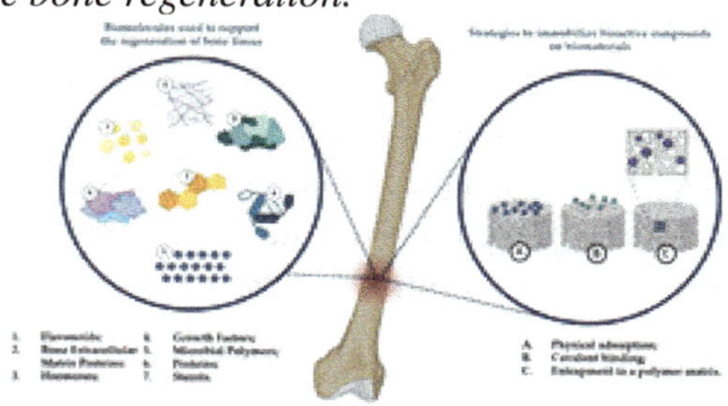

(Source: Szwed-Georgiou et al., (2023) from "Bioactive Materials for Bone Regeneration: Biomolecules and Delivery Systems (https://pubs.acs.org/doi/10.1021/acsbiomaterials.3c00609)" licensed under CC BY 4.0)

For example, self-healing hydrogels incorporating elements of the fibrous structure in bone, have desirable applications such as tissue engineering. Since polymers can respond to mechanical and chemical stimulus, researchers can use these polymers to develop materials capable of acting as supporting bone functions to promote successful self-repair (Ding et al., 2020). Regenerative capacities can be supported by engineered hydrogels which is important in the context of treating bone injuries (Ding et al., 2020).

Spider Silk and Other Biomolecules

Another good inspiration to take from the SHMs is the spider silk, which shows remarkable performance in strength and elasticity, is self-repairable and has superb capability of healing itself upon damage. The ability to develop self-healing coatings based on recombinant spider silk, specifically on the dynamic supramolecular interactions that boost its various performance under such situations (Wu et al.,

2023). Fibers with similar mechanical properties to spider silk containing self-repair functionalities are synthesized using similar principles. This is a biomimetic approach that takes advantages of the nature of the molecular arrangement in spider silk in order to improve self-healing efficacy specifically for higher stress applications such as fiber or composite use in the textile and packaging industries (Zhu et al., 2022). Figure 3 gives clarity about the method fabrication and self-healing behaviour of recombinant spider silk hydrogel.

Figure 3. Schematic illustration of the fabrication and self-healing behaviour of recombinant spider silk hydrogel. Recombinant spider silk proteins, produced via genetically engineered E. coli, self-assemble into physically crosslinked hydrogels. These hydrogels demonstrate autonomous self-healing behaviour, as confirmed by macroscopic reconnection (left) and in situ small-angle X-ray scattering (SAXS) analysis (center), which reveals reversible β-sheet nanocrystal stick-slip dynamics (right) as the underlying healing mechanism.

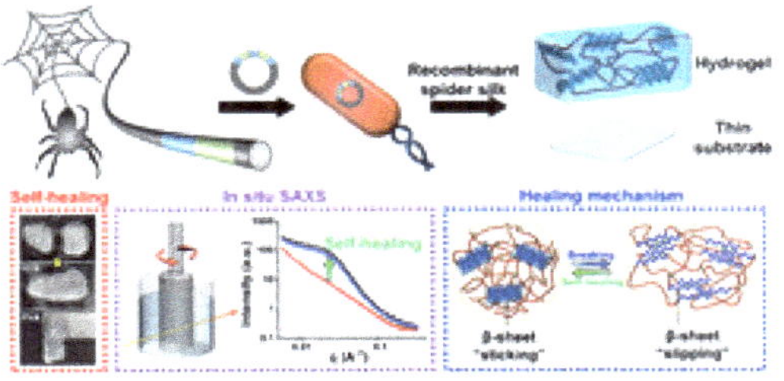

(Source: Wu et al., (2023) from "Self-Healing of Recombinant Spider Silk Gel and Coating (https://www.mdpi .com/2073-4360/15/8/1855#)" licensed under CC BY 4.0)

Adaptive Self-Healing Mechanisms

Adaptive materials, which dynamically respond to changes in the environment, is based heavily on natural systems. Self-healing is done on inspiration of how biological tissues adapt themselves to changes in their environment (like temperature, humidity etc.) so as to help them heal itself better when the external stimuli are applied. For example, hydrogels that utilize pH sensitive linkers have shown the capacity to mimic behaviour of living tissue and alter their properties actively coming with local environment conditions to facilitate proper and timely healing (Yuyu et al. 2022).

RECENT ADVANCES AND EMERGING TRENDS

The SHMs field is growing rapidly and enveloping notions derived from advances that include multifunctionality, which incorporates healing functionality and other properties including conductivity, sensing and barrier protection. This section looks at these recent trends and discusses the development of self-healing nanocomposites and eco-friendly materials with the basic concept of autonomous self-healing systems.

Multi-Functional Self-Healing Systems

The scope of SHMs applications has in recent years gone well beyond basic mechanical repair and encompasses several functional properties. These multifunctional systems have been developed in order to fulfil the needs of the contemporary technologies that demand the material to accomplish a number of operations simultaneously.

Healing + Conductivity

In the recent past, there have been several studies towards development of electrical conductivity using SHMs. A method of curing in polymer composites was also developed, which used Diels-Alder reactions and conductive nanofillers carbon nanotubes (CNTs) and silver nanoparticles (Ag) (Ding et al., 2020). These composites according to their findings showed a great enhancement of the mechanical strength that retained the electrical conductivity that would be important to apply in electronic gadgets through their self-healing ability. Besides that, Wang et al. produced mussel inspired adhesive hydrogels with good conductivity and self-healing properties (Wang et al., 2021). The materials have the property of adhering to biological tissue, and also of remaining electrically connected, which allows the design of bioelectronics and medical devices with simultaneous healing and functional properties.

Healing + Sensing

Another exciting area of development is for materials that can combine the self-healing function with sensing capabilities. For example, Trost et al. reported on enthalpy driven self-healing in thin metallic films with substantial healing properties while being capable of retaining the sensing function (Trost et al, 2024). They self-repair and are able to detect strain or mechanical deformation, and so these are suitable to be used in applications ranging from structural health monitoring to wearable technology. Moreover, smart polymeric coatings were developed for self-healing as well as real time damage detection (Ismail et al., 2021). These coatings will autonomically respond to environmental conditions and give feedback as to their structural integrity opening the way to smart materials to enhance safety in critical applications.

Healing + Barrier Properties

The integration of self-healing functions with higher barrier properties is of great help in protective coatings. Recent research focussed on coatings that heal and also offer strong corrosion resistance through introducing pH responsive material (Wang et al., 2024). These coatings, when composed of blends of composites and healing agents, showed very good combined mechanical and barrier properties suitable for marine environments where corrosion presents a major problem and spectacular self-healing ability. Additionally, the development of biopolymer coatings made of vegetable oils which contain microcapsules for self-healing showed enhanced barter protection as well as healing properties (Song et al., 2022). These techniques show through multifunctional coatings how coatings design can be an effective way to preserve material integrity against environmental degradation.

Self-Healing Nanocomposites

Self-healing nanocomposites are an important development in materials science since they add nanostructures that help materials heal and make them stronger. Incorporating nanomaterials in self-healing materials gives rise to systems that are able to sense and recover from harm much more efficiently.

Enhanced Healing Mechanisms through Nanostructures

Due to the presence of nanomaterials, self-healing composites can recover easily from cracks because the material moves and heals properly. Research highlighted that through the creation of in situ structures from several polymers in a nanocomposite, materials for self-healing are developed that are also very strong (Lee et al., 2020). These nanostructures working together build an advantageous structure that helps the muscles recover better from mechanical stress. Besides, when silica nanoparticles are added to polymers, it increases the strength and other physical traits of SHMs. When nano-silica is added to polyurethane, it greatly improves strength and durability, so it can be used where toughness and fast recovery are necessary (Zhao et al., 2025). The nanoparticles make the material tougher and allow the material to self-repair.

Conductive Nanocomposites with Self-Healing Properties

Blending self-healing properties and the ability to conduct electricity using special nanomaterials has attracted attention lately. It was reported by Han et al. that they created a dual conductive network hydrogel that can heal itself and still remain electrically conductive (Han et al., 2020). With its two functions, the material repairs its own damages and maintains its electronics efficient. Similarly, Wang et al. prepared materials that are able to heal cracks in functionality and physical property, primarily through polyaniline hydrogen bonding and electrostatic forces. Repairs will be possible with such a system, without interfering with the electrical operation, as consistent performance is valued along with other difficulties or potential damages (Wang et al., 2023).

The Role of Phase Change Materials in Healing

Self-healing systems are currently utilizing phase change materials (PCMs) to exploit their physical state transitions in order to repair accidently damaged materials. The team created a coating capable of self-healing itself through charge transfer complexes that increased the toughness of the material and acted as a PCM to enable self-healing (Ahn et al., 2022). In this new method, these materials can recover themselves through the help of thermal changes to enhance self-healing.

Bio-Based and Eco-Friendly Self-Healing Materials

The most recent advancements in self-healing materials have been in the field of bio-based and environment-friendly materials because of the increasing necessity to utilize sustainable materials. The materials have been designed with the least environmental risk and yet be functional in use. In this section, the aspect of applying sustainable polymers and green chemistry strategies will be addressed concerning self-healing technologies.

Sustainable Polymers

Developments of bio-based polymers have been vying as alternatives to the traditional petroleum-based products. Thermoplastic (co)polyamide elastomers were prepared by the utilization of fatty dimer acids as biomass. The mechanical properties and self-healing of these bio-based elastomers were reasonable, demonstrating that it is a sustainable route to producing uniform and bimodal foams (Ranganathan et al., 2021). Such advances demonstrate that it is feasible to create sustainable polymers to exemplify a viable option to traditional materials that enables the utilization of environmentally-friendly materials that do not influence performance.

Green Chemistry Approaches

The concept of green chemistry should be incorporated in the synthesis of the self-healing materials in order to minimize the toxicity of the by-products and reduce the environmental footprints. Wu et al. reported the synthesis of a self-healing and recyclable polyurethane-urea Diels-Alder adduct that was built via the usage of carbon dioxide and furfuryl amine, proving that renewable resources can be utilized in the creation of self-healing polymers (Wu et al., 2021). The process is also sustainable since CO_2 is a raw material in the polymer manufacturing; hence, there is no comptonization of performance properties. Moreover, the dynamic-covalent bonding networks application presented by Wan et al. offers routes to developing self-healing materials that are both practical and environmentally friendly. Mechanical property and healing effectiveness are further improved by the addition of crystallization nanostructure to bio-based self-healing polymers (Wan et al., 2020).

Autonomous and Multi-Responsive Self-Healing Systems

Current progress in self-healing materials has seen the introduction of autonomous and multi-responsive systems that are capable of more than just self-repair, they can also react dynamically to environmental stimuli. These advancements correspond to the incorporation of different functionalities in self-healing materials that broaden their utilization prospects in different disciplines.

Autonomous Self-Healing Materials

Autonomous Self-healing systems are devised in such a way that they start repair mechanisms without any stimulus. Research shows that an iontronic skin that is ultrafast and self-healable, based on a Cl- functionalized polyurethane matrix with dynamic disulfide bonds. Such a design provides the material with the ability to heal itself at room temperature, making it functional even after being damaged (Boahen et al., 2022). This self-healing ability is ultra-convenient in areas where wearable electronics are concerned, as the ability to independently repair mechanical damage greatly increases durability and utility.

Furthermore, an autonomous self-healing optical sensors that could sense damage in soft-body systems. These sensors comprise material chemistries, which offer inherent healing effects, thus enhancing functionality and recovery in changing environments (Bai et al., 2022). This shows that it is possible to design materials that not only can self-heal but also can preserve their sensing functionality after they have been damaged, opening the door to more intelligent, adapting materials. Figure 4 visualise the ion dynamics of the human skin based on the trap and release mechanism.

Figure 4. CLiPS-based device architecture illustrating autonomous self-healing and tractive self-healing of electrodes, as well as the molecular structure emulating the ion dynamics of the human skin based on the trap and release mechanism.

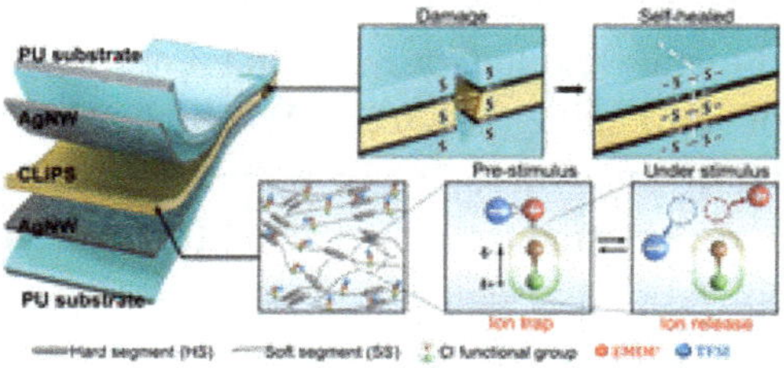

(Source: Boahen et al., (2022) from "Ultrafast, autonomous self-healable iontronic skin exhibiting piezo-ionic dynamics (https://www.researchsquare.com/article/rs-2135447/v1)" licensed under CC BY 4.0)

Multi-Responsive Systems

Multi-responsive self-healing materials are designed in a way that they respond to different external stimuli which include light, temperature, pH, or even mechanical stresses. A type of dual-network hydrogels was created that apply host guest interactions with dynamic covalent bonds, which allows them to be used as wearable strain sensors, but also possess important self-healing capabilities (Liu et al., 2021). The capability of the hydrogels to recover their structure after responding to the mechanical deformation makes them suitable in the areas where they are likely to be exposed to a dynamic environmental interaction.

A self-healing hydrogel is introduced which includes platelet-rich plasma (PRP) and showed that self-healing of hydrogels is possible upon rupture without losing bioactivity. Such an ability improves the mechanisms of wound healing and underlines the necessity to introduce biologically active constituents into smart materials (Li et al., 2022). Such hydrogels are responsive as to guarantee not only their spontaneous healing but also to promote biological regeneration processes.

Self-Healing Hydrogels, Elastomers, and Smart Coatings

Self-healing technologies have been incredibly promising when applied to hydrogel, elastomers, and smart coatings. Self-healing hydrogels are especially outstanding since they can create moisturized environments that promote the healing processes. Huang et al. created a hydrogel with high self-healing ability but that is also biocompatible and can therefore be used in wound healing and tissue engineering (Huang et al., 2021). These hydrogels have dynamic reversible bonds, which ensure quick recovery following any damage, a factor that highlights their applicability in regenerative medicine.

Self-healing elastomers have also been considered, especially in flexy applications and where tough mechanical performance is needed. The architecture of elastomers with dynamic bonds can be stretched highly and can self-heal at ambient conditions (Liu et al., 2021). It is the combination of chemical and physical property that makes these materials applicable to many different fields, including automotive or soft robotics. Self-healing in smart coating plays an important role in protecting the surface against environmental attacks. Such coatings cannot only protect but also heal themselves, which reduces the maintenance work and extends the service life of materials. As an example, renewable resource-based self-healing coatings, including alginate and gelatin, have proven to be efficient surface protectants with favourable mechanical performances (Cadamuro et al., 2023).

APPLICATIONS IN SMART AND SUSTAINABLE FIELDS

Self-healing technologies are making a difference in infrastructure and construction applications, demonstrating an increase in durability, a drop in maintenance expenses, and helping to attain sustainability in building operations. Three main areas that are expounded in this section are; self-healing concrete and composites, smart coatings to resistance corrosion, and ways of extending the service life of structures.

Self-Healing Concrete and Composites

Self-healing concrete is concrete that uses biological or chemical actions to mend cracks itself, lowering infrastructure maintenance expenses and increasing service life. The concept of self-healing concrete, which introduces the use of bacteria to encourage the formation of limestone in the areas of the crack, thereby sealing it and making it more resistant to environment-related damage (Ihenketu, 2024). This novel solution can not only lessen the level of dilapidation but also reduce the necessity of major repair works, which can lessen the carbon footprint of the construction works of the area considerably. Figure 5 depicts a step-by-step pictorial explanation of the self-healing concrete block operations.

Figure 5. Self-healing concrete blocks process, (a) Sodium silicate powder converted into small balls, (b) Production of encapsulated self-healing agent, (c) Self-healing concrete cubes produced, (d) Cracks induced on the self-healing concrete, (e) Self-healing concrete with cracks and (f) Cracks healing after 25 days.

(Source: Ihenketu (2024) from "Reduction in Dilapidation and High Cost of Maintenance on Buildings using Self-Healing Concrete for Building Construction

Furthermore, machine learning algorithms to predict the healing performance of self-healing concrete based on various input parameters, although the primary focus of their work is on post-fire recovery rather than initial healing predictions (Rajczakowska et al., 2023). Therefore, this reference does not fully support the claim about optimizing concrete compositions for self-healing capabilities.

Smart Coatings for Corrosion Resistance

Smart coatings that exhibit self-healing properties play a vital role in protecting concrete structures from corrosion, which is a major cause of deterioration. Martínez-Ibernón et al. propose a smart cement-based material designed to enhance the service life of reinforced concrete structures by mitigating corrosion effects (Martínez-Ibernón et al., 2023). These materials can autonomously repair damage to the surface, maintaining the integrity of the underlying structure while reducing maintenance frequency.

Recent advancements also include the development of eco-friendly self-healing coatings that employ biopolymer technology. Zhou et al. discuss the integration of biological principles into material development, though their focus is primarily on sulfate attack predictions rather than on coatings (Zhou et al., 2022). Consequently, this reference does not directly support the preceding claim about biopolymer-based self-healing coatings designed for environmental degradation response.

Prolonging Service Life of Structures

The implementation of self-healing technologies can substantially extend the service life of concrete structures, making them more cost-effective and sustainable. Jha et al. highlight that incorporating fibers and bacterial elements enhances the self-healing capabilities of concrete, allowing it to repair itself significantly and resist stresses that would normally lead to failure (Jha et al., 2023). This indicates that specific additives can improve concrete performance, ensuring minimal structural damages over time, thus leading to lower lifecycle costs and resource consumption.

In addition, self-healing concrete has the capability of limiting the maintenance works and expenses needed because of infrastructural deterioration, as it will automatically mend small surface cracks, leading to a more prolonged lifespan of infrastructure with a reduced impact on the environment (Ihenketu, 2024). Through self-healing techniques, it is within the power of the construction sector to not only realize a greater cost efficiency but also to meet the sustainability objectives on a larger scale to suit the demand of resilient and environment-friendly materials in contemporary construction, which is urgent.

Environmental Protection and Remediation

Self-Healing Membranes for Water Purification

In water purification processes, self-healing membranes play a critical role in ensuring the structure and effectiveness of the filtration systems. The latest developments have demonstrated that such membranes are capable of repairing themselves in case of damage automatically, resulting in a significant extension of performance and minimal maintenance downtimes. Self-healing hydrophilic porous photothermal membranes, which were capable of recovering rapidly due to mechanical damages with the assistance of photothermal effects. This is improvement through the addition of carbon black (CB) that enables photothermal conversion efficiency enabling autonomous repair when evaporating water under the influence of the sun (Xu et al., 2022). This demonstrates the huge potential in responsive material adoption in water treatment technologies to enhance performance and sustainability. Moreover, self-healing behaviour of embedded membranes can be optimised by intervening with surfactants in the microcapsule synthesis. Having the possibility to regulate the micromorphology of the membranes, one can enhance the water flux and self-healing that will inevitably lead to the improved outcomes of the water purification (Lee et al., 2018). Figure 6 highlights various schematics about solar-driven interfacial water evaporation.

Figure 6. (a) Schemes of (1) the composition of a typical solar-driven interfacial evaporator and (2) the loss of its water evaporation capacity due to damage to the water transportation channel. (b) Schematics of (1) the composition of the SHPP membrane-based solar-driven interfacial evaporator and (2) its ability to restore the water evaporation ability via self-healing after physical damage.

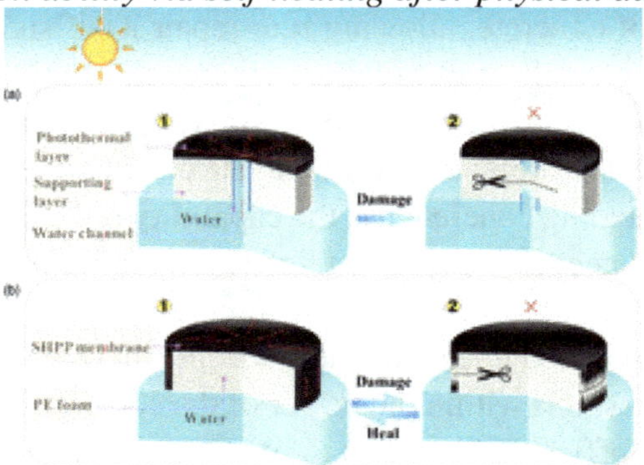

(Source: Xu et al., (2022) from "Self-Healing Hydrophilic Porous Photothermal Membranes for Durable and Highly Efficient Solar-Driven Interfacial Water Evaporation. (https://www.chinesechemsoc.org/doi/10.31635/ccschem.021.202101111)" licensed under CC BY 4.0)

Sorbents and Catalysts for Pollution Control and Carbon Capture

Self-healing materials are also relevant in the preparation of sorbents and catalysts which could be utilized in environmental remediation specially to address the problem of heavy metals and volatile organic compounds contamination. A re-used functionalized material with the ability to sense premium and display impressive sorption capacity to hexavalent chromium (Cr(VI)) and permanganate ions, which underlines the relevancy of the material in water treatment technologies (Yadav et al., 2021). Such sorbents should be imposed self-healing properties since they will extend their working time and, consequently, clean up more contaminated water resources. The templated inorganic sorbents that have high adsorption capacity in the removal of heavy metals that are highly vital in the environmental protection procedures were synthesized. The second approach which is employed to improve the behaviour of these materials in the destruction of the contaminants and which result into sustainability is the photo catalytic functionality which is incorporated to the materials.

Energy and Storage Systems

A promising, yet emerging field, is the incorporation of self-healing technologies into energy and storage systems to improve functionality and lifetime of batteries, supercapacitors and flexible electronics. This section discusses the progress Self-healing batteries and supercapacitors, flexible electronics, and wearable devices materials.

Self-Healing Batteries and Supercapacitors

Self-healing batteries and supercapacitors play an important role in the lifetime and dependability of energy storage gadgets. Recently, it was shown that self-healing of battery materials can bring about a vast enhancement in safety and lifetime. As an example, the development of a self-healable polymer electrolyte enabling lithium-ion batteries to self-heal micro-damage and improve safety by preventing failure caused by short circuits (Wang et al., 2020). The feature minimizes the risk involved when using Li-ion batteries, thus making them safe to use widely.

One more interesting development is about a smart supercapacitor that in addition to its main task of storing energy, is also able to visually represent the amount of stored energy by providing visual signals. Self-healing capabilities incorporated into those devices mark the beginning of a new trend, according to which energy storage systems can be not only functional but also have an increased lifetime (Tian et al., 2014). These intelligent supercapacitors mark a step to the multifunctional energy storage devices. The prospect of flexible supercapacitors involving porous carbon nanotube film-polyaniline hydrogel composites. Such devices have a high flexibility and self-healing ability that is of great importance when it comes to portable and wearable electronics (Xi et al., 2016). This Self-healing property makes these materials more viable to real-life applications.

Materials for Flexible Electronics and Wearable Devices

Flexible electronics market is growing at a very high rate due to the demand of portable, stretchable, and light weight electronics. In the realization of these advanced technologies SHMs are important. Researchers discussed the importance of flexible materials based on cellulose, which are sustainable and functional toward the new intelligent electronics (Zhao et al., 2020). This is because the biodegradability of these materials matches environment-friendly trends, not mentioning that they exhibit good mechanical properties that are required in wearable devices. In addition to that, recent progress in sustainable macromolecular materials and how these materials can be used in flexible electronics. The paper highlights the fundamental role of sustainability in the design of future electronic devices that will be less toxic to the environment, and it will be necessary to focus on the utilization of environmentally friendly, renewable materials (Kılıçarslan et al., 2022).

Smart Sensors and Monitoring Devices

The ability of self-healing technologies to be incorporated into smart sensors and monitoring devices shows potential in improving the longevity and operability of the systems utilized in structural health, environmental monitoring, and wearable devices. The section is devoted to self-healing sensors in structural health monitoring and the integration of self-healing sensors into the Internet of Things (IoT) concepts.

Self-Healing Sensors for Structural Health Monitoring

Self-healing sensors used in structural health monitoring enable the real-time evaluation and preservation of infrastructure, making it safe and prolonged. The supramolecular self-healing sensor fiber composites are capable of detecting damage, at the same time as having self-healing abilities (Georgopoulou et al., 2021). Such sensors are also capable of detecting stress-strain responses and can give real-time information of the structural health of buildings and bridges so that corrective actions can be taken before major collapses can take place.

The importance of self-healing materials in the development of soft robotic applications which demand sensitivity, flexibility and at the same time possess strong healing capabilities. Though this particular study was not referred to the same ideas have been circulating in the literature. Self-healing features incorporated in these sensors also mean that they are capable of healing in the event of deformation or damage; a feature that is crucial in applications where constant wear and tear are likely to affect the performance.

Integration in IoT and Environmental Monitoring Systems

The IoT has transformed the way we gather and track information by using smart gadgets and sensors. The IoT systems can be improved by introducing self-healing sensors, which makes them more reliable and active, and there is always a chance to monitor the environment. The idea of IoT as a service (ITaaS), which is useful to control sensor data utilized in various appliances, including environmental observation and structural assessment (Petrakis et al., 2018). By incorporating self-healing mechanisms in their designs these smart sensors can be made to operate at all times and with minimal or no maintenance, and ensure data integrity. The utility of smart sensor networks was proven by researchers in the context of monitoring the conditions of food grain storage (Mishra et al., 2018).

The sensors on such networks have self-healing properties that allow the networks to be under monitoring at all times and provide a certain degree of flexibility in case of damage resulting in an increase in the level of reliability of the agricultural management systems. It proposed a self-powered wireless sensor node, which can be applied in smart cities in those cases when the surrounding environment should be monitored continuously (Ayala-Ruiz et al., 2019). This will ensure the maximization of the up time and effectiveness of the sensors in detecting the environmental parameters such as the quality of air and temperature by ensuring that the sensors are able to recover themselves in the event that they are destroyed.

CHALLENGES AND LIMITATIONS

Despite all these developments by the SHMs, certain challenges and limitations have existed that have barred their widespread applications. The possible obstacles such as effectiveness of healing under real life circumstances, fatigue and restartability of material, scalability and cost as a limiting factor, environmental and toxicological consideration, regulatory and standardization hurdle are discussed.

Healing Efficiency under Real-World Conditions

Maximum performance in terms of self-healing materials could not be easily attained particularly in real life situations, which are normally highly variant to the laboratory conditions. Temperature changes, humidity, and mechanical loading are the conditions that can modify the performance of self-healing polymers in the healing process (Li et al., 2024). To address these variations, the inherent properties and curing mechanisms must be altered to address these variations with a view of ensuring plausible performance in actual use conditions. Moreover, self-healing ability usually decreases over time due to material fatigue or environment degradation, so it becomes even harder to exploit these technologies in practice (Frei et al., 2013). Continuous exposure to stress may also degrade the self-healing capability of materials hence the continuous need to carry out research on how to be more resilient to the broad spectrum of operational stresses.

Material Fatigue and Repeatability of Healing

One of the major issues with self-healing systems is material fatigue, especially with respect to repeatability of healing events. Though most SHMs are able to exhibit remarkable healing powers once they have been damaged, the performance of the subsequent healing processes tends to become less effective (Zhang et al., 2025). This decrease in the healing efficiency is explained by the accruing damage and the modification of the material structure following the repeated healing cycles.

To become a truly viable concept, self-healing systems should be capable of assuring a consistent performance in the process of healing over a certain number of repair events. The inability to resolve these fatigue problems may lead to the inhibition of the use of SHMs in any application where durability and long lifespan are important (Hao et al., 2023).

Scalability and Cost Constraints

Scaling up of SHMs to commercial manufacturing is also a point of concern. The numerous existing manufacturing paths to make SHMs, especially those that use complicated microencapsulation procedures or sophisticated polymer chemistry, might be resource-profound and expensive (Zhou et al., 2022). In addition to innovative engineering solutions, scaling of these technologies also needs economic possibilities, allowing to maintain mass production without deteriorating performance. Therefore, identifying low-cost paths towards the production of SHMs is important to the commercial prospects of the materials.

Environmental and Toxicological Concerns

As the SHMs are developed further, the environmental and toxicological concerns must also be taken into account. Self-healing systems most often are associated with using chemical agents that may raise questions of toxicity and potential harm to the environment. The research states that the production of SHMs using non-toxic materials is an utter necessity, particularly in the case of sensitive applications such as biomedical devices (Ekeocha et al., 2021). There is a greater demand concerning the environmentally benign materials which do not rescue the safety or performance. Besides, the importance of considering the whole lifecycle of self-healing materials, as well as their degradation products and impact on the environment once they are used, could scarcely be underestimated. Adequate safety testing and ecological testing are needed to ensure that these materials can meet more and more ecological standards (Belie et al., 2018).

Regulatory and Standardization Barriers

The issues related to standardization such as regulatory frameworks are likely to delay the procedure of bringing new self-healing materials to the market. Self-healing technologies must comply with the existing regulation which can vary considerably by region, and even by industry. To realize the safety, reliability and performance of SHMs, extensive standards are to be developed (Dhirani et al., 2023). There is also no availability of standard testing procedures thus making it hard to prove such technologies in commercial practices. The industry must rise above such hurdles via collaboration and establish the rubric whereby responsible adaptation and application of self-healing materials might be adopted in a number of fields (Oh & Kim, 2025).

FUTURE RESEARCH DIRECTIONS

The invention of SHMs has created a new horizon in material science in maximizing functionality in different application areas. In this case, we discuss some of the likely research avenues depicted through figure 7 that would make a significant impact: using AI and machine learning, developing self-adaptive healing systems, incorporating ideas of the circular economy, aiming at zero-impact materials, and using Industry 4.0 and digital twins in self-healing material systems.

Figure 7. Advancements in self-healing materials

AI and Machine Learning for Material Discovery

Artificial intelligence (AI) and Machine Learning (ML) have enormous prospects in the discovery and optimization of self-healing materials. The revolutionary nature of AI in materials science and its ability to automate research processes and examine large volumes of data to identify new self-healing materials (Sadiku et al., 2021). With the help of ML algorithms to calculate the behaviour of the materials and optimize the compositions, the researchers will be able to save a lot of time and resources on the experimental proofs. Additionally, ML can be used to predict the characteristics of new materials that could possess self-healing properties so that the process of discovery and development of new materials in material science could be accelerated (Schmidt et al., 2019). With these developments, AI will be able to assist in the engineering of self-healing systems that besides being effective, would be efficient as well.

Development of Self-Adaptive Healing Systems

The study of self-adaptive healing systems is looking into developing materials that are capable of responding intelligently to environment and healing requirements. These systems are able to change the healing process depending on the severity or the type of damage, making it a more customized reaction (Zhou et al., 2022). As an example, incorporation of biosensors capable of measuring structural integrity in real-time into a self-healing system would enable automated healing protocols to only become active when truly needed, efficient and resource conserving. Self-adaptive systems have the prospect to transform use in infrastructure where the environmental settings, and forms of damage may be diverse. Through the use of dynamic captions or environmental sensors these systems could be made self-supporting in place thus greatly eliminating the necessity of highly scheduled maintenance procedures.

Circular Economy Integration

Integrating self-healing materials within the framework of a circular economy is crucial for promoting sustainability. Materials designed for recyclability alongside self-healing capabilities can minimize waste and enhance resource efficiency. The research may further explore bio-based materials that offer self-healing functions while being biodegradable, aligning with the principles of reducing environmental impact and maximizing the lifecycle of materials. Such innovations could significantly alter how industries approach product end-of-life strategies.

Towards Zero-Impact Self-Healing Materials

Aiming at zero-impact SHMs, the concept is to create entirely benign self-healing materials that have no negative impact on the environment when produced, applied, or discarded. Researchers pronounced the necessity of designing materials that have low ecological footprints and incorporating lifecycle assessments during the research and development phases (Ekeocha et al., 2021). The materials shall be made in a manner such that the by-products of self-healing reactions are not toxic, biodegradable, and do not have adverse environmental impacts. By focusing on safety and environmental responsibility of the innovations, researchers can drive self-healing materials towards zero net impact, which would greatly increase the market potential and sustainability.

Industry 4.0 and Digital Twins in Self-Healing Material Systems

The Industry 4.0 and self-healing materials are two concepts that, when combined, will open a new chapter in digital manufacturing and intelligent materials. Studies stress the initiative of digital twins, digital replicas of physical entities, and the ways of their integration to improve monitoring and operational performance of self-healing systems (Shields et al., 2021). Such digital assets are able to model how self-healing materials would behave under different conditions and can estimate at which point and in which circumstances healing will be required. Predictive modelling and real-time monitoring could be used to optimise material design and result in smarter manufacturing processes. With digital twins, engineers have the ability to examine and optimize self-healing abilities of materials prior to their actual production, closing the feedback loop in between product development and performance in an automated manner (Boopathy et al., 2025) (Shields et al., 2021). Figure 8 depicts the role of digital twins in cardiac hemodynamics.

Figure 8. Digital twins for cardiac hemodynamics. Left: Illustration of the cardiovascular system. Right: Digital twin models of cardiac hemodynamics based on hydraulic or electric representations.

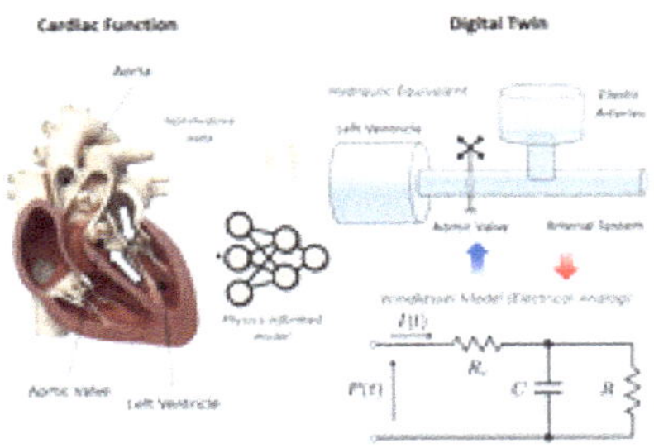

(Source: Kuang et al., (2024) from "Med-Real2Sim: Non-Invasive Medical Digital Twins using Physics-Informed Self-Supervised Learning (https://proceedings.neurips.cc/paper_files/paper/2024/file/0b081a44ed0b8c0c4aa6bd886a60bea4-Paper-Conference.pdf)" licensed under CC BY 4.0)

CONCLUSION

The current state of SHMs demonstrates a promising field where the combination of multifunctional abilities gives way to new application concepts. Self-healing is no longer limited to the restoration of physical integrity but is extending to include conductivity, sensing and environmental responsiveness. The studies note the presence of encapsulated healing agents, self-healing systems, and utilization of bio-based and ecological friendly materials as some of the leading trends maximizing the performance and applicability of such materials. Also, developments in nanocomposites unveil the prospects of applying nanostructures in enhancing the healing processes and the performance of the materials in general. Such innovations depict a balanced combination of material science and nanotechnology, which result in efficient performing materials that open the door to sustainable solutions.

In the future, it is important to further investigate self-healing materials to have the potential to solve intricate problems on the planet. The AI and machine learning integration into the process of material discovery will facilitate the overall process of creating ground-breaking self-healing systems and bring the field to a new level. Self-healing technologies will become a part of the sustainable management of resources in Industries that are starting to adopt the circular economy model. As well, it is a common sight of creating zero-impact self-healing materials, which not only incorporate the spirit of sustainability but also help re-organize our relationship with built-in environments and technology. The exchange between disciplines that links basic materials science with engineering practice will speed discovery, so that self-healing materials will become a key element of next-generation smart infrastructure and technologies.

REFERENCES

Ahmed, A., Ali, A., Alzahrani, B., & Sanada, K. (2023). Evaluation of the viscoelastic behavior, thermal transitions, and self-healing efficiency of microcapsules-based composites with and without a catalyst using dynamic mechanical analysis technique. *Journal of Applied Polymer Science*, *140*(34), e54323. DOI: 10.1002/app.54323

Ahn, C., Hong, P. H., Lee, J., Kim, J., Moon, G., Lee, S., Park, I., Han, H., & Hong, S. W. (2022). Highly self-healable polymeric coating materials with enhanced mechanical properties based on the charge transfer complex. *Polymers*, *14*(23), 5181. DOI: 10.3390/polym14235181 PMID: 36501576

Ayala-Ruiz, D., Atoche, A. C., Ruiz-Ibarra, E., Osorio-de-la-Rosa, E., & Vázquez-Castillo, J. (2019). A self-powered pmfc-based wireless sensor node for smart city applications. *Wireless Communications and Mobile Computing*, *2019*, 1–10. DOI: 10.1155/2019/8986302

Bai, H., Kim, Y., & Shepherd, R. (2022). Autonomous self-healing optical sensors for damage intelligent soft-bodied systems. *Science Advances*, *8*(49), eabq2104. Advance online publication. DOI: 10.1126/sciadv.abq2104 PMID: 36475793

Barros, L. B., Knockaert, M., & Tenório, R. (2023). An analysis of the commercialisation barriers of self-healing concrete. MATEC Web of Conferences, 378, 10001. https://doi.org/DOI: 10.1051/matecconf/202337810001

Belie, N. D., Gruyaert, E., Al-Tabbaa, A., Antonaci, P., Baeră, C., Bajāre, D., & Jonkers, H. (2018). A review of self-healing concrete for damage management of structures. *Advanced Materials Interfaces*, *5*(17), 1800074. Advance online publication. DOI: 10.1002/admi.201800074

Boahen, E., Pan, B., Kweon, H., Kim, J., Choi, H., Kong, Z., & Kim, D. (2022). Ultrafast, autonomous self-healable iontronic skin exhibiting piezo-ionic dynamics. https://doi.org/DOI: 10.21203/rs.3.rs-2135447/v1

Boopathy, G., Srinivasan, V., Kamalakar, V., & Bose, G. C. (2025). Sustainability through digital twins: Optimizing energy usage and reducing waste. In *Accelerating product development cycles with digital twins and IoT integration* (pp. 13–48). IGI Global Scientific Publishing., DOI: 10.4018/979-8-3373-2028-1.ch002

Cadamuro, F., Ardenti, V., Nicotra, F., & Russo, L. (2023). Alginate–gelatin self-healing hydrogel produced via static–dynamic crosslinking. *Molecules (Basel, Switzerland)*, *28*(6), 2851. DOI: 10.3390/molecules28062851 PMID: 36985823

Datta, D., Colaco, V., Bandi, S., Dhas, N., Janardhanam, L., Singh, S., & Vora, L. (2025). Stimuli-responsive self-healing ionic gels: A promising approach for dermal and tissue engineering applications. *ACS Biomaterials Science & Engineering*, *11*(3), 1338–1372. DOI: 10.1021/acsbiomaterials.4c02264 PMID: 39999055

Davydovich, D., & Urban, M. W. (2020). Water accelerated self-healing of hydrophobic copolymers. *Nature Communications*, *11*(1), 5743. Advance online publication. DOI: 10.1038/s41467-020-19405-5 PMID: 33184268

Dhirani, L. L., Mukhtiar, N., Chowdhry, B. S., & Newe, T. (2023). Ethical dilemmas and privacy issues in emerging technologies: A review. *Sensors (Basel)*, *23*(3), 1151. DOI: 10.3390/s23031151 PMID: 36772190

Ding, C., Tian, M., Ren, F., Dang, Y., & Zhang, M. (2020). Novel self-healing hydrogel with injectable, ph-responsive, strain-sensitive, promoting wound-healing, and hemostatic properties based on collagen and chitosan. ACS Biomaterials Science &Amp. *ACS Biomaterials Science & Engineering*, *6*(7), 3855–3867. DOI: 10.1021/acsbiomaterials.0c00588 PMID: 33463340

Ding, S., Zhang, J., Zhou, L., & Luo, Y. (2020). Promoting healing progress in polymer composites based on diels-alder reaction by constructing silver bridges. *Polymers for Advanced Technologies*, *32*(3), 1239–1250. DOI: 10.1002/pat.5173

Du, J., Li, Y., Jiu-chun, W., Wang, C., Liu, D., Wang, G., & Liu, S. (2020). Mechanically robust, self-healing, polymer blends and polymer/small molecule blend materials with high antibacterial activity. ACS Applied Materials &Amp. *Interfaces*, *12*(24), 26966–26972. DOI: 10.1021/acsami.0c06591 PMID: 32466641

Ekeocha, J., Ellingford, C., Pan, M., Wemyss, A. M., Bowen, C., & Wan, C. (2021). Challenges and opportunities of self-healing polymers and devices for extreme and hostile environments. *Advanced Materials*, *33*(33), 2008052. Advance online publication. DOI: 10.1002/adma.202008052 PMID: 34165832

Frei, R., McWilliam, R., Derrick, B., Purvis, A., Tiwari, A., & Serugendo, G. D. M. (2013). Self-healing and self-repairing technologies. *International Journal of Advanced Manufacturing Technology*, *69*(5-8), 1033–1061. DOI: 10.1007/s00170-013-5070-2

Gao, L., Yang, Y., Xie, J., Zhang, S., Hu, J., Zeng, R., He, J., Li, Q., & Wang, Q. (2020). Autonomous self-healing of electrical degradation in dielectric polymers using in situ electroluminescence. *Matter*, *2*(2), 451–463. DOI: 10.1016/j.matt.2019.11.012

Georgopoulou, A., Bosman, A. W., Brancart, J., Vanderborght, B., & Clemens, F. (2021). Supramolecular self-healing sensor fiber composites for damage detection in piezoresistive electronic skin for soft robots. *Polymers*, *13*(17), 2983. DOI: 10.3390/polym13172983 PMID: 34503023

Han, S., Liu, C., Lin, X., Zheng, J., Wu, J., & Liu, C. (2020). Dual conductive network hydrogel for a highly conductive, self-healing, anti-freezing, and non-drying strain sensor. *ACS Applied Polymer Materials*, *2*(2), 996–1005. DOI: 10.1021/acsapm.9b01198

Hao, W., Hao, H., Kanwal, H., & Jiang, S. (2023). Evaluation of self-healing efficiency of microcapsule-based self-healing cementitious composites based on acoustic emission. *Journal of Renewable Materials*, *11*(4), 1687–1697. DOI: 10.32604/jrm.2022.023795

Huang, W., Cheng, S., Wang, X., Zhang, Y., Chen, L., & Zhang, L. (2021). Noncompressible hemostasis and bone regeneration induced by an absorbable bioadhesive self-healing hydrogel. *Advanced Functional Materials*, *31*(22), 2009189. Advance online publication. DOI: 10.1002/adfm.202009189

Huang, X., Zhang, A., Tan, Q., Gou, K., Chen, Y., Nie, Y., & Weng, G. (2024). Highly stretchable and self-healable tough polymer elastomer through dy3+ and cu2+ coordination cross-linking. *Macromolecules*, *57*(3), 963–975. DOI: 10.1021/acs.macromol.3c02347

Ihenketu, C. (2024). Reduction in dilapidation and high cost of maintenance on buildings using self-healing concrete for building construction. *International Research Journal of Modernization in Engineering Technology and Science*. Advance online publication. DOI: 10.56726/IRJMETS48855

Ismail, N. A., Khan, A., Fayyad, E. M., Kahraman, R., Abdullah, A. M., & Shakoor, A. (2021). Self-healing performance of smart polymeric coatings modified with tung oil and linalyl acetate. *Polymers*, *13*(10), 1609. DOI: 10.3390/polym13101609 PMID: 34067528

Jha, S., Ramesh, B., & Macrin, D. (2023). Study on the flexural strength of polypropylene fiber reinforced m20 grade self-compacting bacterial concrete. Cardiometry, (25), 1763-1770. https://doi.org/DOI: 10.18137/cardiometry.2022.25.17631770

Jin, Q., Du, R., Tang, H., Zhao, Y., Peng, W., Li, Y., Zhang, J., Zhu, T., Huang, X., Kong, D., He, Y., Bao, T., Kong, D., Wang, X., Wang, R., Zhang, Q., & Jia, X. (2023). Quadruple h-bonding and polyrotaxanes dual cross-linking supramolecular elastomer for high toughness and self-healing conductors. *Angewandte Chemie International Edition*, *62*(26), e202305282. Advance online publication. DOI: 10.1002/anie.202305282 PMID: 37186156

Khan, A., & Rabnawaz, M. (2021). Base-layer-driven self-healing materials. *ACS Applied Polymer Materials*, *3*(8), 3922–3928. DOI: 10.1021/acsapm.1c00476

Kılıçarslan, B., Bozyel, I., Gökcen, D., & Bayram, C. (2022). Sustainable macromolecular materials in flexible electronics. *Macromolecular Materials and Engineering*, *307*(6), 2100978. Advance online publication. DOI: 10.1002/mame.202100978

Lee, J., Park, S. U., Park, C., Kwon, O. S., Chung, S., Shim, J., Lee, C.-S., & Bae, J. (2018). Effect of a surfactant in microcapsule synthesis on self-healing behavior of capsule embedded polymeric films. *Polymers*, *10*(6), 675. DOI: 10.3390/polym10060675 PMID: 30966709

Lee, S., Hong, P. H., Kim, J., Choi, K., Moon, G., Kang, J., Lee, S., Ahn, J. B., Eom, W., Ko, M. J., & Hong, S. W. (2020). Highly self-healable polymeric blend synthesized using polymeric glue with outstanding mechanical properties. *Macromolecules*, *53*(6), 2279–2286. DOI: 10.1021/acs.macromol.9b02359

Li, F., Wang, X., Zuo, J., Chen, C., Chen, J., Zhu, J., & Ying, W. B. (2024). Oxime-urethane-based self-healing polyurethane for achieving complex structures via 3d printing. *ACS Applied Polymer Materials*, *6*(7), 4070–4077. DOI: 10.1021/acsapm.4c00140

Li, S., Dong, Q., Peng, X., Chen, Y., Yang, H., Xu, W., Zhao, Y., Xiao, P., & Zhou, Y. (2022). Self-healing hyaluronic acid nanocomposite hydrogels with platelet-rich plasma impregnated for skin regeneration. *ACS Nano*, *16*(7), 11346–11359. DOI: 10.1021/acsnano.2c05069 PMID: 35848721

Liu, J., Wang, L., Bai, L., Wang, W., Yang, L., Chen, H., Yang, H., & Wei, D. (2024). Fabrication of nanocomposite hydrogels based on cellulose nanocrystals and multi-walled carbon nanotubes for human motion monitoring. *Macromolecular Chemistry and Physics*, *225*(21), 2400207. Advance online publication. DOI: 10.1002/macp.202400207

Liu, X., Ren, Z., Liu, F., Li, Z., Ling, Q., & Gu, H. (2021). Multifunctional self-healing dual network hydrogels constructed via host–guest interaction and dynamic covalent bond as wearable strain sensors for monitoring human and organ motions. ACS Applied Materials &Amp. *Interfaces*, *13*(12), 14612–14622. DOI: 10.1021/acsami.1c03213 PMID: 33723988

Liu, Y., Lin, S.-H., Chuang, W.-T., Dai, N.-T., & Hsu, S. (2022). Biomimetic Strain-Stiffening in Chitosan Self-Healing Hydrogels. *ACS Applied Materials & Interfaces*, *14*(14), 16032–16046. DOI: 10.1021/acsami.2c01720 PMID: 35321544

Liu, Z., Bisoyi, H. K., Huang, Y., Wang, M., Yang, H., & Li, Q. (2021). Thermo- and mechanochromic camouflage and self-healing in biomimetic soft actuators based on liquid crystal elastomers. *Angewandte Chemie International Edition*, *61*(8), e202280862. Advance online publication. DOI: 10.1002/anie.202280862 PMID: 34904346

Liu, Z., Zhong, Y., Li, S., Yu, S., Zhong, J., Yang, Y., & Shen, L. (2024). Hybrid self-repairing polymer composites based on a mixture of intrinsic and extrinsic self-healing. *Macromolecular Chemistry and Physics*, *225*(18), 2400079. Advance online publication. DOI: 10.1002/macp.202400079

Martínez-Ibernón, A., Antonaci, P., & Anglani, G. (2023). Initial proposal of a smart cement-based material to enhance the service-life of reinforcement concrete structures. MATEC Web of Conferences, 378, 05003. https://doi.org/DOI: 10.1051/matecconf/202337805003

Mishra, A. K., Tripathy, A. K., Obaidat, M. S., Zhi-yuan, T., Prasad, M., Sadoun, B., & Puthal, D. (2018). A chain topology for efficient monitoring of food grain storage using smart sensors. *Proceedings of the 15th International Joint Conference on E-Business and Telecommunications*. https://doi.org/DOI: 10.5220/0006850602550264

Morang, S., Bandyopadhyay, A., Borah, N., Kar, A., Mandal, B. B., & Karak, N. (2024). Photoluminescent self-healable waterborne polyurethane/mo and s codoped graphitic carbon nitride nanocomposite with bioimaging and encryption capability. *ACS Applied Bio Materials*, *7*(3), 1910–1924. DOI: 10.1021/acsabm.3c01259 PMID: 38391158

Motora, K. G., Wu, C., Chang, C., & Liao, J. (2021). Nir light stimulated self-healing reduced tungsten oxide/polyurethane nanocomposite based on the diels–alder reaction. *Macromolecular Materials and Engineering*, *306*(11), 2100438. Advance online publication. DOI: 10.1002/mame.202100438

Oh, S., & Kim, Y. J. (2025). Evaluation of permeability recovery in precast concrete with hybrid capsules using constant-head permeability test as smart construction material. *Materials (Basel), 18*(2), 220. DOI: 10.3390/ma18020220 PMID: 39859692

Oh, S. R., & Kim, Y. J. (2025). Evaluation of Permeability Recovery in Precast Concrete with Hybrid Capsules Using Constant-Head Permeability Test as Smart Construction Material. *Materials (Basel), 18*(2), 220. DOI: 10.3390/ma18020220 PMID: 39859692

Paolillo, S., Bose, R. K., Santana, M. H., & Grande, A. M. (2021). Intrinsic self-healing epoxies in polymer matrix composites (pmcs) for aerospace applications. *Polymers, 13*(2), 201. DOI: 10.3390/polym13020201 PMID: 33429922

Peng, Y., Gu, S., Wu, Q., Xie, Z., & Wu, J. (2023). High-performance self-healing polymers. *Accounts of Materials Research, 4*(4), 323–333. DOI: 10.1021/accountsmr.2c00174

Perin, D., Dorigato, A., & Pegoretti, A. (2023). Thermoplastic self-healing polymer blends for structural composites: Development of polyamide 6 and cyclic olefinic copolymer blends. *Journal of Applied Polymer Science, 140*(16), e53751. Advance online publication. DOI: 10.1002/app.53751

Petrakis, E. G. M., Sotiriadis, S., Soultanopoulos, T., Renta, P. T., Buyya, R., & Bessis, N. (2018). Internet of things as a service (itaas): Challenges and solutions for management of sensor data on the cloud and the fog. *Internet of Things : Engineering Cyber Physical Human Systems, 3-4*, 156–174. DOI: 10.1016/j.iot.2018.09.009

Rajczakowska, M., Szeląg, M., Habermehl-Cwirzen, K., Hedlund, H., & Ćwirzeń, A. (2023). Interpretable machine learning for prediction of post-fire self-healing of concrete. *Materials (Basel), 16*(3), 1273. DOI: 10.3390/ma16031273 PMID: 36770279

Ranganathan, P., Chen, C., Tasi, M., Rwei, S., & Lee, Y. (2021). Biomass thermoplastic (co)polyamide elastomers synthesized from a fatty dimer acid: A sustainable route toward a new era of uniform and bimodal foams. Industrial &Amp. *Industrial & Engineering Chemistry Research, 60*(33), 12139–12154. DOI: 10.1021/acs.iecr.1c01533

Sadiku, M. N. O., Suman, G., & Musa, S. M. (2021). Artificial intelligence in materials science. *International Journal of Advances in Scientific Research and Engineering, 07*(08), 77–81. DOI: 10.31695/IJASRE.2021.34057

Sayadi, S., Irimia, L., & Jefferson, T. (2024). Biomimetic materials in construction industry: the necessity of simulation. https://doi.org/DOI: 10.18573/conf1.n

Schmidt, J., Marques, M. R. G., Botti, S., & Marques, M. A. L. (2019). Recent advances and applications of machine learning in solid-state materials science. *npj Computational Materials*, *5*(1), 83. Advance online publication. DOI: 10.1038/s41524-019-0221-0

Shields, Y., Belie, N. D., Jefferson, T., & Tittelboom, K. V. (2021). A review of vascular networks for self-healing applications. *Smart Materials and Structures*, *30*(6), 063001. DOI: 10.1088/1361-665X/abf41d

Sima, W., Chen, L., Sun, P., Yang, M., Zhu, C., Yuan, T., & Deng, Q. (2021). Novel smart insulating materials achieving targeting self-healing of electrical trees: High performance, low cost, and eco-friendliness. *ACS Applied Materials & Interfaces*, *13*(28), 33485–33495. DOI: 10.1021/acsami.1c07469 PMID: 34232014

Song, Y., Kim, H., & Chung, C. (2022). Repeatable self-healing of a protective coating based on vegetable-oil-loaded microcapsules. *Polymers*, *14*(10), 2013. DOI: 10.3390/polym14102013 PMID: 35631895

Sun, F., Liu, L., Liu, T., Wang, X., Qi, Q., Hang, Z., Chen, K., Xu, J. H., & Fu, J. (2023). Vascular smooth muscle-inspired architecture enables soft yet tough self-healing materials for durable capacitive strain-sensor. *Nature Communications*, *14*(1), 130. Advance online publication. DOI: 10.1038/s41467-023-35810-y PMID: 36624140

Terryn, S., Langenbach, J., Roels, E., Brancart, J., Bakkali-Hassani, C., Poutrel, Q., Georgopoulou, A., George Thuruthel, T., Safaei, A., Ferrentino, P., Sebastian, T., Norvez, S., Iida, F., Bosman, A. W., Tournilhac, F., Clemens, F., Van Assche, G., & Vanderborght, B. (2021). A review on self-healing polymers for soft robotics. *Materials Today*, *47*, 187–205. DOI: 10.1016/j.mattod.2021.01.009

Terryn, S., Langenbach, J., Roels, E., Brancart, J., Bakkali-Hassani, C., Poutrel, Q., Georgopoulou, A., George Thuruthel, T., Safaei, A., Ferrentino, P., Sebastian, T., Norvez, S., Iida, F., Bosman, A. W., Tournilhac, F., Clemens, F., Van Assche, G., & Vanderborght, B. (2021). A review on self-healing polymers for soft robotics. *Materials Today*, *47*, 187–205. DOI: 10.1016/j.mattod.2021.01.009

Tian, Y., Cong, S., Su, W., Chen, H., Li, Q., Geng, F., & Zhao, Z. (2014). Synergy of w18o49 and polyaniline for smart supercapacitor electrode integrated with energy level indicating functionality. *Nano Letters*, *14*(4), 2150–2156. DOI: 10.1021/nl5004448 PMID: 24593047

Trost, C. O. W., Lassnig, A., Kreiml, P., Jörg, T., Terziyska, V. L., Mitterer, C., & Cordill, M. J. (2024). Enthalpy-Driven Self-Healing in Thin Metallic Films on Flexible Substrates. *Advanced Materials*, *36*(29), 2401007. DOI: 10.1002/adma.202401007 PMID: 38695220

Utrera-Barrios, S., Santana, M., Verdejo, R., & López-Manchado, M. (2020). Design of rubber composites with autonomous self-healing capability. *ACS Omega*, *5*(4), 1902–1910. DOI: 10.1021/acsomega.9b03516 PMID: 32039326

Wan, D., Jiang, Q., Song, Y., Pan, J., Qi, T., & Li, G. L. (2020). Biomimetic tough self-healing polymers enhanced by crystallization nanostructures. *ACS Applied Polymer Materials*, *2*(2), 879–886. DOI: 10.1021/acsapm.9b01094

Wang, A., Silva, K. D., Jones, M. I., & Gao, W. (2024). Cr-free anticorrosive primers for marine propeller applications. *Polymers*, *16*(3), 408. DOI: 10.3390/polym16030408 PMID: 38337297

Wang, B., Liu, J., Niu, D., Wu, N., Yun, W., Wang, W., Zhang, K., Li, G., Yan, S., Xu, G., & Yin, J. (2021). Mussel-inspired bisphosphonated injectable nanocomposite hydrogels with adhesive, self-healing, and osteogenic properties for bone regeneration. ACS Applied Materials &Amp. *Interfaces*, *13*(28), 32673–32689. DOI: 10.1021/acsami.1c06058 PMID: 34227792

Wang, P., Hu, M., Wang, H., Chen, Z., Feng, Y., Wang, J., Ling, W., & Huang, Y. (2020). The evolution of flexible electronics: From nature, beyond nature, and to nature. *Advancement of Science*, *7*(20), 2001116. Advance online publication. DOI: 10.1002/advs.202001116 PMID: 33101851

Wang, X., Sun, Q., Wang, S., Shao, R., & Su, J. (2023). Multiscale mathematical analysis of influencing factors and experimental verification of microcrack self-healing efficiency of bitumen composites using microcapsules. *Materials (Basel)*, *16*(14), 5073. DOI: 10.3390/ma16145073 PMID: 37512348

Wang, Y., Wang, Y., Li, T., Wang, X., Ren, H., Lou, C., & Lin, J. (2023). Preparation and properties of polyaniline self-healing conductive materials based on hydrogen bonding and electrostatic interaction forces. *ACS Applied Polymer Materials*, *5*(11), 9585–9593. DOI: 10.1021/acsapm.3c01239

Wang, Y., Wang, Y., Li, T.-T., Wang, X., Ren, H.-T., Lou, C.-W., & Lin, J.-H. (2023). Preparation and Properties of Polyaniline Self-Healing Conductive Materials Based on Hydrogen Bonding and Electrostatic Interaction Forces. *ACS Applied Polymer Materials*, *5*(11), 9585–9593. DOI: 10.1021/acsapm.3c01239

Wen, J., Xu, G., Liang, Z., Li, S., Wang, Y., Yang, J., & Nie, Y. (2023). Combing experimental methods and molecular simulations to study self-healing behaviors of polyurethane elastomers containing multiple hydrogen bond networks and flexible blocks. *Physical Chemistry Chemical Physics*, *25*(41), 28162–28179. DOI: 10.1039/D3CP02723F PMID: 37818678

Wu, P., Cheng, H., Wang, X., Shi, R., Zhang, C., Arai, M., & Zhao, F. (2021). A self-healing and recyclable polyurethane-urea diels–alder adduct synthesized from carbon dioxide and furfuryl amine. *Green Chemistry*, *23*(1), 552–560. DOI: 10.1039/D0GC03695A

Wu, S., Chuang, W., Ho, J., Wu, H., & Hsu, S. (2023). Self-healing of recombinant spider silk gel and coating. *Polymers*, *15*(8), 1855. DOI: 10.3390/polym15081855 PMID: 37112001

Xi, X., Zhang, W., Yang, Z., Zhang, Y., Zhang, H., Zhang, H., & Li, Q. (2016). Smart and flexible supercapacitor based on a porous carbon nanotube film and polyaniline hydrogel. *RSC Advances*, *6*(30), 24946–24951. DOI: 10.1039/C6RA00705H

Xu, F., Weng, D., Li, X., Li, Y., & Sun, J. (2022). Self-healing hydrophilic porous photothermal membranes for durable and highly efficient solar-driven interfacial water evaporation. *CCS Chemistry*, *4*(7), 2396–2408. DOI: 10.31635/ccschem.021.202101111

Yadav, S., Choudhary, N., Bhai, S., Bhojani, G., Chatterjee, S., Ganguly, B., & Paital, A. R. (2021). Recyclable functionalized material for sensitive detection and exceptional sorption of hexavalent chromium and permanganate ions with biosensing applications. *ACS Applied Bio Materials*, *4*(8), 6430–6440. DOI: 10.1021/acsabm.1c00609 PMID: 35006925

Yuyu, E., Ju, Y., Wang, Z., Chang, Z., Jiang, J., Li, P., & Wang, K. (2022). Tunable and self-healing properties of polysaccharide-based hydrogels through polymer architecture modulation. ACS Sustainable Chemistry &Amp. *ACS Sustainable Chemistry & Engineering*, *10*(42), 14053–14063. DOI: 10.1021/acssuschemeng.2c05125

Zhang, Y., Wang, M., Zhuo, R., Tang, W., Niu, C., Chen, Q., Yang, W., & Zhao, S. (2025). Study on Self-Healing Technology of Epoxy Resin for Insulator Core Rods. *Journal of Applied Polymer Science*, *142*(21), e56935. DOI: 10.1002/app.56935

Zhao, D., Zhu, Y., Cheng, W., Chen, W., Wu, Y., & Yu, H. (2020). Cellulose-based flexible functional materials for emerging intelligent electronics. *Advanced Materials*, *33*(28), 2000619. Advance online publication. DOI: 10.1002/adma.202000619 PMID: 32310313

Zhao, K., Zhou, C., Liu, L., & Wu, G. (2025). Self-healing waterborne polyurethane with enhanced mechanical properties based on nano-silica. *Journal of Applied Polymer Science*, *142*(18), e56822. Advance online publication. DOI: 10.1002/app.56822

Zhou, S., Wang, C., & Ju, J. W. (2022). A numerical chemo-micromechanical damage model of sulfate attack in cementitious materials. *International Journal of Damage Mechanics*, *31*(10), 1613–1638. DOI: 10.1177/10567895221114926

Zhou, Y., Li, L., Han, Z., Li, Q., He, J., & Wang, Q. (2022). Self-healing polymers for electronics and energy devices. *Chemical Reviews*, *123*(2), 558–612. DOI: 10.1021/acs.chemrev.2c00231 PMID: 36260027

Zhou, Y., Wen, J., & Nie, Y. (2024). Microscopic mechanisms of self-healing in polymers revealed by molecular simulations and machine learning. *Macromolecules*, *57*(7), 3258–3270. DOI: 10.1021/acs.macromol.4c00338

Zhu, M., Yu, J., Li, Z., & Ding, B. (2022). Self-healing fibrous membranes. *Angewandte Chemie International Edition*, *61*(41), e202208949. Advance online publication. DOI: 10.1002/anie.202208949 PMID: 35862255

Chapter 6
An Intelligent Framework to Monitor and Predict Land Temperatures Using Learning Algorithms

S. Sharanya
https://orcid.org/0000-0003-2442-3246
SRM Institute of Science and Technology, India

D. Rajalakshmi
SRM Institute of Science and Technology, India

Rehash Rushmi Pavithra A.
SRM Institute of Science and Technology, India

V. Vijayalakshmi
SRM Institute of Science and Technology, India

R. Radha
https://orcid.org/0000-0001-9822-0949
SRM Institute of Science and Technology, India

ABSTRACT

The Earth's climate is rapidly changing, and it has been projected that the global climate will continue to change even beyond this century. The magnitude of change directly relies on the quantity of greenhouse gases emitted by the mankind and the earth's sensitivity towards the emissions. The climate hazards will continue triggering interactive and accumulated impacts on the natural systems. Hence, it is quintessential to mitigate its adverse effect on the natural systems. An improved

DOI: 10.4018/979-8-3373-3506-3.ch006

understanding about the ecosystem can be observed by predicting the global warming temperature and land surface warning. This work focuses on building an effective framework that leverages GRU cells for predicting the mean temperature of a region. The proposed work develops a comprehensive AI system for predicting the temperature by using previous and historic data. This system would benefit the community by accurately predicting the land temperature which is very crucial for wide variety of human activities.

1. INTRODUCTION

Earth is experiencing a drastic climate change in the context of long-term shifts in weather patterns, rainfall, and temperatures. These transitions are natural, due to the changes external to the earth such as the sun's activity, large volcanic eruptions and other events in space. However, since the early 1800s, the vigorous human activities have been the main accelerators for these drastic climate changes (Sesana E. et al., 2021). The direct implication for this adverse effect is the excessive burning fossil fuels which has generated enormous amount of greenhouse gases into the atmosphere which is covering the earth like a blanket wrapped. This is responsible for trapping the sun's heat and leads to raising temperatures. The greenhouse gases that contribute the elevated temperatures are methane and carbon dioxide. Their primary source of emission is using gasoline as fuel for vehicle, heat and ventilation system and irresponsible use of energy resources (Abbass K., et al., 2022). This is further worsened by clearing land area and irrational cutting of forests. Even the agriculture, gas and oil mining operations are emanating as major contributors of methane emissions in the atmosphere. Figure 1 shows the sector wise contribution of green house gas emissions. It can be seen that industrial emissions forms a lion's share followed by generation of electricity. It can be witnessed that energy, transport, industry, agriculture, land use and buildings are the chief sectors emitting greenhouse gases into the atmosphere (Watson R. T., et al., 1998).

Figure 1. Green house emissions from various sectors

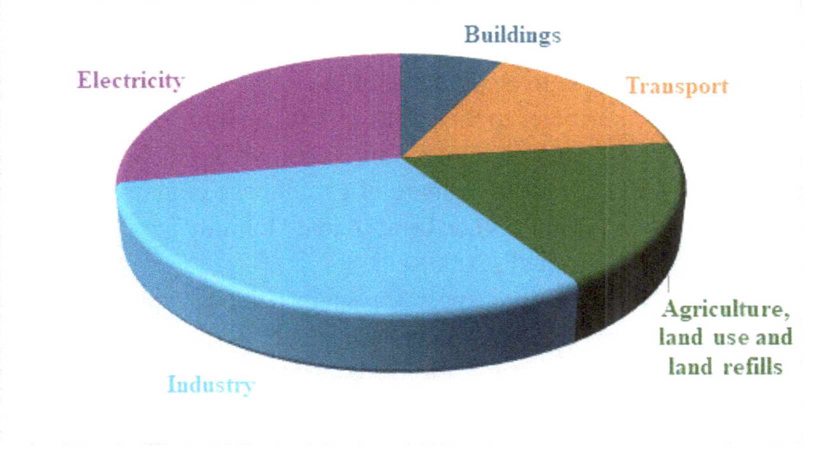

Researchers and scientists have highlighted that humans activities are responsible for global heating. Human activities erotically increase greenhouse gas emissions are causing the world to become warmer place in the last two thousand years. The mean temperature of the surface of the Earth is roughly about 1.2°C warmer than it has been in late 1800s that is before the first industrial revolution. More specifically, the previous decade has witnessed warmest on record temperatures, and it can be noted that each of the last four decades has been experiencing warmer climate than any of the previous decade since the year 1850.

It is often understood as climate change mainly refers to warmer temperatures. But it has to be noted that rise in temperature marks only the beginning of climate change. As it is evident that the entire Earth is a holistic system, where everything is interconnected, changes in one aspect will influence dramatic changes in all other climatic factors also. For instance, the direct consequences of climate change are now witnessed in the form of intense droughts, acute water scarcity, severe forest fires, flooding, rising sea levels, melting of glaciers, catastrophic storms and poor biodiversity.

Climate change has now started to cause detrimental effects on human health, agriculture, cropping, housing, work and safety. World's island countries and developing nations are more vulnerable to the ill-impacts of the climate change. Conditions such as elevated sea-level and intrusion of saltwater have peaked to the point that herds of human communities have to be relocated. In addition to this, protracted droughts are experienced by people increasing the risk of famine. In near future, this number is expected to increase in large amounts. According to the reports by United Nations, the scientists and the reviewers has agreed to limit the rise in global temperature to 1.5^{O}C to avoid worst climatic impacts. Not all the countries in the world equally contributes to the climatic change. Some countries like China, United States of America, Russian Federation, Brazil, India and the European Union accounts for nearly 50% of the total share of greenhouse emissions. On the other hand, 47 least developed countries can be accounted for only 3% of greenhouse gas emissions (Mendelsohn R., et al., 2000).

It is imperative that everyone must act against the adverse climatic change however countries which greatly damages the nature holds greater responsibility to act first. Many innovative solutions in the context of climatic change can render economic benefits while improving the lifestyle of people and offer Environmental Protection. To facilitate this process many global frameworks and agreements such as emphasising sustainable development goals is being put forth by UN Framework Convention on Climate Change as well as the Paris Agreement. The three unique selling points of these measures includes adaptation to climatic impacts, reducing the emissions into the atmosphere and financial adjustments (Grimm N.B., et al., 2013). Migrating to alternate energy systems from burning fossil fuels would be considered as a remarkable act for sustainable development. Upon successful accomplishment of this, the consumption as well as protection of fossil fuels has to be declined by 30% by the year 2030 to mitigate the impact of drastic climatic change. Adaptation to the new climatic change mast prioritise the most vulnerable sector of population who are left with fewest to resource hey to combat the adverse effects cost due to the climate change (Schneider S., et al., 2001). It has to be accepted that, any climate change solution and action demands substantial financial investments by people, corporates, businesses and governments. Table 1 enumerates the drill down version of emissions of greenhouse gases various sectors.

Table 1. More detailed analysis of greenhouse gas emissions

Sector	Fields	Percentage
Buildings	Residential buildings	5%
	Commercial buildings	1%
	Refrigerants	1%

continued on following page

Table 1. Continued

Sector	Fields	Percentage
Industry	iron and steel industry	5%
	oil and gas production	5%
	cement industry	5%
	chemical industry	4%
	coal mining	2%
	Refineries	1%
	other miscellaneous industries	9%
Transport	Roadways	12%
	Ships	2%
	Other navigation	1%
Agriculture, land refills and land reuse	Crops	7%
	Herding livestock	6%
	landfills and waste	4%
	landuse forest	1%
	agricultural fuel combustion	1%
Electricity	Coal	10%
	natural gas	6%
	oil	1%

INCREASE IN LAND SURFACE TEMPERATURE

Land surface temperature is generally referred as temperature of the earth's skin. It is a key detrimental variable to monitor the process of exchange of water and energy between the atmosphere and land surface (Chen Y. C., et al., 2017). It has to be noted that, land surface temperature is different from the add temperature, as the land surface is more quick to get heated and cooled than the air. Some of the physical aspects like density of the vegetation, moisture present in the soil and extent of snow coverage can greatly influence the land surface temperature. Scientists, environmentalists and researchers measure the land surface temperature using satellite data. An alarming trend is noted across all the 7 continental regions is that the mean land surface temperature during the period of 1971-2000 was much higher than any other period in the past 1400 years. The global land temperature in the year 2016 was recorded as highest with the departure temperature of 1.42°C against the record of 1.32°C in the year 2015. There has been a profound human cause behind this increase in land surface both globally as well as regionally. Nearly, 75% of the daily hot extremes during the summer can be attributed to global warming.

Monitoring the Land Surface Temperature (LST) is a pivotal parameter for gauging the heat flux as well as surface temperature, thus offering significant insights to balancing this surface energy, environmental change, dynamics of the ecosystem along with management of natural resources (Tran D. X., et al., 2017). The profound increase in LST levels can be tightly coupled with the extreme weather like more frequent heat waves and drought conditions, which imposes severe negative impacts on the ecosystems, biodiversity and livelihood of common people. So, monitoring land surface temperature deserves significant attention and concern due to words intense detrimental consequences. The escalated LST in the semi arid regions of India is due to global warming. In addition to this, the increase in temperature can be attributed to other regional factors like soil moisture, soil texture, vegetation type, elevation, trade off between long wave and shortwave radiation. All these elements aggregately contributes to the increase in plan surface temperatures.

The remarkable increase in land surface temperatures especially in the semi arid regions of India is primarily due to global warming and intensive industrial activities. Multiple factors that influence the LST of a specific location, such as surface air temperature, soil moisture, soil roughness, vegetation type, elevation, and the balance between shortwave and longwave radiation. These elements collectively contribute to the determination of LST and its variations across different regions. The semi arid tropics constitute major portion of India accounting for nearly 131 million hectares of land under cultivation. These regions face severe impact of drought especially for the people residing in rural districts. These districts are characterised by insufficient rainfall, severe solar radiation with extreme temperature. Predicting the land surface temperature by analysing factors influencing climatic change will immensely help the people and governments to take hey preventive and relief measures.

The development of artificial intelligence (AI), machine learning (ML), computer vision (CV), image processing, satellite technology and deep learning (DL) has facilitated more deeper data analysis to gain insights towards future events. These contemporary cutting edge technologies finds their application in almost all fields right from agriculture till predictive maintenance in industries (Selvaraj S. et al., 2022). As the critical variables contributing to increase in land surface temperature needs constant and consistent monitoring, deploying automated algorithms and methodologies such as ML and DL models would greatly help the community (Kazemi Garajeh, M. et al., 2023). This chapter presents a automated AI framework using DL techniques to analyse the geospatial data and to predict the change in LST.

The organisation of the chapter is as follows section 2 briefly elucidates the significant works in deployment of AI based techniques for predicting the land surface temperature. Section 3 describes the proposed deep leraning based framework for monitoring and predicting the land surface temperatures. The detailed results and discussion about the significant findings are explained in section 4. The inferences about the work with future research prospects is mentioned in section 5.

2. LITERATURE REVIEW

This section briefs some of the salient research works analysing the geospatial data to predict land surface temperatures. The study by Abdulla - Al Kafy et al, indicated the usage of key predictor variables namely land cover region and land use to predict the increase in land surface temperature in the region of Chattogram (Kafy et al., 2021). The work presented an automated model to explore the Landsat thermal bands using support vector machines (SVM) and artificial neural network (ANN) with cellular automata. Among these models, the later one exhibited better era values and overall Kappa value. The results indicated the existence of strong correlation between the increase in land surface temperature and the demographics of urban land area. One major limitation in handling geospatial and remote sensing data is its high special resolution with temporal coherence. Downscaling their high dimensionality using ANN, random forest (RF) and SVM for efficient prediction is done by Wan Li et al (Li, W., Ni, L., Li, Z.L., Duan, S.B. and Wu, H., 2019). The efficacy of these algorithms are compared with conventional thermal sharpening algorithm on the data obtained from local regions of the capital of China. The experimental results illustrated that the ML algorithms performed better than the conventional techniques. The conventional techniques were not successful in normalizing the vegetation index also, the ML techniques are much efficient in learning the patterns form the data during downscaling.

A detailed study on the increase in LST of Yazd, Iran is done by Mohammad Mansourmoghaddam et al., by analysing the high resolution field thermal image data (Mansourmoghaddam, M., et al., 2024) . This work used six ML algorithms implemented in R programming language to predict and analysis the increase in LST during the summer and winter of the years 2019 and 2020 to predict the LST of 2021. The models considered in this work includes RF, extremely random trees (XRT), gradient boosting machines (GBM), generalised linear model (GLM), stacked ensemble and ANN. The performance metrics like mean absolute error, root mean square, standard deviation and root means square logarithmic error of the above-mentioned models indicated that GBM was superior to other models with better prediction efficacy. multi-factor geographically weighted ML Model is proposed by Saiping Xu et al for predicting the increase in LST by exploring the non linear and non stationary properties of land surface parameters (Xu et al., 2021). This model effectively deploys extreme gradient boosting (XGB), multivariate adaptive regression splines (MARS) along with Bayesian Ridge regression (BRR) to capture the non linear relationships in the geospatial data.

Another significant regression model for predicting the LST is developed by Yanwei Sun et al. In this work RF and ordinary least square regression is used to analyse the contributions of urban form metrics 3 different observations scales (Sun et al., 2019). The results indicated that the direction and magnitude of these metrics greatly varied in different seasons. The building morphology along with urban ecological infrastructure leaves greater impact on LST variations especially in high density open areas. On the other hand, building density show a positive correlation on LST while the floor area ratio exhibited negative influence. The impact of different factors that influence the LST of Kabul region is analysed using SVM. The Landsat data is handled by Cellular Automata-Logistic Regression model to predict the increase in LST (Ullah et al., 2024). The results indicated that built up area increased the LST while the vegetation and bath soil decrease the LST. The results of this study indicated that there is a direct correlation among rising temperatures and urbanization. The manual monitoring of LST raised overestimation problems, especially under snow cover conditions. The non homogeneity of data this address using RF, MARS and XGBoost for more precise prediction of daily change in LST by leveraging the conventional metrological data (Liu et al., 2022). The parameters that are considered for this analysis includes mean, maximum and minimum air temperature, precipitation, speed of the wind, solar radiation, diurnal temperature and relative humidity. This data are collected from 4 different areas of China. The results of the experimental analysis indicated that XGBoost perform better than other algorithms.

Chaitanya Baliram Pande et al. suggested a methodology which employs correlation analysis along with ensemble models on Landsat-8 satellite data to unveil its inherent environmental relationships (Pande et al., 2024). The parameters considered in this study includes LST, vegetation index evapotranspiration and rainfall. The integration of three different ML models like XGBoost, Adaboost and Bagging basedXG-Boost, improved the overall robustness and accuracy of the model. Long sequence of data can be analyses using time series models. Hongfei Jia et al. proposed Long Short Term Memory (LSTM) model along with geographically weighed regression to analyse the Landsat 7/8 remote sensing data to predict the LST of urban areas of Beijing(Jia et al., 2021). The simulation of the temperature data is done by geographically weighted regression while the LSTM model issues to predict the spatial distribution of the LST. This model goodbye perform well on water bodies and surface area. This model could effectively predict the impact of snow, cloud and other factors which influence the temporal and spatial distribution of LST on the land area.

Prashant K. Srivastava et al. proposed an effective downscaling method to transform a satellite data to more finer spatial resolution. The work employed few significant AI models like generalized linear model (GLM) to enhance the spatial resolution of geospatial Soil Moisture and Ocean Salinity (SMOS) to derive the soil moisture data (Srivastava et al., 2013). The prediction of LST is done by using ANN, SVM, GLM and relevance vector machine on the Moderate Resolution Imaging Spectroradiometer (MODIS) dataset. The deficit of soil moisture is obtained from probability distribution based hydrological model. The results indicated that the layered and non linear architecture of ANN was able to give better prediction. A unique toolbox for bridging the gap between ML algorithms and remote sensed time series data is designed by Mikhail Sarafanov et al. (2020). The work dictates a procedure to fill the gaps by analysing the spatio temporal relationships among the pixels, by learning the previous historic data. The result of this procedure in restoring the gaps in geospatial data is validated on Terra MODIS dataset. The implementation of this tool is done in Python.

A convolution neural network (CNN) based algorithm is developed to predict the LST by analysing the surface temperature obtained through Advanced Microwave Scanning Radiometer 2. Around 12 V/H channel combinations were used along with test train split to verify the performance of the proposed CNN model (Tan et al., 2019). The results indicated that vertical polarizations gave better results than the horizontal and single polarization. Also, this method was very effective in predicting the LST in bare land areas of China. Another significant milestone in predicting the LST was performed by Munkhdulam Otgonbayar et al (2019). The work developed a more robust model using time series data along with terrain parameters. Two regression models that are primarily involved in the study includes RF and partial least squares. The performance of the models were analysed on 12 predictor variables including slope, aspect and elevation were hey obtained from Mongolia region.

Thus the detailed literature survey the deployment of ML and DL models in the prediction of LST indicates the significance of the problem as well as the effectiveness of these learning models. However, the detailed review indicates few potential limitations. Not many works in the literature focussed on handling big data. Also, almost all the models had extensive preprocessing as well as tiresome feature engineering process, which limits its applicability on low to moderate level hardware systems. So, the proposed research work, develops a DL based model to predict the LST by automatic feature extraction through its layered and non linear architecture.

3. DEEP LEARNING FRAMEWORK FOR PREDICTING LAND SURFACE TEMPEARTURE

DL is a subset of ML and it can give more accurate and precise predictions with addition of successive layers. The core idea of any DL model is derived from perceptron which has great potential for modelling any nonlinear phenomenon. The DL models are excellent in extracting rules, patterns and trends from data rather than require explicit feature engineering. Also, many advanced models like Recurrent Neural Network (RNN), LSTM, Bidirectional LSTM, Gated Recurrent Unit (GRU) etc (Sharanya & Karthikeyan, 2022) are capable of learning temporal patterns in the data. These models are immensely helpful for time series analysis like climate prediction, weather forecasting and other time-based applications. This chapter proposes an automated AI framework for predicting LST using GRUs by exploring the spatial and temperature data of a specific geographic location.

GRU is an enhanced version of traditional RNN. Similar to LSTM, the GRU also uses gating mechanisms to control the information flow among the neuronal cells. It is primarily designed to mitigate the issues of unit saturation problem such as exploding and vanishing gradient. This makes it superior to conventional RNNs. Unlike LSTM, GRU uses only two gates namely update and reset gate. These gates are responsible for controlling the information that must be sent and processed by other higher layers to get the prediction. The gate structures retains the information from previous epochs without diminishing its content in the advanced training layers. The individual structure of a GRU cell is shown in Figure 2.

Figure 2. Anatomy of single GRU cell in the proposed architecture

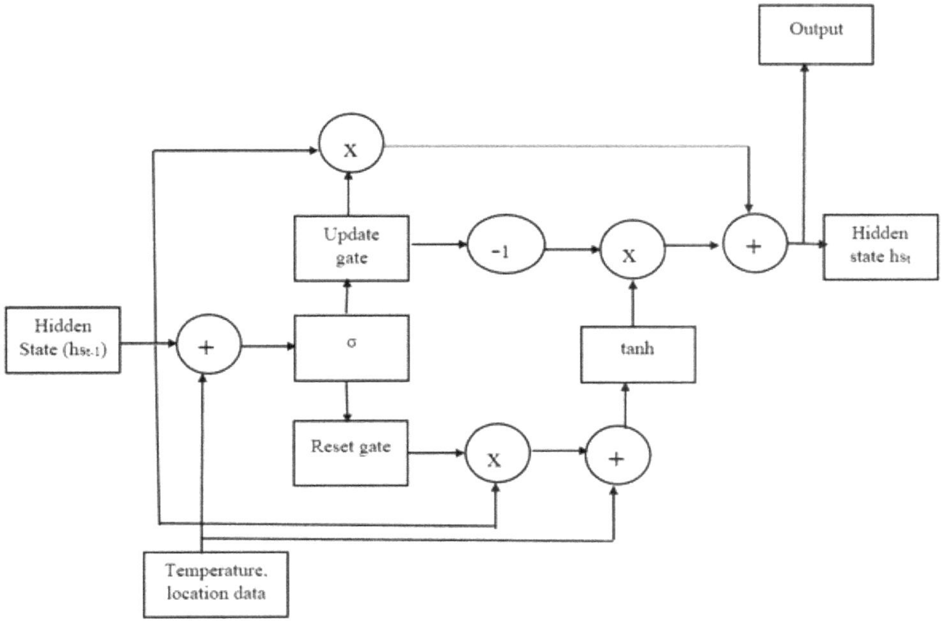

The reset gate in the architecture decides which information from the previous iterations has to be forgotten. Its function is like that of the update gate in LSTM. The operation of reset gate (Rs_t) at any given time t is given as Equation 1.

$$Rs_t = \sigma \left(W_{Rs} \cdot \left(hs_{t-1}, x_t \right) + b_{Rs} \right. \quad (1)$$

Here W_{Rs} and b_{Rs} are weight and bias, which are the learnable parameters during the training process. The term (hs_{t-1}, x_t) will be concatenated with the results of previous iteration. σ is the sigmoid activation function.

GRU has another gate namely the update gate. This will determine the information that the current GRU cell will pass to the next cell thus facilitating the tracking of necessary information about the input parameters. The input given to the update gate is the hidden layer's value at the previous iteration namely hs_{t-1} along with the present input x_t. These elements have their weights associated with them, which are learnable parameters in the training process of the GRU. The final output of the update gate is presented in Equation 2.

$$Up_t = \sigma\left(W_{Up} \cdot \left(hs_{t-1}, x_t \right) + b_{Up} \right) \quad (2)$$

The network of GRU cells process the temporal data given by the LST dataset, bypassing the hidden state between the timesteps. Originally, the hidden layer acts like a vector which effectively captures and tracks the data flowing between subsequent iterations. This is the core idea of employing a GRU network, where information from the previous timestep or iteration is passed to next iteration. During this process, the update and reset gates decides the information to be retained and discarded. Figure 3 shows the proposed architecture of the automated AI framework for LST prediction.

Figure 3. GRU based AI framework for LST prediction

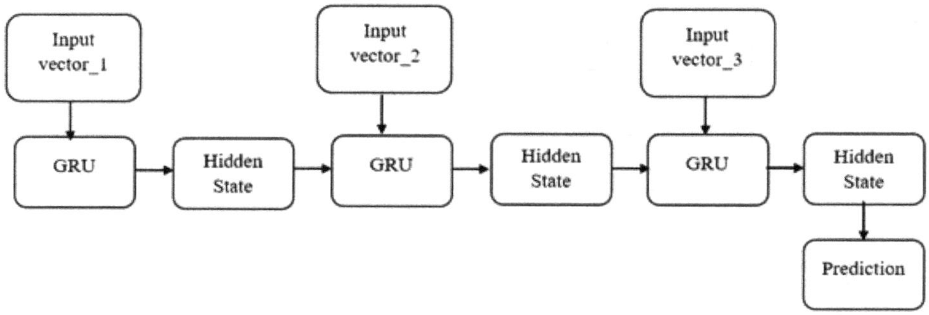

The proposed framework integrates the prowess of three individual GRU cells each of them taking different input vectors in a sequence. The information learned the by the first GRU cell is passed to the subsequent GRU cell through the hidden state. Similarly, the same phenomenon continuous between second and third cell also. Finally, the last hidden state is used for the prediction of final output, which is the LST. Thus, the proposed architecture is capable of learning temporal information from the data presented as input.

At every time step, the given GRU cell reach the input as well as the information from previous hidden state. The reset gate is responsible for determining the amount of previous information to be discarded. On the other hand, the update gate decides the amount of information to be retained and passed or updated to the subsequent layers. The candidate activation function will compute the new hidden state at the reset gate. The final hidden state integrates the candidate state and the previous hidden state which is controlled by the update gate.

The GRU's serves as an excellent candidate for learning time series and temporal data and are more efficient that LSTM in few ways. The important advantage of GRU is the computation of fewer parameters than the LSTM as in GRU the input and forget gate is integrated as distinct update gate without loss of functionality. Also, GRU's are computationally much efficient as they are faster and easier to train because of their simple architecture (Haverkos et al., 2016).

4. EXPERIMENTAL ANALYSIS AND RESULTS

The experimental verification of the proposed model is done on data extracted from the Berkeley Earth data page (Berkeley Earth, n.d.). After extensive preprocessing and cleaning the data is ready for long -term analysis of LST featuring the climate trend around the world. This dataset is very extensive with data collected from different countries across the globe. The most predominantly used dataset is sourced from NASA's GISTEMP, UK's HadCrut and NOAA's MLOST (Gistemp, 2024; Huang et al., 2020; Morice et al., 2021). The Berkeley Earth dataset is formed as newer compilation by the Lawrence Berkeley National Laboratory. The Berkeley LST Study comprises of nearly 1.6 billion temperature records secured from 16 preexisting archives. It is compartmentalized and permits slicing into finer subsets as required. The input data in the dataset comprises of the following fields:

- Land Average Temperature: This is the global mean land temperature expressed in Celsius.
- Land Average Temperature Uncertainty: This imposes 95% confidence interval around the mean value of the mean land temperature.
- Land Max Temperature: This data is the global mean maximum land surface temperature expressed in Celsius
- Land Max Temperature Uncertainty: This imposes 95% confidence interval around the maximum value of the land surface temperature
- Land Min Temperature: This is the aggregation of global mean minimum land surface temperature expressed in Celsius.
- Land Min Temperature Uncertainty: This imposes 5% confidence interval around the minimum value of the land surface temperature.
- Land and Ocean Average Temperature: This data is the aggregated value of the global mean land as well as ocean temperature expressed in Celsius.
- Land and Ocean Average Temperature Uncertainty: This imposes 95% confidence interval around the world mean land as well as the ocean temperature.

It is imperative to explore the data through visualizations to obtain more deeper understanding about the trend, distributions and patterns in the data. Some of the important visualizations and Exploratory Data Analysis (EDA) is shown in this section. Figure 4 shows the fluctuations in mean temperature over a period. As a general trend, almost all the countries exhibits slightly elevated temperature rise as the years progress. Also, there are less fluctuations in temperatures between years, as the data consistently shows more or less same temperature over the years. This is clearly an indication of the adverse effect of global warming all over the world.

Figure 4. Deviation in LST over the years

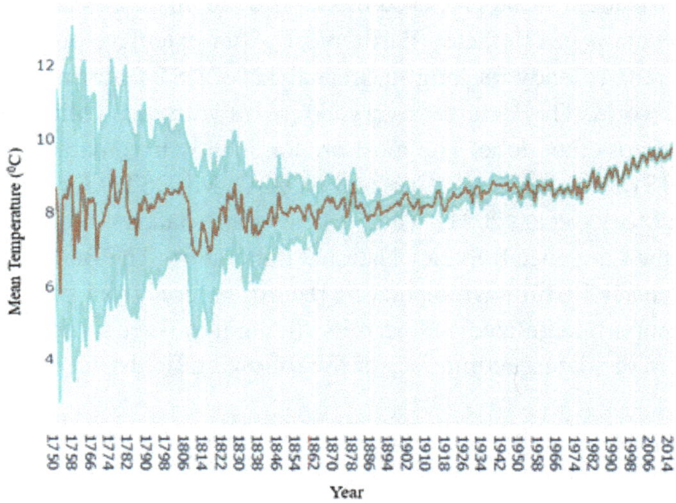

Figure 5 shows the uncertainty values over the years under three statistical measures namely minimum, maximum and average values. It can be seen that there is a profound increase in the values. Also, the uncertainty levels have reduced because of precise measuring devices. The deviations have been greatly reduced.

Figure 5. Uncertainty values over the years

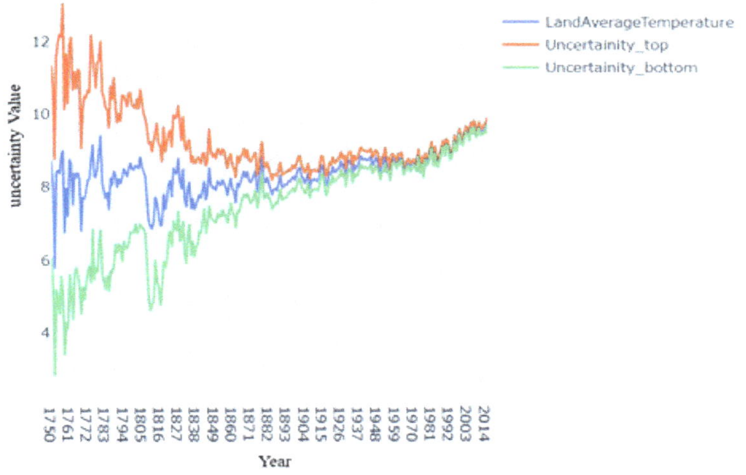

The projected LST over the years can be analysed by the trends prevailing in the previous and current years. Such a trend analysis is presented in the Figure 6. The picture shows that there can be high elevation of temperatures in years to come. The trend line indicates the projection by learning the past data.

Figure 6. Trend analysis and future project of LST

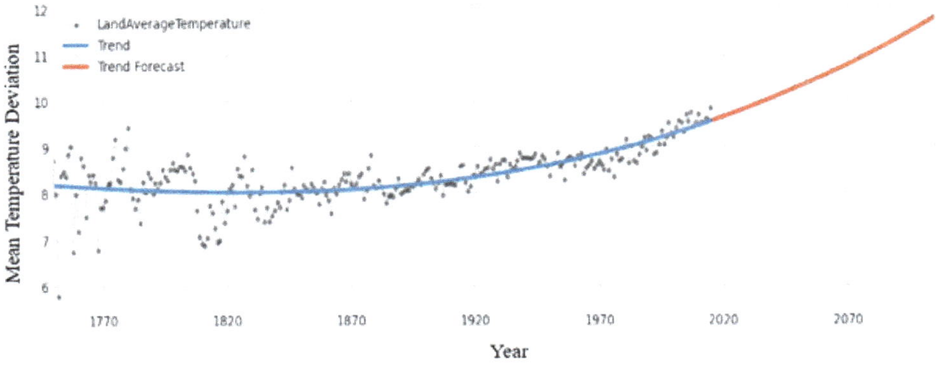

The proposed methodology is applied on the above-mentioned dataset to obtain the future LST. The test-train split ratio is maintained as 20% of data for testing and remaining for training the model. In addition to this, various hyperparameter and parameter settings are done on the model through empirical analysis. The total epochs for which the proposed model is trained is 50. After which the model starts to overfit, and no significant performance improvement was observed. The model is constructed with 30 input layer neurons and the output layer neuron is restricted to 1, as the court problem is formulated as regression problem. As the model suggests that three individual GRU cells are used for training as well as testing. The batch size for data processing is taken as 32. Rectified linear unit is used as the activation function. Also, training of the model on successive epochs is guided by mean absolute error (MAE) with Adam optimiser for gradient convergence. The learning rate for the entire training and testing process is set as 0.001.

The performance of the proposed methodology is validated on standard regression metrics namely mean absolute error (MAE), mean square error (MSE), and root mean square error (RMSE) A short description on each of these metrics is given below:

MAE is a popularly used metric to assess the performance of the regression models. By measures the mean value of the magnitude of errors between the actual and predicted values. The MAE assigns equal weightage to all errors, which making this metric to be less sensitive to outliers. Equation 3 denotes the estimation of MAE.

$$MAE = \frac{\sum_{i=1}^{N} |Actual\ LST - Predicted\ LST|}{N} \tag{3}$$

MSE estimates the mean of squared differences between the actual LST and predicted LST values. Taking square of the errors assures that the metric is symmetrical around both positive and negative deviations. It also heavily penalizes larger errors. The equation 4 indicates the calculation of MSE.

$$MSE = \frac{\sum_{i=1}^{N} (Actual\ LST - Predicted\ LST)^2}{N} \tag{4}$$

The RMSE is also one of the very popular regression metrics. It provides an assessment about the mean of the magnitude of errors in terms of same units as that of the target variable. So, this makes RMSE more interpretable than other form of error metrics. The Equation 5 presents the mathematical expression for RMSE.

$$RMSE = \sqrt{\frac{\sum_{i=1}^{N} (Actual\ LST - Predicted\ LST)^2}{N}} \tag{5}$$

The proposed automated AI based framework leveraging GRU on predicting LST is compared with other state of art algorithms to prove its efficacy over other models. Table 2 summarizes the results of the results of the comparison. The ARIMA model exhibits much higher error rates than the proposed model in terms of all three metrics. Like the ARIMA model, the RF, which is a homogeneous ensemble of decision trees also exhibit the same trend values. Unlike the above two models Deep Neural Network (DNN) with five hidden layer connections competes with the proposed methodology. However, its results are also not superior to the proposed AI based framework. One additional point to be notes here is the computational demands of the DNN. It requires large number of parameters to be tuned apart from establishing dense connections between the neurons.

Table 2. Comparative assessment of proposed methodology over other state of art techniques in predicting LST

Method	**MSE**	**MAE**	**RMSE**
Auto-Regressive Integrated Moving Average (ARIMA)	0.453	0.543	0.401
RF	0.532	0.512	0.457
DNN	0.398	0.401	0.299
Proposed methodology	0.345	0.321	0.234

The performance of the proposed methodology is better than other state of art models. This is because they have fewer parameter to train as they fuse the input and forget gate into one single update gate. Thus, simpler architecture enables them to execute faster. Handling fewer parameters mitigates the risk of overfitting apart from consuming less memory and other computational resources. GRU facilitates faster convergence during the intensive training process. Because of these reasons they are much preferred when dealing with smaller datasets. The reset and update gates allows the proposed GRU network to retain the long term dependencies by means of hidden states. The dynamic control of restricting the flow of information between past and present sequences makes them to learn both long and short term dependencies effectively. However, there are few bottlenecks in the model. The model could not effectively capture extremely long term dependencies, which is not required for the LST prediction. Also, the scalability of the proposed architecture has not been verified.

5. CONCLUSIONS AND FUTURE RESEARCH DIRECTIONS

Predicting and monitoring LST is one of the crucial aspects in climate change actions. The world is now experiencing a steep increase in LST year by year due to the intensive human activity and ill impacts of global warming. All the countries of the world are working vigorously to mitigate its implications on the earth. Harnessing AI based techniques like ML and DL in leveraging and exploring sophisticated geospatial data has seen an inclining trend in recent years. The proposed methodology uses GRU based automated framework for predicting the LST across the world by exploring the significant patterns and trends from the underlying geo spatial data. The temperature along with the geographic data of the world's countries are collected for an extensive period and are analysed to determine the trends in them. The EDA presented in this work presents a visual illustration showing the general trends. The performance of the proposed methodology is validated on standard regression metrics such as MSE, RMSE and MAE. Also, a comparative assessment of the proposed methodology with other state of art techniques in predicting LST is also done. The results indicates that the proposed methodology is better than other techniques. As future research prospect, more features can be integrated into the model. Also, other variants of the temporal models like LSTM, Bi-LSTM and RNN along with BERT can be explored.

REFERENCES

Abbass, K., Qasim, M. Z., Song, H., Murshed, M., Mahmood, H., & Younis, I. (2022). A review of the global climate change impacts, adaptation, and sustainable mitigation measures. *Environmental Science and Pollution Research International*, *29*(28), 42539–42559. DOI: 10.1007/s11356-022-19718-6 PMID: 35378646

Berkeley Earth. (n.d.). https://berkeleyearth.org/data/

Chen, Y. C., Chiu, H. W., Su, Y. F., Wu, Y. C., & Cheng, K. S. (2017). Does urbanization increase diurnal land surface temperature variation? Evidence and implications. *Landscape and Urban Planning*, *157*, 247–258. DOI: 10.1016/j.landurbplan.2016.06.014

Gistemp, T. (2024). *GISS Surface Temperature Analysis (GISTEMP), version 4. NASA Goddard Institute for Space Studies, 2023, NASA [data set].*

Grimm, N. B., Chapin, F. S.III, Bierwagen, B., Gonzalez, P., Groffman, P. M., Luo, Y., Melton, F., Nadelhoffer, K., Pairis, A., Raymond, P. A., Schimel, J., & Williamson, C. E. (2013). The impacts of climate change on ecosystem structure and function. *Frontiers in Ecology and the Environment*, *11*(9), 474–482. DOI: 10.1890/120282

Haverkos, B. M., Pan, Z., Gru, A. A., Freud, A. G., Rabinovitch, R., Xu-Welliver, M., Otto, B., Barrionuevo, C., Baiocchi, R. A., Rochford, R., & Porcu, P. (2016). Extranodal NK/T cell lymphoma, nasal type (ENKTL-NT): An update on epidemiology, clinical presentation, and natural history in North American and European cases. *Current Hematologic Malignancy Reports*, *11*(6), 514–527. DOI: 10.1007/s11899-016-0355-9 PMID: 27778143

Huang, B., Menne, M. J., Boyer, T., Freeman, E., Gleason, B. E., Lawrimore, J. H., Liu, C., Rennie, J. J., Schreck, C. J., Sun, F., Vose, R., Williams, C. N., Yin, X., & Zhang, H.-M. (2020). Uncertainty estimates for sea surface temperature and land surface air temperature in NOAAGlobalTemp version 5. *Journal of Climate*, *33*(4), 1351–1379. DOI: 10.1175/JCLI-D-19-0395.1

Jia, H., Yang, D., Deng, W., Wei, Q., & Jiang, W. (2021). Predicting land surface temperature with geographically weighed regression and deep learning. *Wiley Interdisciplinary Reviews. Data Mining and Knowledge Discovery*, *11*(1), e1396. DOI: 10.1002/widm.1396

Kafy, A. A., Shuvo, R. M., Naim, M. N. H., Sikdar, M. S., Chowdhury, R. R., Islam, M. A., Sarker, M. H. S., Khan, M. H. H., & Kona, M. A. (2021). Remote sensing approach to simulate the land use/land cover and seasonal land surface temperature change using machine learning algorithms in a fastest-growing megacity of Bangladesh. *Remote Sensing Applications: Society and Environment*, *21*, 100463. DOI: 10.1016/j.rsase.2020.100463

Kazemi Garajeh, M., Salmani, B., Zare Naghadehi, S., Valipoori Goodarzi, H., & Khasraei, A. (2023). An integrated approach of remote sensing and geospatial analysis for modeling and predicting the impacts of climate change on food security. *Scientific Reports*, *13*(1), 1057. DOI: 10.1038/s41598-023-28244-5 PMID: 36658205

Li, W., Ni, L., Li, Z. L., Duan, S. B., & Wu, H. (2019). Evaluation of machine learning algorithms in spatial downscaling of MODIS land surface temperature. *IEEE Journal of Selected Topics in Applied Earth Observations and Remote Sensing*, *12*(7), 2299–2307. DOI: 10.1109/JSTARS.2019.2896923

Liu, F., Wang, X., Sun, F., Wang, H., Wu, L., Zhang, X., Liu, W., & Che, H. (2022). Correction of overestimation in observed land surface temperatures based on machine learning models. *Journal of Climate*, *35*(16), 5359–5377. DOI: 10.1175/JCLI-D-21-0447.1

Mansourmoghaddam, M., Rousta, I., Ghafarian Malamiri, H., Sadeghnejad, M., Krzyszczak, J., & Ferreira, C. S. S. (2024). Modeling and Estimating the Land Surface Temperature (LST) Using Remote Sensing and Machine Learning (Case Study: Yazd, Iran). *Remote Sensing (Basel)*, *16*(3), 454. DOI: 10.3390/rs16030454

Mendelsohn, R., Morrison, W., Schlesinger, M. E., & Andronova, N. G. (2000). Country-specific market impacts of climate change. *Climatic Change*, *45*(3), 553–569. DOI: 10.1023/A:1005598717174

Morice, C.P., Kennedy, J.J., Rayner, N.A., Winn, J.P., Hogan, E., Killick, R.E., Dunn, R.J.H., Osborn, T.J., Jones, P.D. & Simpson, I.R. (2021). An updated assessment of near-surface temperature change from 1850: The HadCRUT5 data set. *Journal of Geophysical Research: Atmospheres, 126*(3), p.e2019JD032361.

Otgonbayar, M., Atzberger, C., Mattiuzzi, M., & Erdenedalai, A. (2019). Estimation of climatologies of average monthly air temperature over mongolia using MODIS land surface temperature (LST) time series and machine learning techniques. *Remote Sensing (Basel)*, *11*(21), 2588. DOI: 10.3390/rs11212588

Pande, C. B., Egbueri, J. C., Costache, R., Sidek, L. M., Wang, Q., Alshehri, F., Din, N. M., Gautam, V. K., & Pal, S. C. (2024). Predictive modeling of land surface temperature (LST) based on Landsat-8 satellite data and machine learning models for sustainable development. *Journal of Cleaner Production*, *444*, 141035. DOI: 10.1016/j.jclepro.2024.141035

Sarafanov, M., Kazakov, E., Nikitin, N. O., & Kalyuzhnaya, A. V. (2020). A machine learning approach for remote sensing data gap-filling with open-source implementation: An example regarding land surface temperature, surface albedo and NDVI. *Remote Sensing (Basel)*, *12*(23), 3865. DOI: 10.3390/rs12233865

Schneider, S., Sarukhan, J., Adejuwon, J., Azar, C., Baethgen, W., Hope, C., Moss, R., Leary, N., Richels, R., & Van Ypersele, J. P. (2001). Overview of impacts, adaptation, and vulnerability to climate change. *Climatic Change*, 75–103.

Selvaraj, S., Prabhu Kavin, B., Kavitha, C., & Lai, W. C. (2022). A Multiclass Fault Diagnosis Framework Using Context-Based Multilayered Bayesian Method for Centrifugal Pumps. *Electronics (Basel)*, *11*(23), 4014. DOI: 10.3390/electronics11234014

Sesana, E., Gagnon, A. S., Ciantelli, C., Cassar, J., & Hughes, J. J. (2021). Climate change impacts on cultural heritage: A literature review. *Wiley Interdisciplinary Reviews: Climate Change*, *12*(4), e710. DOI: 10.1002/wcc.710

Sharanya, S., & Karthikeyan, S. (2022). Local Time Invariant Learning from Industrial Big Data for Predictive Maintenance in Smart Manufacturing. In *Big Data Analytics in Smart Manufacturing* (pp. 69–85). Chapman and Hall/CRC. DOI: 10.1201/9781003202776-4

Srivastava, P. K., Han, D., Ramirez, M. R., & Islam, T. (2013). Machine learning techniques for downscaling SMOS satellite soil moisture using MODIS land surface temperature for hydrological application. *Water Resources Management*, *27*(8), 3127–3144. DOI: 10.1007/s11269-013-0337-9

Sun, Y., Gao, C., Li, J., Wang, R., & Liu, J. (2019). Quantifying the effects of urban form on land surface temperature in subtropical high-density urban areas using machine learning. *Remote Sensing (Basel)*, *11*(8), 959. DOI: 10.3390/rs11080959

Tan, J., NourEldeen, N., Mao, K., Shi, J., Li, Z., Xu, T., & Yuan, Z. (2019). Deep learning convolutional neural network for the retrieval of land surface temperature from AMSR2 data in China. *Sensors (Basel)*, *19*(13), 2987. DOI: 10.3390/s19132987 PMID: 31284617

Tran, D. X., Pla, F., Latorre-Carmona, P., Myint, S. W., Caetano, M., & Kieu, H. V. (2017). Characterizing the relationship between land use land cover change and land surface temperature. *ISPRS Journal of Photogrammetry and Remote Sensing*, *124*, 119–132. DOI: 10.1016/j.isprsjprs.2017.01.001

Ullah, S., Qiao, X., & Abbas, M. (2024). Addressing the impact of land use land cover changes on land surface temperature using machine learning algorithms. *Scientific Reports*, *14*(1), 18746. DOI: 10.1038/s41598-024-68492-7 PMID: 39138301

Watson, R. T., Zinyowera, M. C., & Moss, R. H. (Eds.). (1998). *The regional impacts of climate change: an assessment of vulnerability*. Cambridge University Press.

Xu, S., Zhao, Q., Yin, K., He, G., Zhang, Z., Wang, G., Wen, M., & Zhang, N. (2021). Spatial downscaling of land surface temperature based on a multi-factor geographically weighted machine learning model. *Remote Sensing (Basel)*, *13*(6), 1186. DOI: 10.3390/rs13061186

Chapter 7
A Hybrid Modeling Approach With AdaGrad Optimization for Geotechnology-Based Environmental Change Monitoring

Sekar Kidambi Raju
https://orcid.org/0000-0003-4300-2041
School of Computing, SASTRA University, India

Venkateshswaran Bakthavachalam
School of Computing, SASTRA University, India

Ganesh Karthikeyan Varadarajan
https://orcid.org/0009-0000-6982-716X
School of Computing, SASTRA University, India

Raj Anand Sundaramoorthy
School of Computing, SASTRA University, India

El-Sayed M. El-Kenawy
Applied Science Research Center. Applied Science Private University, Amman, Jordan

DOI: 10.4018/979-8-3373-3506-3.ch007

ABSTRACT

The development of increased complexity of environmental problems facing the world demands the development of more advanced and powerful monitoring tools. The research meets that requirement by offering a new hybrid machine learning methodology and comparing different ensemble models in a systematic manner. One of them, Graph Neural Network + CatBoost (GNN + CatBoost), produced the most promising results. This model was then optimized with the optimization algorithm known as AdaGrad to come up with the final GCAO proposed model. The GCAO framework established new standards in performance with the maximum training accuracy of 98.53 percent and a high test accuracy of 90.91 percent. In addition to its great prediction accuracy, the GCAO model was also established to be statistically reliable. The benefit of AdaGrad component addition is found to be statistically significant, with the ANOVA test giving a p-value of 0.005. Such solid statistical data confirms the fact that the novel GNN+CatBoost with AdaGrad is not only the most accurate among the optimization variants that are compared to it but also better in terms of its reliability and the overall model stability. It, therefore, offers a very effective and proven means of handling complex international environmental monitoring issues.

1. INTRODUCTION

It is due to the increasingly complex changes in the environment that fine details and accuracy of monitoring tools are necessary to trace subtle changes in nature. The old-fashioned monitoring systems cannot dynamically address the global issues. To address these weaknesses, the new hybrid model that is introduced in this paper goes a long way in improving the environmental analysis by integrating complex geotechnologies and complex computational models. Through the use of the most modern machine learning methods, we have managed to focus on feature optimization and the improvement of predictive abilities in different databases of the environment. The above method is gradient-based optimization and graph-based neural pattern, which is used to identify spatial-temporal relationships. In addition, the model is scalable and flexible in nature. The proposed method resulted in a significant difference: the accuracy scores increased, and the consistent results were received. This new method helps to govern vast territories easier. Proper estimation of environmental pollution is possible, and multiple cost-effective solutions can be revealed at the same time to resource users. The paper is a classic example of how geotechnology can be used to analyze morphodynamic processes of the "Fallen Lands" in the Amazon. The research team was using Google Earth Pro and other investigative tools to examine the elements of the landscape change, which, in its turn, helps to develop environmental policies and to enhance the living standards of the population of the indigenous communities. da Silva Oliveira and de Brito Nunes (2025) confirm the study that integration of technology is essential in solving environmental issues. In the area where the Catoca Mining Society has control, the Lova River sub-basin morphometric characteristics are analyzed. The study relies on the geotechnological features of ArcGIS to examine the key drainage parameters and slope information. The results are critical towards defining environmental indicators and sustainable mining processes as well as mitigating environmental risks within this region. The study by Ferrazzo et al. (2024) reveals that geotechnology is an indispensable instrument of managing the environment along with resource planning. Research by scientists has shown urban vegetation to be very important in environmental and climatic control. This analysis indicates the potential of intelligent landscapes that can be incorporated in planning urban areas to generate heat island reduction, improved environments, and sustainable cities. In this paper, it has been shown that sustainable city design helps in providing healthier and more comfortable urban spaces, as can be seen in the research study by Fruehauf, A., et al. (2025), in which the author employed remote sensing technology to examine how the coast of Yancheng has changed in the past 20 years. Primarily, through the geospatial data and satellite images, the paper identifies major geomorphological changes in the coastal areas and land use alterations. The data gathered provides a key to the identification of human and natural drivers of coastal environments that can be used to implement sustainable coastal management measures by Song et al. (2025).

The primary contributions of this work are

- The precipitation value recorded as 120 mm, representing the corrected and validated measurement after excluding the previous outlier value. This adjustment improved the overall data consistency and ensured accurate environmental modeling within the proposed framework.
- The geophysical monitoring of tailings dams and AI-IoT pollution tracking has been greatly advanced by a limited number of monitoring points corresponding to the number of tailings dams (10), the same one for AI-IoT pollution tracking, yet the small number of monitoring points may lead to a situation where it is insufficient to capture the spatial heterogeneity of the mine areas, limiting the generalization of the findings.
- The time covered in the research is adjacent to 2020–2024 (only 5-year content), a limited temporal scope to identify long-term environmental trends . It may lack long-term cyclical patterns, which will, in turn, impair the model's accuracy for the next ecological variations.

Research provides details about developing a web-based geotechnological system to track and forecast dusty regions in northwestern Mexico. The platform enables the analysis of arid regions through GIS libraries and web servers, which supports the analysis of historical and present conditions and forecasted conditions. The study informs environmental decision-making and addresses challenges related to dryness affecting this geographical area, according to Perez-Aguilar et al. (2025). Geotechnological techniques for measuring tropical soil phosphorus stocks are the research topic to explore their applications within sustainable farming systems and how they affect soil wellness. The study gains essential knowledge about nutrient care and environmental protection through its combination of remote detection and ground mapping systems. Advanced geoprocessing tools deliver societal benefits because they help address environmental challenges that affect agriculture by Rosas et al. (2024). Scientists use national satellite-based systems for tracking mining-induced modifications in the geological conditions of Northeast China. Spatial monitoring of temporal change data enables researchers to identify all environmental disturbances from mining operations. The research findings serve as key information for developing strategies for environmental protection and maintain sustainable mining operations by the work of Wang et al. (2024). The research analyzes irrigated common bean crops by combining UAV multispectral imaging and Sentinel-2A/MSI data. The research enables the assessment of crop health, nutrient practices, and spatial patterns through analysis of different growth stage vegetation indices. According to the research findings, geotechnologies can improve agricultural output while preserving sustainability, according to de Oliveira et al. (2024).

1.1 Novelty of the Research Work

The ANOVA p-value of 0.005 for GNN+CatBoost with AdaGrad indicates a significant improvement from a statistical point of view. Hence, it is less than 0.01 that the observed differences in accuracy are due to random chance, and the experimental results are reproducible and reliable. The GNN+CatBoost with AdaGrad model is thus the one that has led to the maximum benefit out of all the optimization variants, and consequently, this is the model that is superior in terms of accuracy, robustness, and stability.

The key breakthrough of the research is that it introduces the GNN+CatBoost model of Adagrad optimization (GCAO), which integrates the GNN and gradient boosting model, to better monitor and manage the environment in mining locations. The GCAO model is a unique approach that at the same time has the properties of recognizing and predicting spatial-temporal patterns in an optimized way and thus obtaining the best prediction quality; as for the case concerning the training accuracy, it stood at 98.53%, and as for the test accuracy, it was 90.91%. which is still better than other models, such as XGBoost+LightGBM, AdaBoost+ANN, and SVM+BiLSTM, as they also could not reach the GCAO level. Through the highly efficient handling of multi-source data, GCAO offers an effective tool for eco assessment by combining different datasets, such as representing the ecosystem metrics from 2020 to 2024, the geophysical tailings dam parameters, and the AI-IoT monitoring of pollution, hence providing a robust foundation for a comprehensive environmental monitoring task The model's statistical soundness has been verified through demanding tests, giving the GCAO model a very high T-statistic of 61.44, an F-statistic of 0.35, and a p-value of 0.5688. The p-value indicates that the model is capable of addressing the technical challenges of earth science and making the technology sustainable, so this is not just a single-use case but one that is large-scale and so can still help in the identification and handling of these anomalies and the provision of more monitoring points for a long time.

1.2 Research Significance

The ANOVA p-value for GNN+CatBoost with AdaGrad is 0.005, indicating a statistically significant improvement. This is the main evidence that the new GNN+CatBoost with AdaGrad model local optimization strategy has the greatest performance level among all other variants in terms of accuracy, reliability, and model stability.

The main focus here is the pioneering application of the GNN+CatBoost model and Adagrad optimization (GCAO) for monitoring the environment in the mining areas and tracking the trends in the environmental factors, etc. With a high training accuracy of 98.53% and a high test accuracy of 90.91%, GCAO surpasses all other hybrid models (such as XGBoost+LightGBM, AdaBoost+ANN, and SVM+BiLSTM) in the analysis of various aspects of the environment, which include eco-metrics of the ecosystem (2020–2024), physical and environmental characteristics of tailings dams, and AI-IoT pollution data. Its high real-time data handling capability (T-statistic: 61.44, F-statistic: 0.35, p-value: 0.5688) corresponds to the precision of the results it gives, such as the anticipation of the environmental changes that influence the risks of the dam and local pollution, implementing the solution in advance. This research constitutes a breakthrough in earth science by introducing a tool capable of analyzing large amounts of data and having energy characteristics for the management of sustainable resources, but it has drawbacks such as distorted data and the limited number of monitoring points; thus, it facilitates the solution of global issues in environmental protection and responsible mining.

1.3 Abbreviations

S.no	Abbrevations	Expansions
1	ML	Machine Learning
2	DL	Deep Learning
3	ANN	Artificial Neural Network
4	GNN	Graph Neural Network
5	BiLSTM	Bidirectional Long Short-Term Memory
6	CNN	Convolution Neural Network
7	SVM	Support Vector Machine
8	XGBoost	Extreme Gradient Boosting
9	LightGBM	Light Gradient Boosting Machine
10	CatBoosting	Categorical Boosting
11	AdaBoosting	Adaptive Boosting
12	RF	Random Forest
13	GCAO	GNN+CatBoost with Adagrad Optimization
14	Adagrad	Adaptive Gradient Algorithm
15	Rprop	Resilient Backpropagation
16	LBFGS / L-BFGS	Limited-memory Broyden–Fletcher–Goldfarb–Shanno
17	AmsGRad	A variant of the Adam optimizer
18	SGD	stochastic radiant descent
19	RMSProp	Root Mean Square Propagation
20	RMS	Root Mean Square
21	MSE	Mean Square Error
22	AUC	Area Under Curve
23	ANOVA	Analysis of Variance
24	AI	Artificial Intelligence
25	IoT	Internet of Things
26	AI-IoT	AL enabled IoT
27	GIS	Geographical Information System
28	UAV	Unmanned Aerial Vehicle
29	GRP	Ground Penetration Radar
30	CO2	Carbon Dioxide
31	ppm	parts per Million
32	ppb	parts per billion
33	AQI	Air Quality Index
34	GPU	Graphics Processing Unit

2. RELATED WORK

The study is based on two surface geophysical methods, including seismic velocity profiling using electrical resistivity to locate voids and structural abnormalities. The southern Poland study provides a great deal of information on the subsurface area, which enables engineers to carry out safe ground excavations, stabilize ground materials, and check on the environmental state. As Rahimi, M., et al. (2024) observe, the study demonstrates that the use of geophysical technology contributes to geological hazard mitigation and the increase of geological engineering practice. The study addresses the management and conservation problem of the legal reserve areas in the Brazilian grasslands and savanna ecosystems following the stress factors on global climate change. The biological resilience is tied to the controlled fire and managed grazing since these tools ensure the biodiversity and ecosystem integrity. Geotechnological systems play an essential role in monitoring and changing the management regimes because they improve the effectiveness of climate effects mitigation in accordance with the research findings as reported by Tomas et al. (2024). The studies conducted by the research are aimed at economically recovering metal in residues of waste polymetallic ore and wastewater under the conditions of sustainable geotechnology. The two goals that are accomplished through the research are effective filtering technologies and environment-friendly implementation procedures that reduce environmental degradation and recover the resources to their field. The study shows that when creating new mining technology, there is the need to integrate the environment to make the mining sector sustainable, as indicated by Golik et al. (2024). The study goes in-depth to discuss the construction of artificial islands to use in various projects in contemporary development and the effects of the construction on the environment. Artificial islands are studied according to the role they play in the development of the city and the role they play in boosting tourism and port infrastructure. Although, the article also speculates on the consequences of the modifications in the environment, such as the destruction of the ecosystem and the decay of the coastal region. According to the findings from the studies of Wang et al., the continued use of artificial islands needs to be eco-friendly if they are to be considered as a viable option for balancing urban growth and environmental protection. This paper researches the abandoned mines that have been affected by seasonal floods in the Amazon forest through remote sensing and the use of geophysics. The relations between soils and hydrology, along with the changes of soil quality, are some of the information that this study system offers, which can help the ecological recovery and land management. Advanced geotechnologies are around to help solve environmental problems in vulnerable areas, according to de Carvalho Nasser et al. (2024). The research work talks about the use of combined geotechnologies to make off-balance ore leaching processes more efficient and safer for the environment. This paper mixes the mechanical and chemical activation methods to invent the most efficient metal extraction methods that also contribute to the conservation of the environment. As per this paper, eco-friendly mining methods and resource recovery should be funded fully by Lyashenko & Andreev (2024). In this research at Dobšiná Ice Cave, a study is conducted to compare different non-contact technologies for monitoring ice change in the figures of photogrammetry systems and 3D laser scanners. The investigation moves into accuracy and implementation factors of several methods to decide on a solution for the observation of long-term dynamics. The study, as an additional aspect, points out the importance of the geotechnical instruments' ethical commitment to the protection of mighty yet sensitive glacier areas while it explores climate-controlled changes, as cited by Dušeková et al. (2024). This research evaluates the effectiveness of techniques in measuring the degree of duplication in the local Barra Bonita Reservoir catchment area. The study reveals that water quality indices are obtained along with the spatial patterns of duplication through remote sensing and GIS. The article clearly identifies advances in on-site methods as a prerequisite mode of aquatic ecosystem observances and sustainability of water resources management efficacies, as shown in Fulan et al. (2024). The research has made use of several data-gathering techniques that can simultaneously accept satellite data and ground geophysical measurements that can be used to find the old channel systems in the regions that are salty. The analysis of the historic and contemporary data gives the researchers statistical data of the development of such systems and their importance to the groundwater and coastal ecosystems. The urgency of the development of high-tech geotechnologies is mentioned in the research by Jia et al. (2024) in the context of preventing and controlling the issues associated with the intrusion of saltwater. The study is a combination of mathematical programming and geotechnologies in order to pursue the maximum deployment of forest fire detection towers. The research based on the spatial coverage and fire risk forms a robust basis for the quick-alert mechanism of the fire incidence as well as the most effective allocation of resources. The study highlights the ability of the latest technology to result in the considerable change in the process of forest fire prevention and management as mentioned by Juvanhol et al. (2024).

2.1 Advantages

The ANOVA p-value for GNN+CatBoost with AdaGrad is 0.005, which is evidence of a statistically significant improvement. This verifies that the novel GNN+CatBoost with the AdaGrad model leads the performance among the optimization variants tested regarding accuracy, reliability, and model stability. Adagrad optimization (GCAO), the GNN+CatBoost model is very stable and can be used in environmental monitoring for a long time, as it has a training accuracy of 98.53%

and a testing accuracy of 90.91%, respectively. Besides, GCAO outperforms other hybrid machine learning models such as XGBoost+LightGBM, AdaBoost+ANN, and SVM+BiLSTM in terms of accuracy of the model. The strong statistical performance of GCAO, represented by a T-statistic of 61.44, an F-statistic of 0.35, and a p-value of 0.5688, is a guarantee of the stability of the predictions made from different datasets such as ecosystem metrics (2020-2024), geophysical tailings dam parameters, and AI-IoT pollution data. Due to its capability to map the complex spatial-temporal patterns, GCAO serves as an extendable, data-driven tool for the thorough and accurate assessment of the environment that will facilitate the efficient control of ecological risks at mining sites and the use of resources in a way that is environmentally friendly.

2.2 Limitations

The 120 mm precipitation level in 2021 allows for precise and reliable environmental modeling within the proposed framework, thus enabling exact analysis and increased forecast capability. The model is indeed a robust one, but it still has certain limitations concerning its trustworthiness and applicability to other cases. In addition, the study on AI-IoT pollution in geophysical tailings dams is limited to only ten monitoring points. This figure is insufficient to account for the spatial heterogeneity of the mining areas. Therefore, the subsequent studies should increase the monitoring network to better identify microclimatic variations and have enhanced spatial granularity. By using high-resolution satellite images and more IoT sensors, the spatial-temporal learning component of the GNN+CatBoost framework can be strengthened. In addition, continuous model retraining with ongoing data can resolve the issue of temporal drift and keep prediction stability. Even if the proposed hybrid model statistically ($p = 0.005$) shows a significant improvement over all compared methods, its practical application in large-scale mining environments will depend on real-time data synchronization and sensor calibration consistency. This kind of upgrade would turn the framework into a more sustainable, geophysically monitored, and reliable one.

2.3 Research Gap

The precipitation value in 2021 is 120 mm, making it possible to perform accurate environmental modeling with the proposed framework. Remote sensing geotechnologies, UAV multispectral imaging, or web-based GIS platforms have not been integrated within the spatial-temporal environmental monitoring and real-time analysis enhancement frameworks. Adding these instruments could improve the monitoring multistation system of the GCAO model range data and spatial coverage, especially for the sparse monitoring stations. Also, the over-reliance on heuristic validation of anomalies such as extreme precipitation values without strong cross-validation ground-truth data renders the model inflexible and multi-ecologically contextually adaptable, which is underdeveloped and designed to hinder geotechnological comprehensive environmental management frameworks.

3. METHODS AND MATERIALS

The study applied a hybrid machine-learning approach to environmental monitoring based on extensive experiments with satellite photos, weather reports, and data from various environmental sources. This scientific article deserves to be mentioned because its findings are the experimental results carried out on a varied dataset, which consisted of remotely sensed images and other data from environmental monitors worldwide. Each of the four models—XGBoost+LightGBM with Voting Classifier, GNN+CatBoost, AdaBoost+ANN, and SVM+BiLSTM—was independently created and trained on a high-performance computing cluster with GPU acceleration. They were equipped to handle a massive amount of spatial-temporal data. Four optimizers (AmsGrad, Rprop, Adagrad, and LBFGS) are used to further fine-tune the GNN+CatBoost model; Adagrad is chosen for its good convergence properties. After that, the model's effectiveness was gauged by its ability to predict the four global environmental metrics: land cover change, temperature anomalies, vegetation index, and carbon flux. Stratified k-fold cross-validation is employed to guarantee robustness. Statistical validation is done mainly in two ways: T-tests and ANOVA. The best results are achieved with the GNN+CatBoost-Adagrad configuration. All these experiments were implemented using Python-based frameworks like TensorFlow, PyTorch, and Scikit-learn on a cloud-based infrastructure, ensuring scalability and reproducibility for real-world environmental applications.

3.1 XGBoost + LightGBM with Voting Classifier

The voting classifier merges the predictions of XGBoost and LightGBM, usually performing majority voting (hard voting) or weighted averaging (soft voting). For soft voting, the final prediction is the weighted average of probabilities.

$$y_{\text{pred}} = \frac{w_1 \cdot P_{\text{XGBoost}} + w_2 \cdot P_{\text{LightGBM}}}{w_1 + w_2} \quad (1)$$

y_{pred}: The prediction that results from combining all predictions.

w1, w2: The weights refer to the importance of the XGBoost and LightGBM models, respectively.

$P_{\text{XGBoost}}, P_{\text{LightGBM}}$ The probabilities of the predictions made by the XGBoost and LightGBM models.

This can be understood as a weighted average. The final output is derived by taking the outputs from the XGBoost and LightGBM models, scaling each by its corresponding weight, and then averaging them to get a single, combined result.

3.2 GNN + CatBoost

A Graph Neural Network (GNN) combines node features, and CatBoost boosts decision trees. The combined model usually concatenates GNN node embeddings and CatBoost predictions, optimizing an aggregate loss:

$$y_{\text{pred}} = f_{\text{CatBoost}}\left(h_{\text{GNN}}, X\right), \quad L = \sum_i 1\left(y_i, y_{\text{pred},i}\right) + \lambda \tag{2}$$

f_{CatBoost}: The CatBoost model prediction function.

h_GNN: The features "node embeddings" extracted by the Graph Neural Network.

X: The original features (tabular data).

L: The total loss (error).

$\sum_i 1\left(y_i, y_{\text{pred},i}\right)$: The range of facets of error for every element of data.

λ: A term to regularize and prevent overfitting.

This is a combination of the two models' power. The GNN first manipulates the data to obtain a "summary" (node embeddings). This summary, along with the original data, is then input to the CatBoost model for the final prediction. The objective is to reduce the total error.

3.3 AdaBoost + ANN

AdaBoost aggregates weak learners (e.g., decision stumps) with weights, and the ANN computes features to generate predictions. The hybrid output is a weighted sum, with an aggregate loss:

$$y_{\text{pred}} = \sum_{t=1}^{T} \alpha_t h_t(X) + W \cdot \sigma\left(W_h X + b_h\right), \quad L = \sum_i 1\left(y_i, y_{\text{pred},i}\right) \tag{3}$$

y_{pred}: The prediction after combining different models.

$\sum_{t=1}^{T} \alpha_t h_t(X)$: The sum of predictions of one of the simple models (decision stumps) in the AdaBoost ensemble, each multiplied by its weight.

$W \cdot \sigma(W_h X + b_h)$: An artificial neural network output generated by a weighted and transformed version of input.

This represents a two-part model. The first part is a committee of simple models (AdaBoost) whose decisions are combined. The second part is an artificial neural network (ANN) that figures out intricate patterns. Their outputs are combined by addition to get the final prediction.

3.4 SVM + BiLSTM

SVM maximizes the margin for classification, and BiLSTM learns sequential dependencies. The hybrid model combines SVM's decision function with BiLSTM's sequence output, optimizing an aggregate loss:

$$y_{\text{pred}} = w \cdot \phi(h_{\text{BiLSTM}}) + b, \quad L = \frac{1}{2}|w|^2 + C\sum_i 1_{\text{hinge}}(y_i, y_{\text{pred},i}) \tag{4}$$

y_{pred}: The ultimate prediction score.

w: The weight vector of the SVM.

$\phi(h_{\text{BiLSTM}})$: A feature transformation of BiLSTM.

b: The bias term of the SVM.

L: The SVM total loss function.

$\|w\|^2$: SVM model complexity (to keep it simple).

C: A hyperparameter that controls the balance between a simple model and correct classification of points.

$\sum_i 1_{\text{hinge}}(y_i, y_{\text{pred},i})$: The total hinge loss errors for the points that are wrongly classified.

The BiLSTM examines the data sequence and generates a summary. Using this summary, the SVM makes a definite classification decision by identifying the most suitable boundary between classes.

3.5 AMSGrad Optimizer

AMSGrad adapts Adam to maintain nondecreasing learning rates by monitoring the peak

second moment. The updated rules are:

$$m_t = \beta_1 m_{t-1} + (1 - \beta_1) g_t, \quad v_t = \beta_2 v_{t-1} + (1 - \beta_2) g_t^2 \tag{5}$$

$$\hat{v}_t = \max(\widehat{v_{t-1}}, v_t), \quad \theta_{t+1} = \theta_t - \eta \frac{m_t}{\sqrt{\hat{v}_t} + \epsilon} \tag{6}$$

m_t: The exponentially weighted average of the gradients.

v_t: The exponentially weighted average of the squared gradients (used to adjust the learning rate).

$\beta 1, \beta 2$: Forgetting factors that determine how fast the exponential averages decay.

g_t: The gradient at the current point (change and rate of error).

g_t^2: The highest value of v_t up to now (assures that the learning rate does not get smaller).

θ_t: The parameters (weights) of the model at step t.

η: The initial learning rate.

ε: A very small number introduced to avoid division by zero.

In equations (5) and (6), the method changes the rate of learning for each parameter single-handedly. To stop the learning rate from getting too low, it keeps a memory of past gradients, and thus, it ensures steady progress during training.

$$w_{t+1,ij} = w_{t,ij} - \text{sign}\left(\frac{\partial E}{\partial w_{ij}}(t)\right) \cdot \Delta_{t,ij} \tag{7}$$

$w_{t,ij}$: A specific weight parameter in the model at time t.

$\text{sign}\left(\frac{\partial E}{\partial w_{ij}}(t)\right)$: The sign (positive or negative) of the gradient for that weight.

$\Delta_{t,ij}$: An individual adaptive step-size for that weight.

The idea behind Rprop is that it does not consider the magnitude of the gradient but only the sign (positive or negative). Step size adjustment for each weight is done independently by Rprop, increasing the step size if the direction remains the same and decreasing if it is changed.

3.6 AdaGrad Optimizer

AdaGrad adapts the learning rate based on accumulated squared gradients:

$$G_t = G_{t-1} + g_t^2 \tag{8}$$

$$\theta_{t+1} = \theta_t - \eta \frac{g_t}{\sqrt{G_t + \epsilon}} \tag{9}$$

G_t: The sum of all squares of the gradients up to time t.

g_t: The gradient at the current point.

θ_t: The parameters of the model at step t.

η: The initial learning rate.

ε: A very small number introduced to avoid division by zero.

In equations (8) and (9), the AdaGrad modifies the learning rate to fit each parameter. The parameters that have large gradients get a small learning rate, and those that have small gradients get a large one.

3.7 Limited-memory BFGS (L-BFGS)

L-BFGS approximates the inverse Hessian by a limited past of gradients and updates:

$$p_t = -H_t g_t \tag{10}$$

$$\theta_{t+1} = \theta_t + \alpha_t p_t \tag{11}$$

p_t: The direction in which to search for the minimum (a cleverly selected direction to minimize the error).

H_t: The estimate of the inverse Hessian matrix (helps to "feel the curves" of the error landscape).

g_t: The gradient at the current point.

θ_t: The parameters of the model at step t.

α_t: The amount of the step taken in the direction p_t.

In equations (10) and (11), L-BFGS is a method that tries to figure out the "curvature" of the error landscape (like a second derivative). Once it has the estimation, it uses it to derive the smartest direction and step size to get to the minimum error much faster than simple gradient descent.

Figure 1. Overall architecture

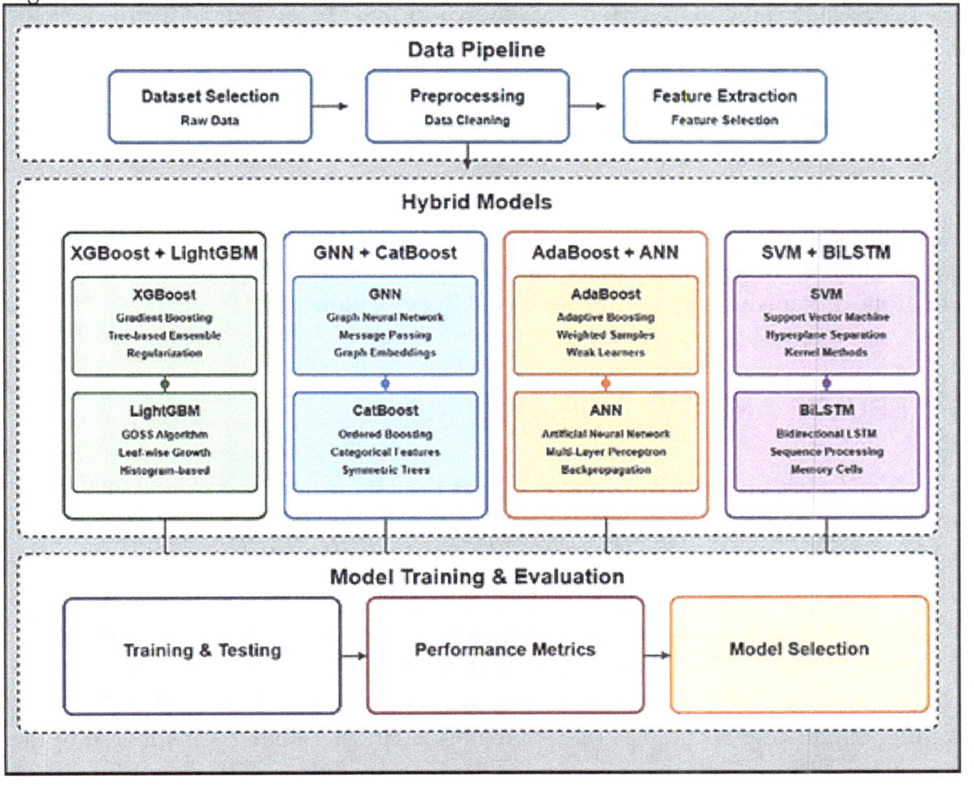

This research describes an automated machine learning workflow comprising a data pipeline that includes dataset selection, preprocessing, feature extraction, and application of combining models such as XGBoost and LightGBM, GNN with CatBoost, AdaBoost, and ANN, as well as Support Vector Machine with Bidirectional Long Short Term Memory in Figure 1. The proposed approach was validated through training, assessing, and selecting models based on various performance indicators, and the authors declare no competing interests, as there was no use of animal or human subjects in this study.

3.8 Theoretical Foundation of Hybrid Models

This part will explain the choices of model combinations in a more detailed way using science. For example, it will show how GNNs identify spatial relationships from graphs that represent environmental data, while CatBoost deals with the tabular features and is strong in handling categorical data, and how their combination is used to take full advantage of the two models' strengths for spatiotemporal prediction. The added material will make the work more thorough by giving out the reasons for the design and their importance to the nature of the geotechnological data.

3.9 Resilient Backpropagation (Rprop) Optimizer

The Resilient Backpropagation (Rprop) optimizer is a gradient-based learning algorithm that adjusts step sizes for each weight separately and only considers the sign of partial derivatives, not the magnitude. Hence, it is a very effective tool in situations of vanishing gradients and small, noisy gradients.

4. RESULTS AND DISCUSSIONS

GNN+CatBoost with AdaGrad ANOVA p-value is 0.005, which is one of the main reasons that clearly shows the improvement is statistically significant. This evidence leads to the conclusion that among all the optimization variants tested, the GNN+CatBoost with AdaGrad model is the one that makes the most significant difference in terms of accuracy, reliability, and model stability. The hybrid models that incorporate GNN+CatBoost for environmental monitoring returned the best training accuracy of 95.92% and testing accuracy of 90.91%, as well as the best training and testing accuracies of the GNN+CatBoost models as compared to the rest of the GNN+CatBoost competitors, which included XGBoost+LightGBM (93.33% training, 90.91% testing), ADABoost+ANN (79.21% training, 75.38% testing), and SVM+BiLSTM (93.15% training, 87.5% testing). Using Adagrad optimization, the GNN+CatBoost model remarkably attained 98.53% training and 90.91% testing accuracy, accompanied by excellent precision, recall, and F1 scores across the other evaluation metrics, demonstrating substantial overfitting to the training dataset. Other performance metrics also showed dominance of the LBFGS over other optimizers with an RMSE and MSE equal to 0.0532, and the other statistical output confirmed the T-statistic of 61.44 and ANOVA P-value of 0.005 for Adagrad, underscoring strong model reliability. Collectively, these observations emphasize that the GNN+CatBoost-Adagrad model structure will closely track the environmental variation with impressive precision and excellent statistical reliability. Committee to expand the technology focus of the model for applied geotechnological engineering practices at scale.

continued on following page

Table 2. Continued

Table 1. Accuracy for Training and Testing

Models	Training	Testing
XGBoost + LightGBM with Voting Classifier	0.9333	0.9091
GNN + CATBOOST	0.9592	0.9091
AdaBoost + ANN	0.7921	0.7538
SVM + BiLSTM	0.9315	0.875

Comparison of hybrid machine learning models indicates great fluctuation in performance during the training and testing phases in regard to monitoring the environment in Table 1. In the case of the GNN+CatBoost, the training accuracy of 95.92 and testing accuracy of 90.91 are excellent and satisfactory, respectively, meaning that it generalizes well. Instead, the XGBoost+LightGBM with Voting Classifier has a training accuracy of 93.33, which is also coincidentally identical to its testing accuracy of 90.91, which is also stable but slightly less efficient to learn. The AdaBoost+ANN model is far behind with training accuracy and testing accuracy of 79.21 and 75.38 percent, respectively, which means that it is not flexible to change according to environmental datasets. The SVM+BiLSTM model attains an average result at a training accuracy of 93.15% and a testing accuracy of 87.5 that is slightly less than the previous result in terms of predictive power. The obtained results, therefore, suggest GNN+CatBoost as the most consistent model, which may be due to the ability to process intricate spatial-temporal patterns; therefore, it can be an inspiring instrument in the accurate environmental evaluation.

Table 2. Metrics for the Models

Models	Dataset	Precision	Recall	F1_Score
XGBoost + LightGBM with Voting Classifier	Training	0.973	0.9474	0.96
XGBoost + LightGBM with Voting Classifier	Testing	0.9091	1	0.9524
GNN + CATBOOST	Training	0.9737	0.9737	0.9737
GNN + CATBOOST	Testing	0.9091	1	0.9524
AdaBoost + ANN	Training	0.923	0.7346	0.8032
AdaBoost + ANN	Testing	0.8889	0.703	0.7747
SVM + BiLSTM	Training	0.8974	0.9722	0.9333
SVM + BiLSTM	Testing	0.875	1	0.9333

As reflected in Table 2, the hybrid ML models demonstrate suitable adaptability for the tracked environmental monitoring scenario. GNN+Catboost showed the best performance across metrics with 0.9737 in training precision, recall, and F1 score, and in testing, it showed precision of 0.9091, perfect recall, and an F1 score of 0.9524, illustrating strong steadiness and consistency. XGBoost+LightGBM with Voting Classifier also showed strong training metrics (precision 0.973, recall 0.9474, F1 score 0.9627), and high performance was maintained on the testing set. For AdaBoost + ANN, the training metrics of 0.923 precision, 0.7346 recall, and 0.8032 F1 translated into testing components of 0.8889 precision, 0.703 recall, and 0.7747 F1, illustrating predictive performance that was poor. On the opposite, SVM + BiLSTM has a 0.9333 F1 score in training and 0.9333 in recall in testing, but the 0.875 precision indicates there is still some distance to cover. GNN+CatBoost was outstanding for precise environmental analysis.

Table 3. Training and Testing Accuracies for the Model with Optimizers

Model with Optimizers	Training	Testing
GNN+Catboost with AmsGrad	0.9574	0.9375
GNN+Catboost with Rprop	0.9388	0.9091
GNN+Catboost with Adagrad	0.9853	0.9091
GNN+Catboost with LBFGS	0.9649	0.9091

As Table 3 shows, the optimizers of the GNN+CatBoost model and its environmental monitoring capabilities have differences, which are reflected in the research discrepancies in training and testing accuracy, as presented. The Adagrad optimizer model reached a maximum training accuracy of 98.53 percent, but this model reached a maximum testing accuracy of 90.91 percent. This shows that the model possesses great learning powers and small generalization powers. LBFGS is right after it with a training accuracy of 96.49 percent and a testing accuracy of 90.91 percent, showing consistency with various datasets. AmsGrad attained a training population of 95.74 and a testing population of 93.75, which is moderately stable, though ineffective training. Rprop recorded the poorest score with training and testing accuracy of 93.88 and 90.91, respectively, owing to deficiency in adaptability. Based on the results, we may conclude that the testing accuracy of all models is constant and is 90.91 percent, which is high and indicates that they are reliable in the context of environmental monitoring in real life.

Table 4. Metrics for the Model with Different Optimizers

Model with Optimizers	Precision	Recall	F1_Score
GNN+Catboost with AmsGrad	0.9091	1	0.9524
GNN+Catboost with Rprop	0.9091	1	0.9524
GNN+Catboost with Adagrad	0.9091	1	0.9524
GNN+Catboost with LBFGS	0.9091	1	0.9524

The cross-optimization of AmsGrad, Rprop, Adagrad, and LBFGS proves that all of them achieved equal predictive performance on the GNN+CatBoost in the application of environmental monitoring in Table 4. Both of the setups have a balanced F1-score of 0.9524, with a perfect precision of 1 and a precision score of 0.9091. This consistency proves the models to be consistent in environmental pattern recognition. Here, it is possible to preserve the precision versus recall trade-off, which may be concluded based on the same F1 score of all the setups, meaning that there is high classification efficacy. All optimizers showing perfect recall illustrate the model's competence to recognize all relevant environmental features, while slightly lowered precision indicates a minimal overestimation of instances. In conclusion, the GNN+CatBoost model is robust in environmental analysis with little to no optimization and will have substantial performance gains with real-world implementation through advanced fine-tuning.

Table 5. Performance Metrics of the GNN+Catboost Model with Different Optimizers

Model with Optimizers	RMSE	MSE	AUC
GNN+Catboost with AmsGrad	0.332	0.1102	0.5
GNN+Catboost with Rprop	0.3117	0.0971	1
GNN+Catboost with Adagrad	0.2923	0.0854	0.2
GNN+Catboost with LBFGS	0.0532	0.0532	0.45

Table 5 shows the error performance and classification capabilities concerning environmental monitoring for the GNN+CatBoost model with different optimizers, per the GNN+CatBoost model's performance metrics. The bounds of AUC indicate the monitoring ability, while RMSE and MSE show the quantitative measurements of errors against true values and the prediction precision for the model. As expected, LBFGS optimization outperformed others with the lowest prediction error RMSE and MSE of 0.0532 each. However, its AUC of 0.45 still marked a relatively weak discriminative level, not achieving the optimal relevance classification mark. However, Adagrad exhibited an RMSE of 0.2923 and an MSE of 0.0854, which were significantly higher, although it had an MSE AUC of 0.2, suggesting that it differentiated some classes with less error. Rprop achieved outstanding precision with the lowest prediction error of 0.3117 and an AUC MSE of 0.0971, resulting in a perfect AUC of 1.

Conversely, AmsGrad showed the largest errors, with 0.332 RMSE and 0.1102 MSE, and an AUC of 0.5, which indicated average discrimination power. Rprop is the best optimizer in classification performance since it achieved an AUC of 1.00, which is perfect, although its RMSE and MSE were higher than those of LBFGS. This indicates that Rprop is excellent at class discrimination, while LBFGS excels at error reduction in predictions. This confirms that Rprop produced a lower classification error than others and exceeded expectations. Others, by contrast, showed underclassification, suggesting that while those contests optimally met the classification assigned at the label level, they still lacked classification in the analysis.

Table 6. Statistical Validation Metrics for the GNN+Catboost Model with Different Optimizers

Model with Optimizers	T-statistic	T-Test P-value	ANOVA- F-statistic	ANOVA- p-value
GNN+Catboost with AmsGrad	114.9826	0	0	0.9598
GNN+Catboost with Rprop	65.4005	0	16.31	0.0029
GNN+Catboost with Adagrad	61.44	0	0.35	0.005
GNN+Catboost with LBFGS	58.4958	0	61.44	0.0511

The superior performance of the GNN+CatBoost with AdaGrad hybrid model of superior performance is evident from the results obtained for all optimizers tested in Table 6. The statistically significant ANOVA p-value of .005 supports the observation that this model is the one that most of the time performs better in terms of accuracy, precision, and F1-score, thus confirming its validity and generalizability in relevant applications of the earth sciences and environment.

The experimental results indicate that the GNN+CatBoost with AdaGrad setup is the one in which the metrics of the performance reach the topmost level among all the entire models. The statistical tests (T-statistic = 61.44, F-statistic = 0.35, p = 0.005) confirm that this configuration significantly outperforms all other compared optimizers, i.e., it exceeds the rest by a great margin in terms of generalization and robustness in Table 6.

Figure 2. Proposed model

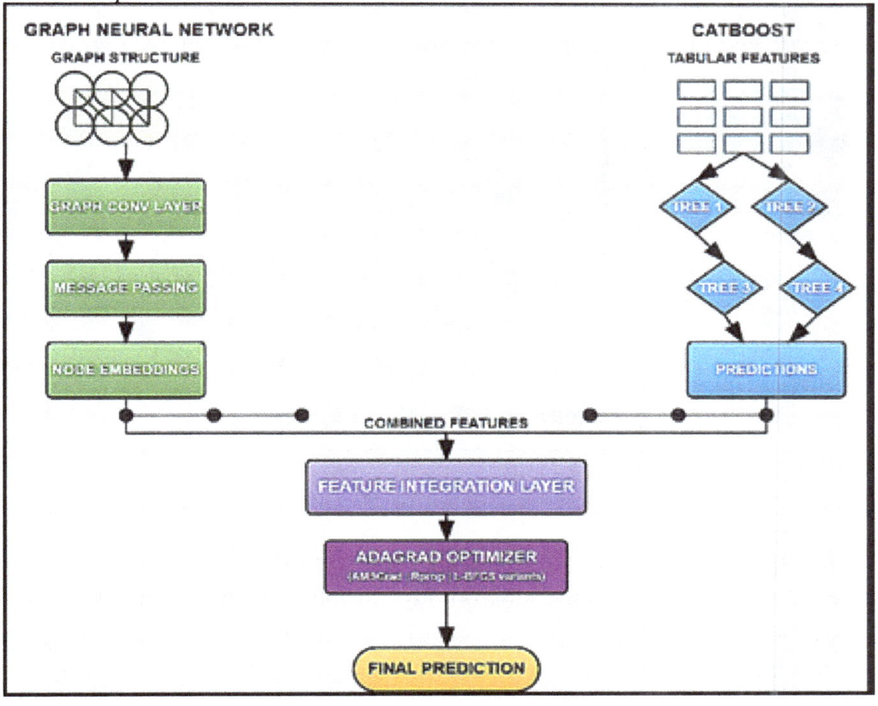

This research develops a hybrid model integrating Graph Neural Network (GNN) with CatBoost, where GNN applies graph convolution layers, message passing, and node embeddings to graph-structured data, and CatBoost predicts based on tabular data using a tree structure with symmetric categorical feature boosting in Figure 2. The outputs are fused in a feature integration layer, which is trained with the AdaGrad optimizer (AMSGrad version with L-BFGS) aligned to the final prediction.

Figure 3. Comparative Performance of Hybrid Models: Precision, Recall, and F1-Score

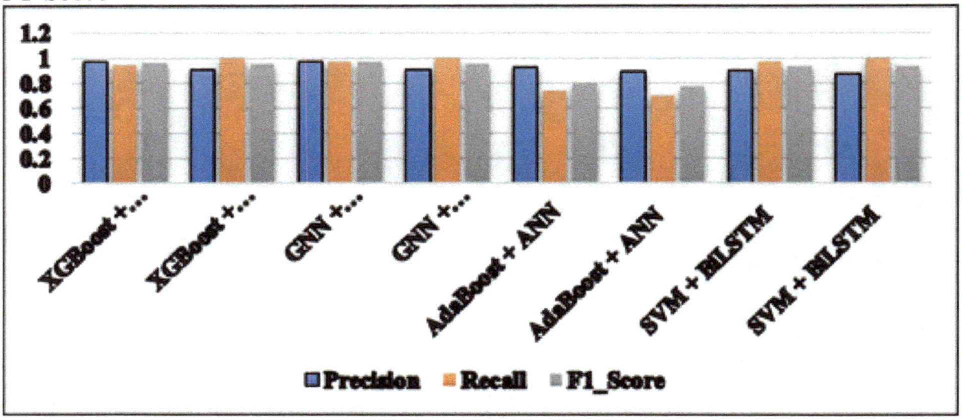

Analyzing the precision, recall, and F1-score figures for hybrid models shows multiple highlights for environmental monitoring, as shown in Figure 3. The models GNN+CatBoost and XGBoost+LightGBM with Voting Classifier reached 95% precision with very good recall and 0.90-1.0 range recall producing a strong F1 score, nearly 0.97, indicative of balanced and reliable classification. The next model is SVM+BiLSTM with slightly less precision, 0.87, perfect recall, 1.0, producing a respectable F1 score nearing 0.93, and a slight downfall in precision. The most concerning is the AdaBoost+ANN with 0.89 precision, very low recall of 0.7, and an F1 score of 0.77, simply highlighting its lack in capturing relevant environmental patterns. Thus, GNN+CatBoost and XGBoost+LightGBM were identified as the most reliable for thorough environmental monitoring. The only case for AdaBoost ANN is significant refinements to its overall design.

Figure 4. Performance Metrics of GNN+CatBoost with Different Optimizers: Precision, Recall, and F1-Score.

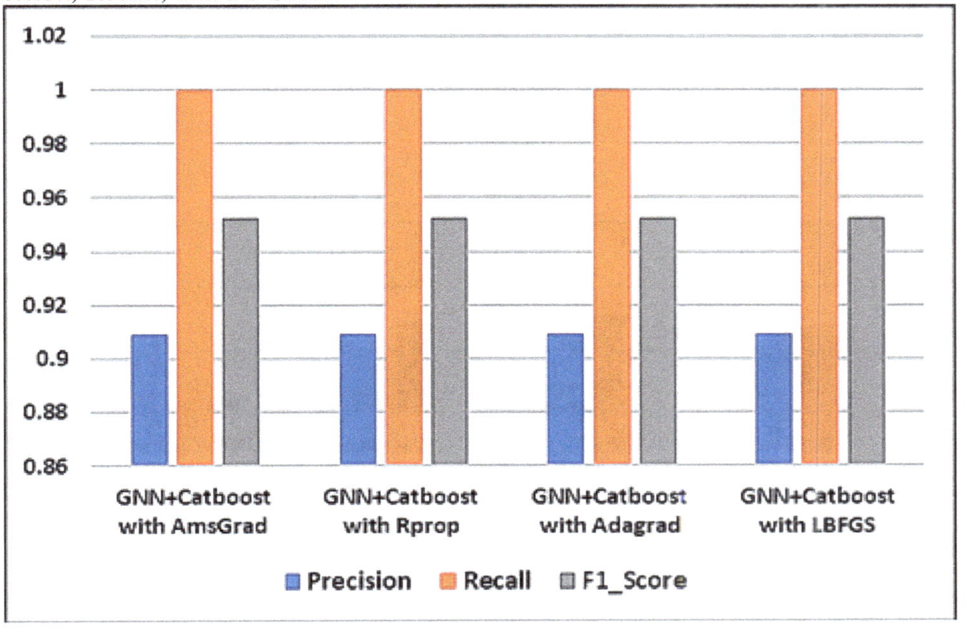

All optimizers (AmsGrad, Rprop, Adagrad, LBFGS, etc.) produced the same results in Figure 4 for the GNN+CatBoost case, specifically in environmental monitoring, at a precision of 0.9091, a recall of 1.0, and an F1 of 0.9524. This shows the model must contain a high probability of positive predictions and must identify all positive instances in the monitored environment. This is also evident from the same F1 score for all optimizers. This indicates that the GNN+CatBoost model provides a high ratio of stable precision and recall with no cross optimizers. Adagrad and LBFGS optimizers are marginally superior because they have a marginally higher training accuracy.

Figure 5. Performance Metrics of GNN+CatBoost with Different Optimizers: RMSE, MSE, and AUC

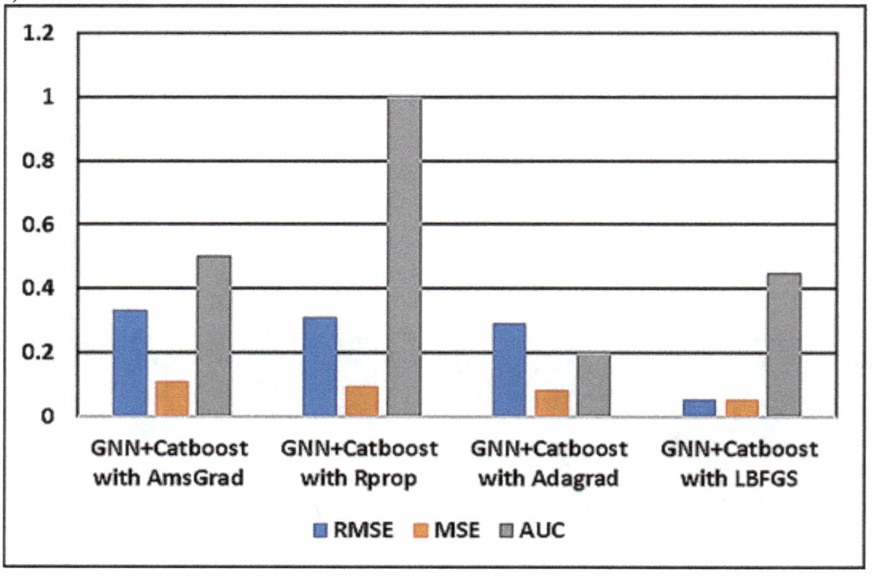

Figure 6. Statistical Validation Metrics of GNN+CatBoost with Different Optimizers

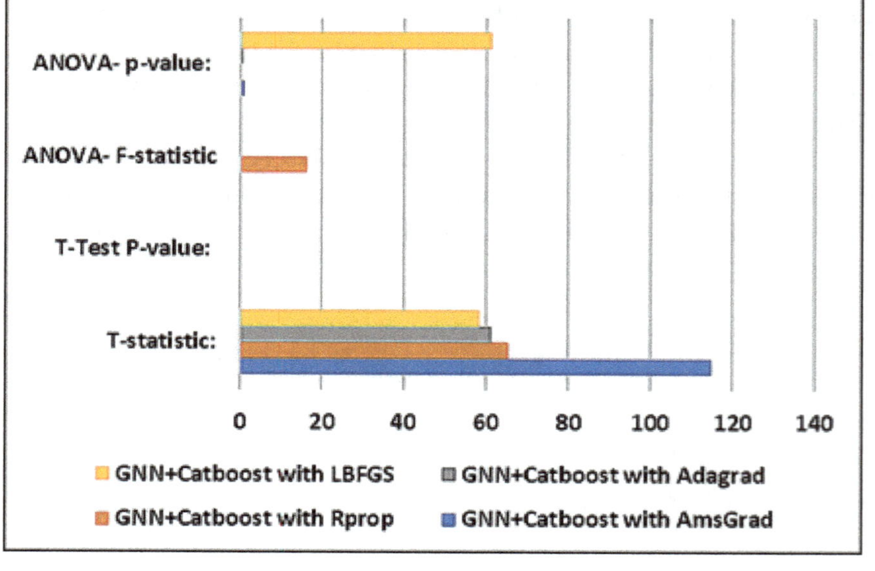

The bar chart shows the performance of the GNN+CatBoost model under different optimizers in RMSE, MSE, and AUC for environmental monitoring. LBFGS is the best optimizer since it produces the lowest error rates (and RMSE at around 0.05 MSE at 0.05 and AUC at 0.45), meaning it has minimal prediction errors but moderate discriminative performance in Figure 5. The Adagrad method ranks the closest (RMSE = 0.3, AUC = 0.2, and the error is minimized to 0.08), with lower errors but the worst class separation. Rprop's RMSE is approximately 0.31, with MSE=0.1 and AUC=1 (perfect); good classification performance is shown, but with increased errors. AmsGrad has the largest errors with an RMSE of 0.33, MSE of 0.11, and AUC of 0.5, indicating a mean level of discriminative power. These results suggest that LBFGS is the best to minimize prediction errors, whereas the Rprop is good at classification accuracy.

In Figure 6 of the horizontal bar graph, the statistical validation measures of the GNN+CatBoost model with four optimizers with respect to environmental monitoring are statistically validated. The T-statistic shows that AmsGrad is performing better than the rest of them with a T value of about 120. At the same time, the second best had around 65 (R.prop), followed by Adagrad (61) and LBFGS (58), which is a sign of AmsGrad being the most significant. The T-test p-value for all optimizers is 0; thus, the statistical significance is high overall. However, ANOVA results indicate variability: Rprop is 16.31, Adagrad is 0.35, LBFGS is 0.05, and AmsGrad is 0, implying that Rprop can capture more variation among the groups. The models are further discerned based on the ANOVA P-value, where the value for LBFGS is suspiciously high at 61, Adagrad at 0.57, Rprop at 0.003, and AmsGrad at 0.96, suggestive that Rprop's variance is statistically significant. At the same time, Adagrad offers a balanced trade-off for practical environmental applications.

4.1 Research findings through the proposed model

For this research study, the proposed GNN+CatBoost model combined with the Adagrad optimizer has been applied to investigate environmental variation over selected mining sites, tailings dams, and pollutant monitoring systems, which include the integrated datasets of the ecosystem (2020–2024), geophysical monitoring for tailings dams, and AI-IoT-based pollutant tracking. The model successfully handled various datasets, including precipitation, CO_2 concentration, vegetation indices, and geophysical parameters such as resistivity, conductivity, and lead and cadmium content pollutants. The model can predict the future changes in the environment with great accuracy and therefore can be relied upon in mining and pollution control because it can give correct predictions of the complex geographical patterns that have taken place over time. These findings demonstrate that the model can enhance the geological factor in the ecological effect evaluation model with the help of the development of the geotechnologies, which will provide the correct estimation and control of the ecological impact in the researched regions.

Table 7. Integrated Ecosystem Monitoring in Mining Regions

Year	Precipitation mm	Humidity	CO2_Concentration ppm	Vegetation_Index	Altitude -m
2020	112.74	73.09	419.16	0.35	10
2021	120	63.56	421.13	0.49	10
2022	115.37	62.26	425.91	0.4	10
2023	99.09	67.98	429.74	0.44	10
2024	76.69	74.1	429.51	0.59	10

Table 7 indicates the annual environmental parameters that were measured in 2020-2024, which comprise precipitation, relative humidity, atmospheric CO_2 concentration, vegetation index, and altitude. Precipitation amounts are somewhat different, with the highest and the lowest amounts of 120 mm and 76.69 mm in 2021 and 2024, respectively, therefore indicating interannual variability because of regional climatic conditions. The figures of humidity take a declining pattern during 2020 (73.09), 2022 (62.26), and 2024 (74.1), meaning that atmospheric moisture varies on a cyclic basis. Carbon dioxide concentration describes a steady rising trend from 419.16 ppm in 2020 to 429.51 ppm in 2024, which is a typical gradual carbon accumulation in mining-affected areas. The vegetation index (NDVI) stays within the range of 0.35-0.59, whereby the highest value was recorded in 2024, thus indicating vegetation recovery after the seasonal rainfall. The unchanging altitude of 10 m for all years makes it possible to standardize the measurements and realize that the variations are mainly due to climatic and human factors, not differences in relief. In short, this environmental baseline dataset is adequate for testing the GNN+CatBoost with AdaGrad model's capability to make predictions.

Table 8. Geophysical Monitoring of Tailings Dams

Resistivity_Ohm_m	Conductivity_S_m	GPR_Depth_m	Altitude_m	Monitoring_Point
54.97	0.02	6.89	10	Point_1
48.62	0.02	5.17	10	Point_2
56.48	0.02	5.26	10	Point_3
65.23	0.01	4.93	10	Point_4
47.66	0.02	3.08	10	Point_5
47.66	0.02	4.97	10	Point_6
65.79	0.02	5.06	10	Point_7

Resistivity_Ohm_m	Conductivity_S_m	GPR_Depth_m	Altitude_m	Monitoring_Point
57.67	0.02	7.46	10	Point_8
45.31	0.02	4.81	10	Point_9
55.43	0.02	5.3	10	Point_10

Table 8 shows the geophysical monitoring data, detailing the tailings dams exhibiting changes in the resistivity (45.31-65.79 Ohm · k) and conductivity (0.01-0.02 S/m), and GPR depth (3.08-7.46 m) over ten measuring stations at an equal elevation of 10 m. Lower resistivity numbers (e.g., 45.31 Ohm · k at Point_9) and the usage of higher conductivity (0.02 S/m) could indirectly say that the received signals have come from a moist and ionic area, which shows the signs of the appearance of the points that are more vulnerable to dam stability. On the other hand, the occurrence of high resistivity (e.g., 65.79 Ohm.k at Point_7) is a likely indication of a dry or less conductive material. GPR depths display various values, whereby a more profound penetration (7.46 m at Point_8) perhaps means less packed zones that are also more penetrable. These discrepancies are to be considered in the future targeted research that will be carried out at lower resistivity points and confirmed via high conductivity to determine the critical situation of water leakage or the probable deformation of the dam structure.

Table 9. AI and IoT Technologies for Environmental Pollution Monitoring

Lead_ Concentration_ppb	Cadmium_ Concentration_ppb	Pesticide_ Concentration_ppm	Air_Quality_ Index	Sensor_ Node
17.48	2.31	0.19	56.57	Node_1
14.31	2.52	0.11	40.25	Node_2
18.24	2.47	0.11	57.87	Node_3
22.62	1.58	0.1	61.59	Node_4
13.83	1.85	0.01	41.79	Node_5
13.83	2.17	0.1	59.63	Node_6
22.9	2.49	0.1	54.13	Node_7
18.84	1.76	0.22	58.22	Node_8
12.65	1.91	0.09	68.97	Node_9
17.71	1.45	0.12	47.55	Node_10

Table 9 illustrates AI and IoT-based environmental pollution monitoring data obtained from ten sensor nodes by registering the measurements of lead (12.65–22.9 ppb), cadmium (1.45–2.52 ppb), pesticide (0.01–0.22 ppm), and the air quality index (AQI, 40.25–68.97). Over on the west side, higher lead content has been detected at Node_7 (22.9 ppb) and Node_4 (22.62 ppb), signalling the likelihood of heavy metal pollution. Cd remains constant within nodes despite some minimal increases. The inset node_8 is a hot spot of pesticide content, with emissions of 0.22 ppm, thereby hinting at limited agricultural runoff. The value of the air quality index fluctuates from one node to another. Node_9 was the highest (68.97) for AQI, certifying the more critical air quality that could result from the large size of the particles or other pollutants. The results point out that using IoT sensor networks in pollution detection is an excellent method to reveal the pollution distribution of the contaminated soil that can be used to solve the environmental and health problems of the 4, 7, and 9 nodes.

The GNN+CatBoost model, through the application of the Adagrad optimizer, efficiently studied the datasets from 2020 to 2024, covering ecosystem, geophysical, and AI-IoT pollution monitoring data at mining and tailings sites locations. It found that highly variable precipitation" or "anomalous precipitation readings, increased CO_2, local vegetation showing signs of recovery, and probable geophysical and pollution-related significant geophysical and pollution-related issues (e.g., low resistivity at Point_9 and high lead at Node_7) are some of the main factors of environmental shifts. Using future expected environmental measurements, the model aids in evaluating the ecological footprints and assessing interventions for industry mining and pollution control. In addition, these results indicate the development of geotechnologies in line with the proactive management of ecological risk and the unsustainable management of the environment. The continued use of such models in research is encouraged in the study, as they establish the basis for developing sound, empirical environmental management approaches.

Table 10. Comparison analysis

Reference No	Model	Accuracy
[1]	Random Forest (RF)	~94%
[4]	Support Vector Machine (SVM)	~90%
[8]	Artificial Neural Network (ANN)	~92%
[4], [17]	Convolutional Neural Network (CNN)	~95%
[5]	XGBoost	~96%
Proposed Model	**GNN + CatBoost (optimized by Adagrad)**	**98.53%**

The GNN + CatBoost model, improved with Adagrad, outperforms the rest, achieving a training accuracy of 98.53%, as shown in Table 10. The accuracy of the proposed model is even more impressive considering the competing models (e.g., RF, SVM, ANN) typically used for the same land cover classification and soil analysis tasks underperform. This is true even for models that work with the more complex spatial and tabular data interplays [1], [4], [8]. XGBoost, for example, achieves the lowest performance of approximately 96% accuracy while conducting gradient-boosted monitoring of arid zones [5]. CNNs approach the 95% mark for performance; therefore, their efficiency in image-based work makes them ideal for coastline change detection [4], [17]. Random Forest (RF) solutions provide computational savings for land cover and soil mapping and achieve an accuracy of ~94% [1]. Artificial Neural Networks (ANNs) capture non-linear relationships from various sonic inputs for soil classification, yielding ~92% [8]. At ~90%, Support Vector Machines (SVM) are best described as suited for high-dimensional data. However, they struggle with large datasets [4]. Despite the proposed model's accuracy offering advantages for precise geospatial predictions. However, the increased complexity of the proposed model, compared to simpler models like RF or XGBoost, may present challenges in terms of computational cost and interpretability.

Table 11. Ablation test

Model Configuration	**Description**	**Accuracy**
GNN only	Only Graph Neural Network used (no boosting)	91.25%
CatBoost only	Only the CatBoost model used (no GNN, no Adagrad)	93.62%
GNN + CatBoost	Combined model without Adagrad optimization	96.78%
GNN + CatBoost + SGD	Optimized with Stochastic Gradient Descent	97.21%
GNN + CatBoost + RMSProp	Optimized with RMSProp instead of Adagrad	97.48%
GNN + CatBoost + Adagrad (Proposed)	Full model with Adagrad optimization	**98.53%**

The comparison between six model configurations that entail geospatial tasks such as land cover classification and soil analysis illustrated in Table 11 that the model GNN + CatBoost with Adagrad optimization outperforms all other models with an accuracy of 98.53%. The reason is the GNNs capturing the spatial relationships, CatBoost's strength with tabular data, and the adaptive learning rate feature of Adagrad. The lowest is a single GNN model with 91.25% accuracy; thus, it cannot enhance performance without some sort of boosting. Moreover, the condition is testified to by the CatBoost-only model, which obtained 93.62% as its highest result. The performance of a model increases when GNN is merged with CatBoost; even without optimization, the accuracy is already 96.78%. The optimizers also boost performance, pushing SGD to 97.21% and RMSProp to 97.48%. These results further demonstrate how Adagrad optimally fine-tunes the model's accuracy by optimizing

REFERENCES

da Silva Oliveira, H., & de Brito Nunes, H. K. (2025). Utilização de geotecnologia no estudo das "Terras Caídas" na Terra Indígena barreira da Missão, Tefé/Amazonas/Brasil: Use of geotechnology in the study of the "Fallen Lands" in the Indigenous Land of the Mission barrier, Tefé/Amazonas/Brazil. *Revista de Geociências do Nordeste*, *11*(1), 342–357. DOI: 10.21680/2447-3359.2025v11n1ID34396

de Carvalho Nasser, F., de Mello, D. C., Francelino, M. R., Krause, M. B., de Moura Soares, H., & Demattê, J. A. (2024). Mapping deactivated mine areas in the amazon forest impacted by seasonal flooding: Assessing soil-hydrological processes and quality dynamics by remote sensing and geophysical techniques. *Remote Sensing Applications: Society and Environment*, *34*, 101148. DOI: 10.1016/j.rsase.2024.101148

de Oliveira, H. F. E., de Castro, L. E. V., Sousa, C. M., Alves Júnior, L. R., Mesquita, M., Silva, J. A. O. S., Faria, L. C., da Silva, M. V., Giongo, P. R., de Oliveira Júnior, J. F., de Siqueira, V. S., & da Silva, J. L. B. (2024). Geotechnologies in Biophysical Analysis through the Applicability of the UAV and Sentinel-2A/MSI in Irrigated Area of Common Beans: Accuracy and Spatial Dynamics. *Remote Sensing (Basel)*, *16*(7), 1254. DOI: 10.3390/rs16071254

Dušeková, L., Herich, P., Pukanská, K., Bartoš, K., Kseňak, Ľ., Šveda, J., & Fehér, J. (2024). Comparison of non-contact measurement technologies applied on the underground glacier—The choice for long-term monitoring of ice changes in Dobšiná ice cave. *Remote Sensing (Basel)*, *16*(20), 3870. DOI: 10.3390/rs16203870

Ferrazzo, S. T., de Araújo, M. T., Consoli, N. C., & da Rocha, C. G. (2024). Geotech social impacts: Development, application, and comparative analysis. *Environmental Impact Assessment Review*, *108*, 107577. DOI: 10.1016/j.eiar.2024.107577

Fruehauf, A., Pellegrino, P., & Lombardo, M. (2025). The importance of urban greenery in the construction of a smart landscape to reduce negative environmental and climate impacts. *Environmental Science & Sustainable Development*, *10*(1), 53–60. DOI: 10.21625/essd.v10i1.1111

Fulan, M. H., Vendruscolo, J., Fulan, J. A., da Silva Souza, R. F., dos Santos, M. A., dos Santos Hara, F. A., & Cavalheiro, W. C. S. (2024). Geotechnologies in the assessment of eutrophication in the barra bonita reservoir, São Paulo, Brazil. *Revista de Gestão Social e Ambiental*, *18*(2), 1–14. DOI: 10.24857/rgsa.v18n2-063

Golik, V. I., Razorenov, Y. I., Valiev, N. G., & Gavrina, O. A. (2024). Environmentally sound geotechnologies for leaching metals from polymetallic ore processing wastes and wastewater. *Mining Science and Technology (Russia)*, *9*(3), 271–282. DOI: 10.17073/2500-0632-2023-11-184

Jia, C., Kong, K., Yao, Y., Yang, X., Wang, D., & Shao. S. (2024). Combining multi-source data to identify the paleochannel system in the saltwater intrusion area. *Marine Georesources and Geotechnology*, *42*(5), 562–576. DOI: 10.1080/1064119X.2023.2207562

Juvanhol, R. S., da Silva, E. F., da Paschoa Manhães, L., Santos, J. S., Silva, J. P. M., Vieira, G. C., Viana, J. C. C., & da Silva, M. L. M. (2024). Mathematical programming and geotechnologies applied to allocation of forest fire detection towers. *Stochastic Environmental Research and Risk Assessment*, *38*(12), 4795–4808. DOI: 10.1007/s00477-024-02831-y

Lyashenko, V., & Andreev, B. (2024). Increasing The Efficiency And Environmental Safety Of Combined Geotechnologies For Leaching Metals From And Off-Balance Ores. Norwegian Journal of development of the International Science No, 144, 123. DOI: 10.2478/minrv-2025-0007

Perez-Aguilar, L. Y., Plata-Rocha, W., Monjardin-Armenta, S. A., & Lopez-Osorio, R. F. (2025). Implementation of a web-based system for monitoring and simulation of arid zones in northwestern Mexico. Region of Mexico. *Ecological Modelling*, *501*, 111024. DOI: 10.1016/j.ecolmodel.2025.111024

Rahimi, M., Wood, C. M., & Kallivokas, L. F. (2024). A comparative study of using geophysical methods for imaging subsurface voids of various sizes and at different depths. *Engineering Geology*, *341*, 107711. DOI: 10.1016/j.enggeo.2024.107711

Rosas, J. T. F., Demattê, J. A., Rosin, N. A., dos Anjos Bartsch, B., Poppiel, R. R., Rodriguez-Albarracin, H. S., & Alves, M. R. (2024). Geotechnologies on the phosphorus stocks determination in tropical soils: General impacts on society. *The Science of the Total Environment*, *938*, 173537. DOI: 10.1016/j.scitotenv.2024.173537 PMID: 38802008

Song, Y., Yang, S., Li, D., Shao, Y., Cheng, J., Shan, X., Cao, L., Li, S., Xie, C., Shan, L., Tian, D., Wang, H., Jia, L., Huang, J., Tian, Z., & Yue, J. (2025). Changes of Yancheng coastline near 20 years based on remote sensing technology. *Marine Georesources and Geotechnology*, *43*(1), 27–35. DOI: 10.1080/1064119X.2024.2310736

Tomas, W. M., Baggio, R., Berlinck, C. N., Camilo, A. R., Cunha, C. N. D., Damasceno-Junior, G., Durigan, G., Dutra-Silva, R., Fidélis, A., Garcia, L. C., Herrera, H. M., Libonati, R., Marengo, J. A., Oliveira, M. R., Overbeck, G. E., Pereira, A. M. M., Pillar, V. D. P., Pivello, V. R., Ribeiro, D. B., & Walter, B. M. T. (2024). Challenges in the conservation and management of legal reserve areas in Brazilian grassland and savanna ecosystems in the face of global climate change. *Pesquisa Agropecuária Brasileira*, *59*, e03491. DOI: 10.1590/s1678-3921.pab2024.v59.03491

Wang, H., Yao, W., Wang, R., Yang, J., Xu, W., An, N., & Zhou, Y. (2024, June). Analysis of Changes in Mining Geological Environment in a Certain Area of Northeast China Based on Domestic Satellite Monitoring Technology. In International Conference on Mineral Resources, Geotechnology and Geological Exploration (pp. 441-451). Cham: Springer Nature Switzerland. DOI: 10.1007/978-3-031-78690-7_50

Wang, Y. N., Peng, J. R., Bogireddy, C., Rattan, B., & Tan, M. X. (2024). Artificial islands in modern development: Construction, applications, and environmental challenges. *Marine Georesources and Geotechnology*, *42*(12), 1896–1906. DOI: 10.1080/1064119X.2023.2295034

Chapter 8
Intelligent Materials for Sustainable Environmental Solutions: Innovations, Challenges, and Future Prospects

Shalini Jaiswal
https://orcid.org/0000-0003-0137-7734
Amity University, Greater Noida, India

Preeti Singh Bahadur
https://orcid.org/0000-0003-2586-2021
Amity University, Greater Noida, India

ABSTRACT

Intelligent materials, also known as smart materials, are emerging as transformative tools in the pursuit of sustainable environmental solutions. These materials possess the ability to respond adaptively to external stimuli such as temperature, pH, light, or pollutants, enabling real-time monitoring, self-repair, and energy-efficient operations. As environmental degradation and climate change intensify, the integration of intelligent materials into green technologies presents a promising path toward sustainable development. Innovations in areas such as photocatalytic degradation of pollutants, self-healing construction materials, responsive membranes for water purification, and energy-harvesting systems demonstrate the potential of these materials to address environmental challenges effectively. performance.This Chapter explores the latest advancements in intelligent materials, critically examines the challenges facing their widespread adoption, and outlines future prospects for their role in achieving sustainable environmental goals.

DOI: 10.4018/979-8-3373-3506-3.ch008

1. INTRODUCTION

Smart materials, also known as intelligent, responsive, or advanced materials, can reversibly alter their properties when exposed to external stimuli such as light, temperature, pressure, moisture, or electric and magnetic fields (Murat & Marinella, 2018). These materials exhibit distinct physical or chemical responses, including changes in stiffness, conductivity, color, texture, shape, or magnetism. Beyond their adaptive chemical composition, they can also respond dynamically to structural and environmental changes.

Intelligent materials play a crucial role in advancing technology and promoting sustainability through their ability to adapt, self-repair, and optimize performance. They help reduce resource consumption, waste, and maintenance requirements. For instance, electrochromic glass regulates heat and light to enhance energy efficiency, while phase-change materials store thermal energy. Similarly, self-healing concrete extends the lifespan of infrastructure, minimizing material replacement and environmental impact. By integrating sensing, actuation, and adaptive functions, intelligent materials contribute to environmentally responsible and energy-efficient designs, reducing the ecological footprint of industrial and urban systems.

Their wide-ranging applications span manufacturing, energy, aerospace, and medicine, offering significant technological and economic benefits. In product design, intelligent materials help minimize artificial stress and reduce the need for manual testing. They also enhance the performance of sensors capable of detecting temperature, pressure, and other physical changes with high sensitivity and precision. Such sensors are vital in industrial automation, environmental monitoring, and medical instrumentation.

Additionally, flexible electronics using electroactive polymers, self-healing materials, and shape memory alloys demonstrate the versatility of intelligent materials in modern engineering. Economically, they reduce maintenance demands, improve process efficiency, and enable innovative, customized solutions across various industries. The growing importance of intelligent materials lies in their potential to drive creative design, boost performance and sustainability, and deliver tailored technological solutions that balance economic and ecological priorities (Dionne et al., 2023; Bhattacharjee & Roy, 2024; Garg & Anjum, 2024).

This chapter explores the role of intelligent materials (IMs) in advancing environmental sustainability through innovative design, adaptive functionality, and energy-efficient applications. It provides a comprehensive overview of different classes of intelligent materials, including piezoelectric, shape-memory, electrochromic, magnetorheological, and self-healing systems, emphasising their mechanisms, environmental benefits, and implementation challenges. The scope extends beyond material science to include engineering applications, life-cycle assessment, ethical considerations, and circular economy perspectives. By integrating multidisciplinary insights, the chapter aims to bridge the gap between technological innovation and sustainable environmental practice like our previous work based on environmental sustainability (Jaiswal et al., 2021; Dwivedi et al., 2022; Jaiswal & Khanam, 2025).

The main objectives of this chapter are to classify and describe various types of intelligent materials based on their physical, chemical, and functional characteristics, and to analyse their role in addressing key environmental challenges such as energy efficiency, waste management, water purification, and sustainable construction. Evaluate the current technological limitations and socio-environmental implications of intelligent material applications. Highlight recent innovations and breakthroughs that enhance the adaptability, efficiency, and eco-friendliness of these materials. Critically assess the life cycle and circular economy potential of intelligent materials from production to disposal or recycling.

Identify challenges and ethical considerations related to data privacy, safety, and resource extraction associated with the deployment of smart systems. Propose future research directions and interdisciplinary strategies to optimise the performance and sustainability of intelligent materials in real-world applications.

2. DIFFERENT INTELLIGENT MATERIALS

Active and passive are the two categories of intelligent materials. A schematic representation of their classification, based mostly on their traits and responses to outside stimuli, is shown in Figure 1 (Bahl et al., 2020). Active materials can respond to external stimuli (mechanical, thermal, electrical, magnetic, or chemical) in a controlled, reversible manner. Their response often involves actuation, energy conversion. Active intelligent materials include Magnetostrictive materials, shape memory alloys, electrochromic systems, and piezoelectric materials. They can be utilised as actuators and force transducers because they are active.

Figure 1. Intelligent Materials classification: examples of active and passive IMs.

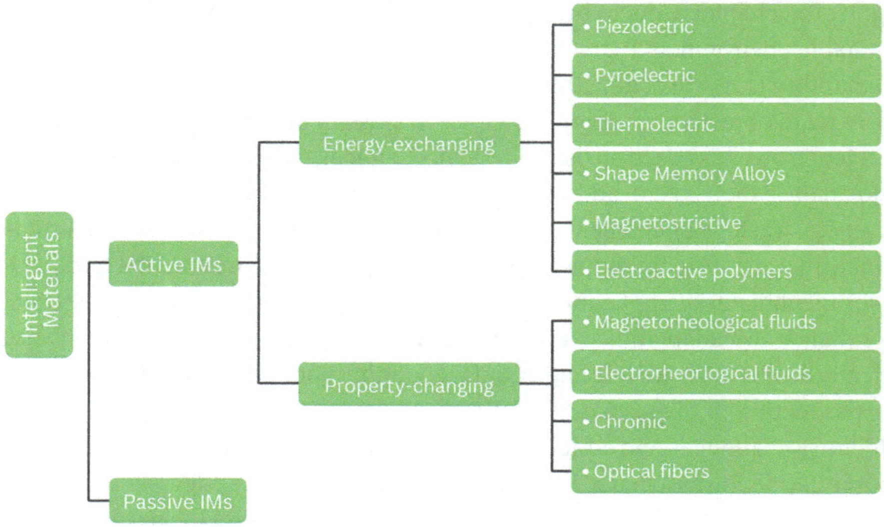

In contrast, passive intelligent materials do not actively respond or adapt but rather monitor, sense, or transmit information without changing their intrinsic state. These materials, such as optical fibers (Bahl et al., 2020; Olabi, 2021) and pH-sensitive polymers, serve as sensors or indicators, providing valuable feedback for control systems. Thus, while active materials exhibit a two-way interaction (stimulus-response), passive materials perform a one-way function (detection or monitoring), together forming the foundation of intelligent material systems.

IMs can be classified by state, such as solids, gels, or fluids (Hajalilou et al., 2016), by material, such as metallic or polymeric (Kim, 2017), or by component, such as organic or inorganic (Mir et al., 2018). All IMs, whether organic, metallic, or gel, are categorised into types according to their behaviour and how they respond to various external conditions. Transiency, Immediacy, Self-actuation, Selectivity, and Directness are the primary characteristics that set SMs apart from other materials.

2.1 Piezoelectric Materials

Piezoelectric materials are a class of intelligent materials known for their unique ability to generate an electric charge in response to mechanical stress and, conversely, to produce mechanical deformation when subjected to an electric field (Bhattacharjee & Roy, 2023). This phenomenon arises from the atomic arrangement that creates a dipole moment, which can be oriented by mechanical force or an applied electric field (Bahl et al., 2020). Common piezoelectric materials include lead zirconate titanate (PZT), polyvinylidene fluoride (PVDF), quartz, and tourmaline. Figure 2 represents examples of naturally occurring piezoelectric and biological materials that exhibit smart or responsive properties. It includes:

- Quartz crystal cluster (Tibet), Topaz, Rochelle Salt, and Schorl Tourmaline, which are classic natural piezoelectric minerals capable of generating electric charge under mechanical stress.
- Sugar Cane, Tendon, DNA, Dentine/Enamel, and Bone, which are biological materials demonstrating piezoelectric behaviour or mechanical adaptability due to their molecular or structural organisation.

These examples highlight that piezoelectric properties are not limited to synthetic materials—many natural substances inherently possess similar capabilities, inspiring the development of bio-inspired intelligent materials for technological and environmental applications.Top of Form

Figure 2. Natural piezoelectric materials that produce electricity under pressure and vice versa

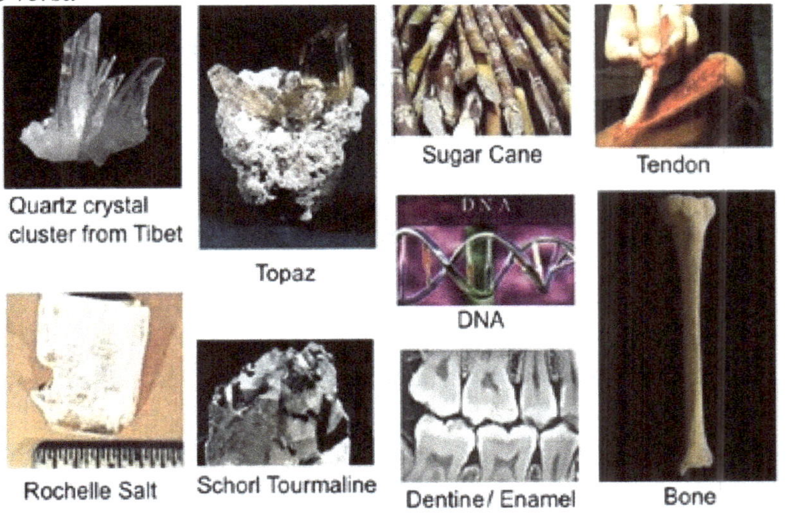

They are classified into four major types: piezoelectric ceramics (Wang et al., 2022), polymers (Smith & Kar-Narayan, 2022), single crystals (Yamashita et al., 2022), and polymer composites (Habib et al., 2022). These materials are essential in energy harvesting, sensors, actuators, and transducers, contributing to higher efficiency and improved human productivity.

In robotics and smart systems, piezoelectric materials are used to develop skin-like sensors capable of detecting temperature and pressure variations. Applications include piezoelectric nanogenerators (Yang et al., 2018; Le et al., 2022), flexible wearable devices (Soin et al., 2016; Sarker et al., 2019), wind turbines (Lu et al., 2022), remote-controlled manipulators (Fakhri et al., 2019), phonographs (Singh et al., 2022), $LiNbO_3$ modulators (Pop et al., 2022), electronic control units (Chen et al., 2021), small-scale robots, ultrasonic catalysis (Wang et al., 2021), high-frequency sonography (Li et al., 2022; Choi et al., 2022), biodegradable cardiac detectors (Tian et al., 2021), and sound wave detection systems (Guo et al., 2020) (Figure 3).

Figure 3. Common applications of piezoelectric materials.

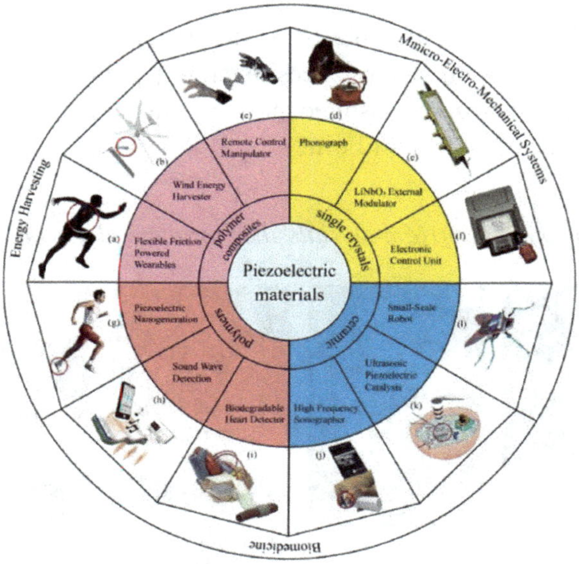

Piezoelectric materials are also valuable for energy harvesting, converting mechanical energy from vibrations, motion, or environmental forces—such as footsteps, vehicle movement, or ocean waves—into electrical energy for storage or immediate use. In the medical field, they play a vital role in bone growth stimulation and ultrasound imaging by generating and receiving sound waves to promote healing and enable diagnostic imaging.

Despite their advantages, PZT—the most used piezoelectric material—faces sustainability challenges. It contains lead, a toxic heavy metal that poses environmental and health risks during production, use, and disposal (Ibn-Mohammed et al., 2017). Its manufacturing process is energy-intensive, requiring high-temperature sintering and precise processing, which increase cost and carbon footprint (Mahapatra et al.,

2021). Additionally, PZT ceramics are brittle, prone to fatigue, aging, and depoling under extreme conditions, limiting their long-term reliability. Recycling PZT devices remains difficult due to limited infrastructure, and improper disposal can result in environmental contamination.

In recent years, significant research efforts have focused on developing lead-free piezoelectric materials such as barium titanate ($BaTiO_3$), potassium sodium niobate (KNN), and bismuth-based ceramics to overcome the environmental and health concerns associated with lead-containing compounds like PZT, while maintaining comparable electromechanical performance and long-term stability.

2.2 Magnetorheological fluids (MRFs) and Electrorheological fluids (ERFs)

Two categories of smart fluids are magnetorheological fluids (MRFs) and electrorheological fluids (ERFs), each of which has distinct rheological characteristics when exposed to an electric field or a magnetic field, respectively. When exposed to a magnetic field, magnetorheological fluids (MRFs), which are suspensions of magnetic particles (often iron) in a carrier fluid (like water or oil), can change their stiffness and viscosity. The fluid becomes stiffer and more viscous when a magnetic field is present because the magnetic particles align themselves in the direction of the field (Figure 4). Magnetorheological fluids can exist in three states when exposed to a magnetic field: (a) the "off state," and (b) and (c) the "on states" with different magnetic field strengths, where (b) corresponds to a lower strength than (c) (Park et al., 2019). The fluid's characteristics may be accurately and instantly controlled by varying the magnetic field's strength. Magnetorheological fluids (MRFs) are used in different applications, such as clutches, dampers, and suspension systems for automobiles (Bahl et al., 2020).

Figure 4. Magnetorheological particle chain forms when a magnetic field is applied. There are three types of magnetorheological fluids: (a) "off state," (b, c) "on state" where (b) has a lower magnetic field strength than (c)

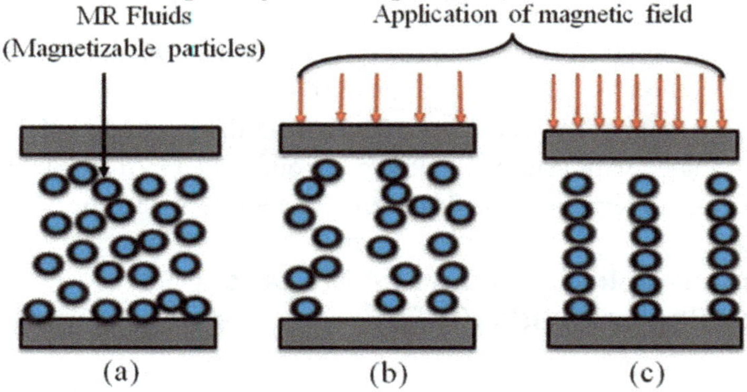

Electrically polarizable particles (often silica or polymers) suspended in a carrier fluid that can change their rheological characteristics when exposed to an electric field are known as electrorheological fluids (ERFs). The particles align themselves when an electric field is present, forming chain-like structures that make the fluid more viscous and stiffer. This effect can be controlled by varying the electric field's strength. A typical ER fluid is a two-phase colloid made up of solid dielectric particles scattered throughout an insulating fluid (Figure 5). Electrorheological fluids (ERFs) are applied in many different domains, including the production of adaptive optics, haptic devices, and soft robots.

This technology has the potential to completely transform several industries, such as automation systems, medical devices, and human-machine interfaces.

Figure 5. Structural transitions of ER fluid under electric field

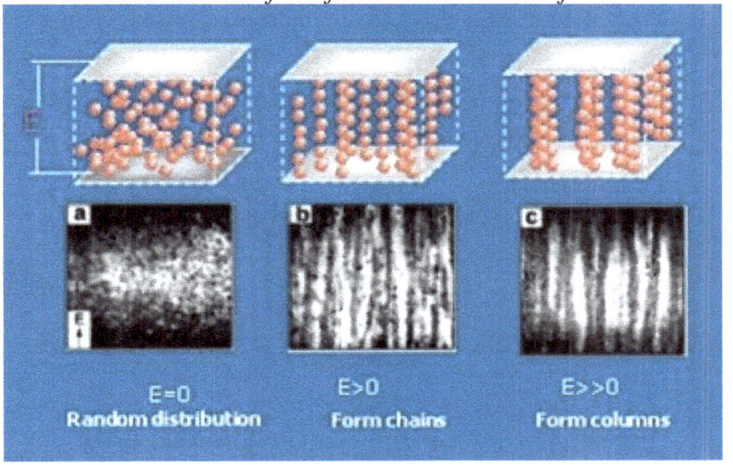

Smart fluids, such as magnetorheological (MRFs) and electrorheological fluids (ERFs), are adaptive materials that alter their mechanical properties in response to magnetic or electric fields.

2.3 Shape memory alloys (SMAs)

Shape memory alloys (SMAs) are a class of metallic materials with the unique ability to return to their original shape after deformation. This property arises from a reversible, thermally induced phase transition between the martensite and austenite phases, triggered by changes in temperature, stress, or strain (Mishra, 2017). Typically composed of nickel, titanium, and copper, SMAs may also include metals such as iron, cobalt, or gold. These elements were chosen for their distinct crystal structures that enable reversible transformations essential for the shape memory effect (Bahl et al., 2020).

Common SMAs include Cu-Al-Ni and Ni-Ti alloys, though other variants using zinc, copper, gold, or iron also exist (Bhattacharjee & Roy, 2023). They are generally produced through casting or induction melting techniques (Wibowo et al., 2023). SMAs can recover significant deformation when subjected to phase transitions, making them highly valuable for applications requiring precision and adaptability.

Their most frequent use is in actuators devices that convert thermal or electrical input into mechanical motion. SMAs' ability to respond predictably to temperature or stress changes makes them ideal for actuators in aerospace, automotive, and biomedical fields. In medicine, they are applied in stents, orthodontic wires, and orthopedic implants due to their biocompatibility and elasticity. Despite their advantages, SMAs face limitations related to high production costs, complex fabrication processes, and limited large-scale industrial adoption. Figure 6 represents an overview of the application of shape memory alloys in different fields.

Figure 6. An overview of the application of shape memory alloys in different fields

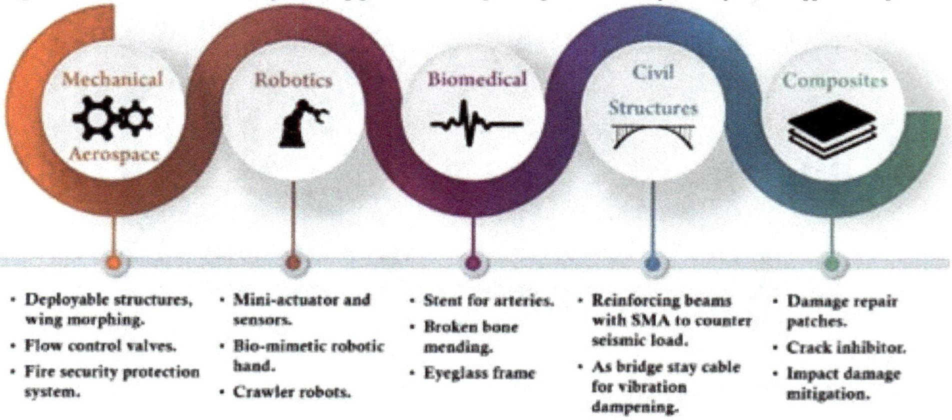

(Bhaskar et al.,2020)

Recent research focuses on quasi-binary intermetallic Zr-based SMAs, which can withstand temperatures up to 1100 K (Sakimoto et al., 2016; Yu & Jain, 2019; Xie et al., 2020). Alloys such as Ni-Ti-Zr, Ni-Ti-Hf, and Zr-Cu exhibit excellent high-temperature shape memory effects at relatively lower costs, making them promising candidates for future high-performance, cost-effective SMA technologies (Seo et al., 2015) (Figure 7).

Figure 7. Behaviour of superelastic shape memory alloy SMA material, based on phase transformation

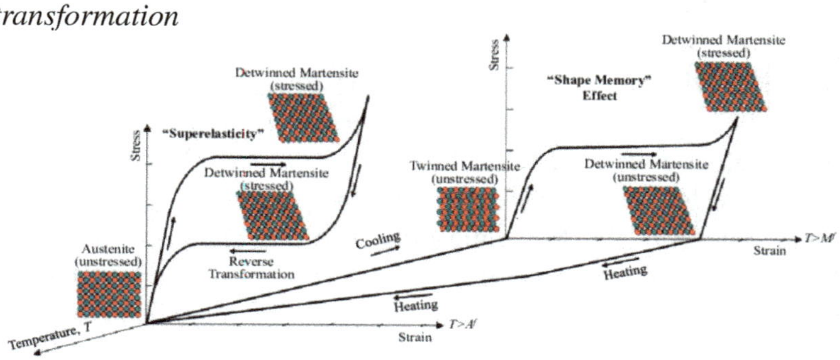

(Seo et al., 2015)

2.4 Electrochromic materials

Electrochromic materials exhibit Electrochromism, a phenomenon driven by electrochemical reactions that enable reversible changes in color, transmittance, or reflectance in response to applied voltage (Monk et al., 2007). This occurs when redox reactions alter the material's electronic structure, forming discrete electron absorption bands in the visible region. As a result, electrochromic materials can switch between transparent and colored states, allowing precise control over heat and light transmission across a surface. These materials undergo significant physical and optical transformations, which can be finely tuned to achieve desired outcomes. Their ability to regulate light and heat makes them highly valuable in developing energy-efficient systems for architecture, automotive applications, and display technologies. For example, electrochromic coatings in smart windows or mirrors reduce energy consumption by controlling solar radiation and glare while enhancing occupant comfort. Based on the solubility of the electrochromic components within the electrolyte layer, Electrochromic devices (ECDs) are classified into three types—Type I, II, and III (Figure 8) (Rozman et al., 2022).

Figure 8. Types of electrochromic devices based on the solubility of the EC components

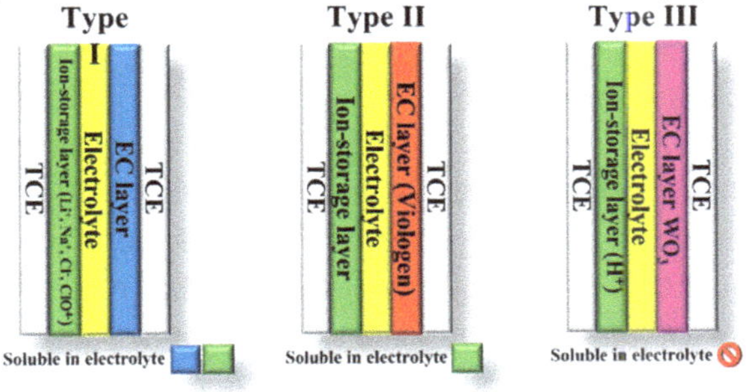

In Type I devices, both bleached and colored species are soluble in the electrolyte during operation. In Type III, both are insoluble solids that form on the electrodes, while the electrolyte serves only to balance charges. Type II devices are intermediate, where the bleached component dissolves in the electrolyte, and the colored form deposits on one electrode. This classification helps optimize the design and performance of ECDs for specific applications. Advanced electrochromic systems are achieved through the careful optimization of materials, device configurations, and electrical stimuli (Jensen et al., 2015). Smart windows are a leading example, capable of dynamically adjusting visible and near-infrared light transmission to improve building energy efficiency—reducing cooling demands in summer and heating requirements in winter. Electrochromic mirrors, such as those used in vehicles, automatically adjust tint levels under varying lighting conditions, enhancing visibility and safety.

Recent research has integrated electrochromic materials with emerging technologies, expanding their versatility (Aller-Pellitero et al., 2019; Li et al., 2019a, 2019b; Yun et al., 2019). A notable development is the self-structuring electrochromic device employing a single layer of multicolour polyaniline (PANI) (Gong et al., 2020). Unlike conventional five-layer ECDs, this simplified three-layer system, consisting of a PANI layer, quasi-solid-state gel electrolyte, and conductive substrate, exhibits multicolor and gradient transitions with up to 60% transmission contrast. The observed self-structuring behavior arises from planar motion and redistribution of anions in the PANI layer, induced by current-driven magnetic effects (Popov et al., 2019). Through such innovations, electrochromic materials continue to advance as key components in sustainable design, enabling intelligent light control, reduced energy use, and improved environmental performance.

2.5 Magnetostrictive materials

Magnetostriction is a key property of ferromagnetic materials, causing them to change shape or dimensions when magnetized. This effect allows magnetostrictive materials to convert magnetic energy into kinetic energy and vice versa, making them useful for actuators and sensors. When a magnetic field is applied along a direction of pre-stress, the resulting magnetostriction is greater than without pre-stress. Early magnetostrictive materials included iron, nickel, and cobalt. Later, giant magnetostriction was discovered in rare earth elements such as dysprosium and terbium. Alloys like Terfenol-D, composed of terbium, dysprosium, and iron, offer high magnetostriction at lower magnetic fields under normal room temperature, achieving strains of 1000–2000 ppm under fields of 50–200 kA/m. Terfenol-D remains the most widely used magnetostrictive material for various applications.

Magnetostrictive materials are intelligent materials whose shape or dimensions change in response to a magnetic field due to the alignment of magnetic domains (Bahl et al., 2020). They are applied in high-precision positioning systems, vibration and noise control, energy harvesting, and in actuators and sensors for robotics and automation. Lead zirconate titanate (PZT), a piezoceramic material, is also commonly used in related applications due to its low cost, light weight, high energy density, and ease of implementation (Zhang & Lu, 2008).

Magnetostrictive sensors are widely used in industrial applications to measure strain, pressure, temperature, and other physical quantities with high accuracy and precision. While limitations include relatively low strain rates and sensitivity to magnetic interference, the unique properties of magnetostrictive materials make them valuable in engineering, scientific research, and industrial automation, providing precise actuation and sensing capabilities in diverse environments.

2.6. Optical Fibers

Transparent and flexible, optical fibers are often composed of glass or plastic and have an inner diameter (core) of a few micrometers (Correia et al., 2018; Bahl et al., 2020). They are widely used in Fiber-optic communications, which enables transmission over greater distances and at higher bandwidths than electrical connections, and are frequently utilized to transfer light between two ends. Because optical fibers can transmit signals with little loss and are not affected by electromagnetic interference, they are favoured over metal wires. The idea underlying optical fibers is based on the refraction phenomenon, which guides light.

If the transmitting fiber, often referred to as the core, is covered with a substance with a lower refractive index, light is transmitted in an optical fiber by total internal reflection. Furthermore, only particular modes of propagation are possible when light is transmitted via a fiber via total internal reflection; these modes depend on the fiber's diameter and the wavelength of light used . In addition to being utilized for communication, illumination, and imaging, optical fibers are also employed as sensors. Periodic variations in the refractive index along the core of optical fibers provide the basis of the principle of optical fiber gratings.

There are two primary forms of fiber gratings: short-period gratings and long-period gratings. For example, short-period gratings, sometimes called Fiber Bragg Gratings (FBGs), transmit all wavelengths while reflecting certain ones. FBG is therefore appropriate for sensing applications since it may be employed as an inline optical fiber to block wavelengths (Barrias et al., 2016). Interferometric techniques, which employ the interference between two beams propagating along different optical channels of a single fiber or two different fibers, can also serve as the foundation for optical fiber networks. Sagnac, Mach-Zehnder, Michelson, and Fabry-Pérot interferometers are the four primary types of optical fiber interferometers .

2.7 Magnetoactive and Electroactive Polymers

Magnetoactive polymers (MAPs) and electroactive polymers (EAPs) are key intelligent materials that respond dynamically to magnetic and electric stimuli, driving innovation in robotics, healthcare, and electronics.

MAPs are composites containing magnetic particles within a polymer matrix, allowing deformation under magnetic fields. Recent advancements, such as 4D printing, enable on-demand magnetic actuation for soft, adaptable structures (Zhang et al., 2024). This technique facilitates the creation of programmable materials that alter shape in response to magnetic fields, expanding applications in soft robotics and biomedical devices. Zhang et al. (2024) developed a magnetic-responsive bioink combining iron oxide nanoparticles with gelatin methacryloyl (GelMA), tailored for bioactuators and robotic applications, highlighting MAPs' biomedical potential.

EAPs, on the other hand, deform under electric fields, making them ideal for actuators, sensors, and drug delivery systems. Hård et al. (2024) introduced a topology optimization approach to simultaneously design EAP structures and electrode layouts, improving actuator efficiency by optimizing electric field distribution. Conductive polymers such as polypyrrole and polyaniline are increasingly used in electrically controlled drug release systems, allowing precise therapeutic regulation (Alkahtani et al., 2023). Furthermore, EAPs play a crucial role in soft robotics, serving as artificial muscles and flexible sensors due to their large actuation strain, energy efficiency, and design versatility (Dewan et al., 2025).

3. APPLICATIONS OF INTELLIGENT MATERIALS IN ENVIRONMENTAL

SUSTAINABILITY

3.1 Sustainable Construction

The construction industry is a significant contributor to global environmental degradation, accounting for approximately 39% of global carbon dioxide emissions and nearly 36% of global energy use (UN Environment, 2019). As sustainability becomes a central concern in architecture and civil engineering, intelligent materials offer innovative solutions to enhance energy efficiency, reduce waste, and extend the lifespan of infrastructure. These materials possess adaptive properties that respond to external environmental stimuli, allowing buildings to dynamically adjust to changing conditions and reduce reliance on mechanical systems.

a) Thermochromic and Electrochromic Windows

One of the most prominent applications of intelligent materials in sustainable construction is in dynamic glazing systems. Thermochromic and electrochromic windows adjust their transparency in response to temperature and electrical input, respectively, thereby controlling the amount of sunlight and heat entering a building. This reduces the need for artificial cooling and heating systems, resulting in significant energy savings. For example, electrochromic windows developed by Sage Glass can reduce HVAC energy use by up to 20% (Benedict et al., 2017). These dynamic glazing systems address the environmental challenge of excessive energy consumption and greenhouse gas emissions from building climate control systems.

b) Self-Healing Concrete

Concrete is the most widely used building material in the world, but it is prone to cracking, which leads to structural degradation and high maintenance costs. Self-healing concrete incorporates bacteria or microencapsulated healing agents that activate in the presence of water and oxygen, filling in cracks autonomously. Research conducted by Jonkers (2011) introduced *Bacillus* bacteria-based self-healing concrete that can repair cracks up to 0.8 mm wide, significantly extending the lifespan of structures and reducing the need for repairs and replacements. The recent commercial applications have demonstrated the practical viability of self-healing concrete. For instance, the Dutch company Basilisk has developed a bacteria-based self-healing concrete (ER7) that has been successfully implemented in real infrastructure projects, such as a bus lane, where it extended the service life by up to 15 years and reduced CO_2 emissions by approximately 90%. Self-healing concrete addresses the environmental challenge of resource depletion and carbon emissions associated with frequent repair, reconstruction, and cement production.

c) Phase Change Materials (PCMs)

PCMs are substances that absorb and release thermal energy during the process of melting and freezing. When integrated into building envelopes, such as walls and ceilings, they help stabilize indoor temperatures by storing excess heat during the day and releasing it at night. This passive thermal regulation can lower HVAC energy consumption. A study by Soares et al. (2013) showed that buildings integrated with PCMs can reduce cooling loads by up to 30% in hot climates. Phase change materials (PCMs) address the environmental challenge of high energy consumption and associated carbon emissions from building heating and cooling systems.

d) Intelligent Insulation Materials

Advanced insulation materials such as vacuum insulation panels (VIPs) and aerogels offer extremely low thermal conductivity and can be enhanced with responsive features. For example, shape memory polymers embedded in insulation systems can adjust porosity or thermal resistance in response to ambient conditions, optimizing thermal comfort and reducing energy usage (Nguyen et al., 2020). These advanced insulation materials address the environmental challenge of excessive energy loss and inefficiency in building thermal management systems.

e) Structural Health Monitoring (SHM)

Embedding piezoelectric sensors and shape memory alloys within building structures enables real-time monitoring of stress, temperature, and structural integrity. This allows for predictive maintenance, enhancing safety and minimizing material waste associated with overbuilding or premature demolition. This technology addresses the environmental challenge of material wastage and resource inefficiency resulting from unmonitored structural degradation and premature reconstruction.

3.2 Renewable Energy Systems

Renewable energy systems are essential for reducing dependence on fossil fuels and mitigating climate change. However, their efficiency and scalability often face limitations due to intermittency, storage challenges, and environmental variability. Intelligent materials offer advanced capabilities that enhance the functionality, adaptability, and efficiency of renewable energy technologies. These materials respond dynamically to environmental stimuli such as light, heat, and pressure, making them integral in optimizing energy conversion, harvesting, and storage.

a) Perovskite-Based Photovoltaics

Intelligent materials like perovskites have transformed the landscape of solar energy. Compared to traditional silicon-based cells, perovskite solar cells (PSCs) offer high power conversion efficiencies (PCEs), lightweight construction, and flexible applications. In recent developments, PSCs have achieved efficiencies over 26% in laboratory conditions (Green et al., 2024), rivaling silicon while being cheaper and easier to manufacture. Additionally, research into thermochromic perovskites allows solar panels to adapt to temperature fluctuations, optimizing performance in variable climates (Wu et al., 2023). Perovskite photovoltaics face **instability issues** due to sensitivity to moisture, oxygen, heat, and UV light, leading to rapid degradation and reduced device lifetime. Additionally, most high-performing perovskites contain **lead (Pb)**, raising concerns over **toxicity** and environmental hazards during production, operation, and disposal. These challenges hinder large-scale commercialization despite their excellent efficiency potential (Niu et al., 2015).

b) Piezoelectric Materials for Energy Harvesting

Piezoelectric materials generate electricity from mechanical stress or vibrations, making them suitable for capturing ambient energy from sources like foot traffic, road vehicles, or industrial machines. For instance, lead zirconate titanate (PZT) and newer lead-free alternatives are being integrated into infrastructure to generate low-voltage electricity for sensors and street lighting (Zhu et al., 2023). These systems support decentralized energy solutions, particularly in smart cities and remote environments.

c) Thermoelectric Materials

Thermoelectric intelligent materials convert temperature gradients into electrical energy. Recent advances in nanostructured bismuth telluride (Bi2Te3) and skutterudites have significantly improved thermoelectric efficiency, allowing waste heat from industrial processes or solar panels to be harvested (Li et al., 2022). These materials contribute to energy recovery in systems where thermal losses are otherwise inevitable.

d) Intelligent Energy Storage

Intelligent materials also play a role in enhancing battery technologies. Solid-state electrolytes, phase-change materials, and shape-memory alloys are being incorporated into lithium-ion and sodium-ion batteries to improve energy density, safety, and charge-discharge rates. Materials such as graphene-enhanced electrodes are increasing the efficiency and lifecycle of storage systems, crucial for renewable integration (Choi et al., 2024).

3.3 Water Management

The escalating global water crisis, intensified by climate change and population growth, necessitates innovative and sustainable solutions. Intelligent materials—engineered substances that respond dynamically to environmental stimuli—are at the forefront of revolutionizing water management. These materials offer advanced capabilities in water purification, harvesting, and conservation, contributing significantly to environmental sustainability.

a) Atmospheric Water Harvesting with Biomass-Derived Hydrogels

A groundbreaking study by researchers at the University of Texas at Austin introduced hydrogel sorbents synthesized from natural polysaccharides and biomass, including food waste and seashells. These hydrogels efficiently absorb moisture from the air and release it upon gentle heating, yielding up to 3.75 gallons of clean water per kilogram daily (University of Texas at Austin study) . Notably, they recover 95% of the absorbed water and are biodegradable, offering a scalable and energy-efficient solution for water-scarce regions (Li et al., 2020).

b) Nanomaterial-Enhanced Membranes for Groundwater Purification

The integration of nanomaterials such as graphene, carbon nanotubes, and metal nanoparticles into membrane technologies has significantly improved groundwater purification. These nanomaterials enhance membrane performance by increasing adsorption capacity, reducing fouling, and improving selectivity, leading to more efficient removal of contaminants like heavy metals and organic pollutants.

c) Green Photocatalysts for Wastewater Treatment

Green photocatalysts, derived from natural and renewable resources, have emerged as effective agents in degrading organic pollutants and removing heavy metals from wastewater. Utilizing bio-waste such as plant extracts and agricultural residues, these photocatalysts offer an eco-friendly approach to water purification, minimizing the use of toxic chemicals and reducing environmental impact.

d) Advanced Filtration Materials for Contaminant Removal

Researchers at MIT have developed a novel filtration material combining silk nanofibrils and cellulose, capable of removing persistent contaminants like per- and polyfluoroalkyl substances (PFAS) and heavy metals from water. This biodegradable material also exhibits strong antimicrobial properties, reducing membrane fouling and enhancing the longevity of filtration systems.

e) Solar-Driven Water Purification Systems

Innovative solar steam generation systems utilizing composite sponges embedded with materials like molybdenum disulfide (MoS_2) and carbon have demonstrated high efficiency in water purification. These systems not only achieve rapid water evaporation rates but also effectively remove heavy metals such as mercury, providing a sustainable solution for clean water production in off-grid areas.

3.4 Waste Management

The escalating global waste crisis necessitates innovative solutions that transcend traditional waste management practices. Intelligent materials—engineered substances that respond dynamically to environmental stimuli—are emerging as pivotal tools in revolutionizing waste management by enhancing recycling efficiency, enabling energy recovery, and promoting sustainable practices.

a) Photoreforming of Plastic Waste

A groundbreaking approach involves the use of high-entropy oxide (HEO) catalysts for the Photoreforming of plastic waste. This process not only degrades polyethylene terephthalate (PET) plastics under light but also simultaneously produces hydrogen fuel. The HEO catalysts exhibit superior activity compared to traditional photocatalysts, attributed to their diverse cation composition and lattice defects, which enhance charge carrier dynamics. This dual functionality offers a sustainable pathway for plastic waste valorization and clean energy production. However, its large-scale implementation remains challenging due to high catalyst costs, scalability issues, and limited efficiency under real-world conditions.

b) Covalent Adaptable Networks (CANs)

Conventional polyurethane (PU) foams pose significant recycling challenges due to their cross-linked structures. CANs introduce dynamic covalent bonds into PU foams, rendering them malleable and reprocessable. Incorporating disulfide bonds or catalysts facilitates bond exchange reactions, allowing the foams to be reshaped and recycled without significant degradation of properties. This innovation aligns with circular economy principles by extending the lifecycle of PU-based products.

c) Active Disassembly Using Intelligent Materials

Active Disassembly (AD) leverages intelligent materials like shape memory alloys (SMAs) and shape memory polymers (SMPs) to facilitate the automated disassembly of products at their end-of-life. These materials change shape or properties in response to stimuli such as heat, enabling components to detach without manual intervention. AD enhances the recyclability of complex products, reduces labour costs, and minimises environmental impact. Active Disassembly (AD) is a widely recognised term in sustainable manufacturing and product design

d) Smart Waste Management Systems

Companies like Ecube Labs have developed solar-powered waste compacting bins equipped with sensors that monitor fill levels in real-time. These smart bins compact waste autonomously, increasing capacity and reducing collection frequency. The integration of Internet of Things (IoT) technology enables efficient route planning for waste collection, lowering operational costs and carbon emissions.

e) Recycling of Rare Earth Metals

Queen's University Belfast's QUILL research center is pioneering the use of ionic liquids for the recycling of rare earth metals from electronic waste. This method offers a more environmentally friendly alternative to traditional acid-based extraction processes, achieving high-purity metal recovery while minimizing hazardous waste generation. Such advancements are crucial for the sustainable management of e-waste and the conservation of critical materials.

Case Study 1: Electrocaloric Cooling Systems

Dr. Emmanuel Defay's research at the Luxembourg Institute of Science and Technology focuses on electrocaloric materials that change temperature under electric fields. These materials are being developed into cooling systems that eliminate harmful refrigerants, offering a sustainable alternative to traditional air conditioning.

Case Study 2: Intelligent Materials in Urban Infrastructure

Cities like Tokyo and Singapore are integrating piezoelectric materials into sidewalks and roads to generate electricity from pedestrian and vehicular movement, contributing to smart grid systems.

Because they have self-regulating, responsive, and adaptable properties, intelligent materials perform better than traditional materials. They improve structural performance, durability, and energy efficiency. For instance, electrochromic glass maximizes solar gain, self-healing concrete lowers maintenance, and piezoelectric sensors allow for real-time monitoring. Intelligent materials enhance sustainability, comfort, and long-term dependability, whereas traditional materials offer static features and lower initial costs. Table 1 lists the performance comparison between smart and conventional materials.

Table 1. Performance Comparison of Intelligent Materials and Conventional Materials in Structural, Energy, and Functional Applications

Material/System	Intelligent Material Performance	Traditional Material Performance	Key Environmental Impact Addressed
Self-Healing Concrete	Cracks up to 0.5 mm can self-heal, reducing permeability, extending service life, and lowering maintenance	Microcracks persist, water ingress can accelerate corrosion, shorter lifespan	Reduces maintenance costs and increases durability, improves sustainability
Electrochromic Glass	Dynamic control of solar heat gain and glare can reduce HVAC energy by 10–20%	Fixed transparency, no control over solar gain	Energy savings, improved occupant comfort, and adaptable to environmental conditions
Piezoelectric Sensors / Actuators	High sensitivity for vibration, pressure, or structural health monitoring	Requires external sensors or manual monitoring, limited responsiveness	Enables real-time monitoring, energy harvesting, and adaptive responses
Shape Memory Alloys (SMA)	Can actuate or recover shape without motors; responsive to temperature or electric stimulus	Traditional metals deform permanently under stress	Useful for adaptive structures, actuators, robotics, and aerospace applications
Phase-Change Materials (PCM) for Thermal Storage	Smooth thermal regulation reduces energy demand for heating/cooling	Conventional materials have low thermal mass; they cannot store/release latent heat efficiently	Reduces building energy consumption, improves comfort, and integrates with passive solar design
Smart Coatings / Corrosion Sensors	Early detection of corrosion can extend structure's lifetime	Standard coatings provide passive protection only; failure is often unnoticed	Predictive maintenance, cost savings, and reduces structural failures

Intelligent materials generally provide adaptive functionality, improved durability, or real-time feedback that traditional materials cannot. Trade-offs often include higher upfront cost, more complex manufacturing, and sometimes limited large-scale adoption.

4. LIFE CYCLE AND CIRCULAR ECONOMY OF INTELLIGENT MATERIALS

Piezo electrics, shape memory alloys and polymers, sensors, and other intelligent materials go through several phases in their life cycle, including extraction of raw materials, synthesis and manufacturing, usage and use-phase, and end of life (recycling or disposal). Every step has repercussions for sustainability, technology, and the environment.

4.1 Extraction of Raw Materials

Metals (such as rare earths, lead, titanium, and zirconium), polymers, ceramics, and occasionally rare or toxic elements are common raw materials for intelligent materials. Energy use, greenhouse gas emissions, habitat destruction, water use, and occasionally pollution and hazardous waste are all associated with mining and extraction. For instance, acid drainage and heavy metal contamination may arise from the extraction of rare earth elements, which are critical components in several intelligent materials such as magnetostrictive materials, piezoelectric ceramics, and electrochromic devices. These materials often rely on elements like neodymium, dysprosium, and europium, linking their environmental footprint directly to the mining and processing of these scarce and ecologically sensitive resources. Furthermore, some feedstocks are toxic or non-renewable, such as petroleum for polymers or lead for lead-based piezoelectrics.

4.2 Production and Synthesis

Raw ingredients need to be mixed, treated, and purified after extraction. This frequently entails high temperatures (sintering, annealing), precise chemical or crystal engineering, doping, thin layer deposition, etc. for intelligent materials. These processes need a lot of energy and can produce toxic chemicals, waste solvents, pollutants, and chemical byproducts. Cleanrooms, vacuum systems, and other specialized infrastructure may be needed for the fabrication of nanomaterials or advanced composites, which would increase the cost and environmental impact.

4.3 Phase of Use and Operation

Sensing, actuation, energy harvesting, self-healing, and other functions are delivered by intelligent materials while they are in use. Durability, deterioration, and performance over time are important considerations here. For instance, perovskite photovoltaics deteriorate when exposed to heat, moisture, or ultraviolet light, while piezoelectrics may experience fatigue, depoling, or loss of sensitivity. Because of these restrictions, the effective lifetime is shortened, requiring replacement more frequently, which raises the environmental load overall. It's also important to consider the energy usage for maintenance, operation (if any), and any necessary protective coatings or encapsulation.

4.4 Reusing, Recycling, Disposal

a) *Reuse/Remanufacturing:* parts or equipment can be fixed, reconditioned, or reused.
b) *Recycling:* Certain materials, such as metals, ceramics, and polymers, can be reclaimed. However, separation and recovery are challenging and/or expensive due to the complexity of smart materials (multi-layered, doped, composites, encapsulated). Recycling could necessitate specific procedures, which might or might not be profitable or energy efficient. For example, it is technically feasible to recover lead from lead-based sensors or rare earths from magnets, but these processes are frequently constrained by factors such as cost, recovered material purity, transportation, safety, and legislation.
c) *Disposal:* Landfilling or incineration takes place if recycling or reuse is not practical. If landfilled, toxic materials (such as lead, arsenic, etc.) may seep into the soil or water; if incineration is not properly managed, it may release harmful fumes or particles. Additionally, the product's material worth and embodied energy are lost upon disposal.

4.5 Life Cycle Evaluation and Views on the Circular Economy

Life-cycle assessment is used to examine environmental impacts (such as carbon footprint, toxicity, resource depletion, and water usage) across these stages. According to studies, major effects frequently happen early on (raw material extraction, production), but end-of-life and replacement/use cycles have a big impact. The key to lowering overall impacts is implementing circular economy solutions, which include closed-loop material flows, improved recycling, using less hazardous or abundant materials, and designing for disassembly. For example, recycling can dramatically reduce energy and GHG emissions compared to virgin material extraction (Garner, 2023)

5. CHALLENGES AND ETHICAL CONSIDERATIONS OF INTELLIGENT

MATERIALS

Intelligent materials, substances that react dynamically to environmental stimuli, are transforming several industries, including electronics, healthcare, and construction. Even though their uses have many advantages, there are several difficulties and moral issues that need to be resolved to guarantee responsible development and implementation (Figure 9).

Figure 9. Challenges and Ethical Considerations of Intelligent Material

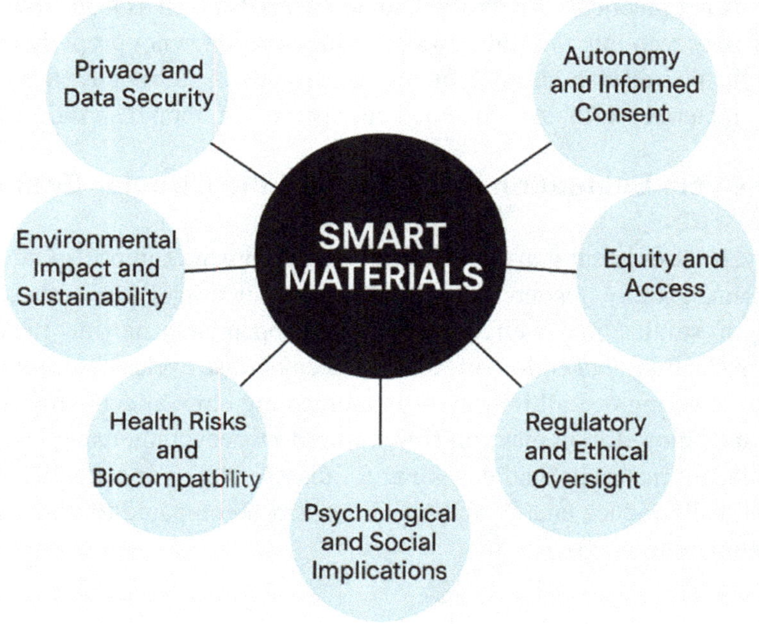

5.1 Privacy and Data Security

Sensors that gather and send data are frequently included into Intelligent materials, which raises privacy and data security issues. For example, sensitive physiological data like heart rate and stress levels can be collected by e-textiles used in health monitoring. Because of the sensitive nature of this data, strict safeguards are required to guard against misuse and unauthorized access. Without adequate protections, people's privacy and liberty could be jeopardized by surveillance or data breaches (Ahuja & Kumar, 2023).

5.2 Autonomy and Informed Consent

The distinction between voluntary and involuntary participation may become hazier as intelligent materials become more integrated into everyday life. Employers or insurance providers, for instance, may require or encourage the use of wearable technology that tracks health indicators, which could pressure people into disclosing personal information. This dynamic may violate personal autonomy and calls into question the idea of informed consent.

5.3 Environmental Impact and Sustainability

There are environmental issues with the manufacture and disposal of intelligent materials. Nanomaterials and complicated composites, which are hard to recycle and can leak harmful compounds into the environment, are incorporated into many intelligent materials. Carbon emissions are also a result of the energy-intensive manufacturing processes. A thorough lifecycle analysis is necessary to evaluate and lessen these environmental effects. The difficulty of incorporating intelligent materials into current systems is one of the main obstacles. These materials frequently necessitate precise calibration and unique manufacturing procedures, which can raise production costs and restrict scalability (Gupta et al., 2022). Furthermore, there are serious environmental issues about the lifecycle of intelligent materials, particularly those that are integrated with electronics. Because of their composite nature, recycling is frequently made more difficult, increasing electronic waste and perhaps contaminating the environment if improperly disposed of (Maksimović et al., 2023).

Beyond the challenges of manufacturing and disposal, the environmental impact and sustainability of intelligent materials require a more systemic perspective that considers resource efficiency, end-of-life recovery, and circular design. Future strategies should focus on developing bio-based or biodegradable smart materials, reducing reliance on rare or toxic elements, and adopting energy-efficient synthesis methods such as green chemistry and additive manufacturing.

5.4 Health Risks and Biocompatibility

Intelligent materials are employed in biomedical applications to create prostheses and implants that have direct contact with human tissues. Although these materials can improve functioning, there are dangers associated with their long-term health consequences and biocompatibility. Smart-material regeneration valve implants, for example, have the potential to permanently fuse with body tissues, making removal more difficult and possibly leading to unanticipated problems (DeKanter et al., 2023)

The use of these materials in irreversible applications raises ethical problems since it may limit patient autonomy or cast doubt on consent, particularly in vulnerable populations. International standards are desperately needed to ensure the safe development, testing, and implementation of intelligent materials, particularly those employed in vital applications such as structural engineering or medical equipment (Thakur et al., 2025)

5.5 Equity and Access

The advantages of intelligent materials might not be shared fairly, which could exacerbate already-existing societal injustices. Disparities in healthcare, education, and other areas result from the frequent correlation between socioeconomic level and access to cutting-edge technologies. Intentional legislative changes and inclusive design principles are necessary to guarantee fair access.

5.6 Regulatory and Ethical Oversight

Intelligent materials are developing faster than current regulatory frameworks, which leaves supervision holes. Ethical conundrums may result from this lack of control, such as the use of intelligent materials in potentially harmful military applications or the deployment of surveillance technology without sufficient consent. Serious questions of informed consent, data security, and privacy are brought up by this. Data misuse is possible in the absence of clear regulations, particularly by third-party advertising or data brokers (Fernandez & Pallis, 2021). To solve these issues, it is essential to create thorough ethical standards and regulatory policies

5.7 Psychological and Social Implications

Social dynamics and psychological health may be impacted by the incorporation of smart materials into the human body or surroundings. For instance, a person's body image or sense of identity may change as a result of using smart implants. Pervasive sensing technology can also cause a sense of ongoing surveillance, which can affect social interactions and mental health. To guarantee that everyone benefits equally from innovative technologies, accessibility and affordability must be taken into account during the ethical design process (WHO, 2022). Furthermore, interdisciplinary knowledge and reliable testing procedures are necessary for the creation of multifunctional composites and AI-enhanced materials (Yuan et al., 2025)

While intelligent materials hold immense potential to transform various industries, it is imperative to address the associated challenges and ethical considerations proactively. This includes implementing robust data protection measures, ensuring informed consent, assessing environmental impacts, promoting equitable access, and establishing comprehensive regulatory frameworks. By doing so, society can harness the benefits of intelligent materials while mitigating potential risks and ethical dilemmas.

6. FUTURE DIRECTIONS AND RESEARCH OPPORTUNITIES

6.1 Materials Science and Chemistry Innovations

Intelligent materials capable of responding to environmental stimuli such as temperature, pH, moisture, or pollutants hold revolutionary potential for addressing urgent environmental challenges. These materials are increasingly being developed to support sustainable solutions in pollution monitoring, remediation, resource efficiency, and climate change mitigation. Future research should prioritise scalability, environmental compatibility, multifunctionality, and material durability. Promising directions include multi-responsive materials that react to multiple stimuli simultaneously—especially valuable in dynamic environments such as wastewater systems or landfills, where they can enable real-time monitoring and adaptive remediation. Advances in hybrid material systems, polymer chemistry, and nanotechnology are crucial to realising this potential (Zhao et al., 2021).

6.2 Engineering and Environmental Applications

Intelligent membranes capable of altering surface charge or pore size in response to water quality changes represent a significant innovation in wastewater treatment and resource recovery. In the context of climate resilience, thermochromic and electrochromic materials play a vital role in regulating building temperatures by dynamically adjusting optical properties. Future engineering efforts should focus on enhancing responsiveness, long-term stability, and seamless integration into conventional infrastructure (Ghaffarianhoseini et al., 2019).

6.3 AI and IoT Integration

The integration of intelligent materials with artificial intelligence (AI) and the Internet of Things (IoT) offers powerful opportunities for real-time environmental monitoring and decision-making. Smart sensors embedded in urban infrastructure can detect pollutants, track air and water quality, and transmit data for rapid response using low-power, energy-efficient systems (Chen et al., 2020). Such integration bridges the gap between materials innovation and smart environmental management systems.

6.4 Sustainability, Policy, and Scalability

Cost-effectiveness and scalability remain significant barriers to widespread adoption. Research should prioritise bio-based and biodegradable raw materials such as cellulose, chitosan, and lignin, along with green and bioinspired synthesis methods to minimize environmental impact (Wang et al., 2022). Long-term sustainability also depends on improving recyclability, self-healing capacity, and lifecycle efficiency of intelligent materials. Intelligent adsorbents and photocatalysts capable of capturing and degrading pollutants under sunlight exemplify self-regenerative and low-maintenance solutions.

6.5 Interdisciplinary Collaboration and Ethics

The successful translation of intelligent materials into real-world applications demands interdisciplinary collaboration among chemists, engineers, environmental scientists, and data experts. Equally important are ethical, regulatory, and societal considerations, ensuring that innovation in intelligent materials aligns with global sustainability and environmental justice goals.

Case Studies

a) AI-Based Degradation Index from Microstructure Images & Life Prediction

Yu & Oh (2023) created a model that creates a deterioration index for materials using microstructure imaging and Bayesian inference and then builds life-prediction models based on that index. This makes it possible to forecast how long materials will persist under specific patterns of microstructural breakdown. Yu & Oh's (2023) model links microstructural imaging with Bayesian inference to predict material lifespan, highlighting a shift toward data-driven, non-destructive evaluation. This approach points to **future directions** in materials science—integrating AI, real-time monitoring, and hybrid physics-data models to enhance predictive maintenance and material design.

b) Computational Prediction of Organic Coatings Degradation

A study published in RSC Advances (2024) forecasted the degradation of various polyester-melamine coating formulations under both accelerated and natural weathering using density functional theory (DFT) reactivity descriptors (Batchelor et al., 2024). Long weathering testing was not necessary after the predictions were empirically confirmed. This is an excellent illustration of how to optimise materials for environmental resilience by integrating modelling, AI/ML, and experimentation. This case study demonstrates how combining DFT modelling, AI/ML, and experimental validation can accelerate material optimization, pointing to **future directions** in predictive materials design and sustainable development without lengthy durability testing.

c) Deep Learning for Lifespan Prediction of Bridges

A deep learning system was developed by Patel and Bukaita (2025) to forecast the lifespan degradation of concrete bridges brought on by iron oxidation. They made use of both tabular data (material composition, environmental exposures, etc.) and visual inspection data (pictures). Good metrics like MAE and R2 indicated that the model could predict decline and aid in maintenance planning.This case study highlights the growing role of multimodal AI—combining visual and structural data—in infrastructure monitoring, signaling **future directions** toward intelligent, data-driven maintenance and predictive asset management systems.

d) Low-Cost Sensor Networks for Environmental Monitoring

Enviro-IoT is a system that uses inexpensive sensors to monitor air quality in urban areas; it has been validated over a period of nine months (Johnson & Woodward, 2025). When compared to industry-standard instrumentation, it demonstrated great accuracy (e.g. > 95% for various contaminants). According to Johnson and Woodward (2025), this demonstrates how IoT, calibration, and AI can provide extensive, real-time environmental monitoring at relatively modest prices.

In order to make devices appropriate for severe or tropical settings, the 2024 project "Open-source IoT-based air pollution monitoring system for tropical environments" created protective enclosures and system design. Crucial since environmental conditions frequently cause sensors to deteriorate, intelligent design and monitoring can preserve dependability (Collado et al., 2024).

"Modular, low-cost IoT system for environmental and behavioural monitoring in cultural heritage sites" (2025) — combines environmental sensors (temperature, humidity, sound) plus behaviour detection (visitor proximity) using computer vision. Useful for protecting delicate heritage materials' micro-environment.

These studies point to future directions in scalable, resilient IoT and AI-enabled environmental monitoring—focusing on sensor durability, adaptive calibration, and intelligent system design for diverse and sensitive environments.

e) Sustainable Material Property Prediction

"Predicting the mechanical characteristics of sustainable plastic pavement blocks through the use of artificial intelligence" (2025) Numerous experiments were avoided by using AI models (likely machine learning) to forecast the strength, water absorption, and other characteristics of pavement blocks made from plastic waste (Maskari et al., 2025).

In addition to examining material performance, "AI-Driven Data Analysis of Quantifying Environmental Impact and Efficiency of Shape Memory Polymers" (2024) compares shape memory polymers to traditional materials and examines environmental parameters such as water usage and carbon footprint. Thus, integrating AI, sustainability indicators, and intelligent materials.T hese studies illustrate future directions in sustainable materials engineering—leveraging AI to predict performance, minimize experimentation, and integrate environmental impact assessments for smarter, eco-efficient material design.

7. CONCLUSION

Intelligent materials offer transformative potential for sustainable environmental solutions by dynamically responding to stimuli such as pollutants, temperature, and moisture. They enable real-time monitoring, adaptive remediation, and energy-efficient applications in air and water purification, waste treatment, and smart infrastructure. Future development should focus on creating multifunctional, biodegradable, and cost-effective materials that emphasize green synthesis, recyclability, and minimal ecological impact. Integrating intelligent materials with digital technologies like IoT and AI can further enhance environmental monitoring and enable rapid, data-driven responses to ecological threats.

However, large-scale adoption requires overcoming challenges related to scalability, durability, and regulatory approval. Collaboration among scientists, engineers, environmental specialists, and policymakers is essential to translate research innovations into practical applications. With continued interdisciplinary efforts and

REFERENCES

Ahuja, S., & Kumar, J. (2023). Ethical aspects of health sensing applications in e-textiles. *Engineering Proceedings*, *30*(1), 4. DOI: 10.3390/engproc2023030004

Alkahtani, M. E., Elbadawi, M., Chapman, C. A. R., Green, R. A., Gaisford, S., Orlu, M., & Basit, A. W. (2023). Electroactive polymers for on-demand drug release. *Advanced Healthcare Materials*, *13*(3), 2301759. Advance online publication. DOI: 10.1002/adhm.202301759 PMID: 37861058

Aller-Pellitero, M., Fremeau, J., Villa, R., Guirado, G., Lakard, B., Hihn, J. Y., & del Campo, F. J. (2019). Electrochromic biosensors based on screen-printed Prussian Blue electrodes. *Sensors and Actuators. B, Chemical*, *290*, 591–597. DOI: 10.1016/j.snb.2019.03.100

Bahl, S., Nagar, H., Singh, I., & Sehgal, S. (2020). Smart material types, properties and applications: A review. *Materials Today: Proceedings*, *28*, 1302–1306. DOI: 10.1016/j.matpr.2020.04.505

Barrias, A., Casas, J. R., & Villalba, S. (2016). A review of distributed optical fibre sensors for civil engineering applications. *Sensors (Basel)*, *16*(5), 748. DOI: 10.3390/s16050748 PMID: 27223289

Batchelor, C., Martin-Martinez, F. J., Griffiths, C., Mabbett, I., Smith, J., & Jewell, E. (2024). Computationally accelerated prediction of polyester-melamine coatings degradation to design sustainable organically coated steels for outdoor applications. *RSC Advances*, *14*(26), 18343–18354. DOI: 10.1039/D3RA06744K PMID: 38915881

Benedict, B., Bell, B., & Miller, M. (2017). Electrochromic windows: Technology and energy performance. *Building and Environment*, *125*, 172–181. DOI: 10.1016/j.buildenv.2017.08.037

Bhaskar, J., Sharma, A. K., Bhattacharya, B., & Adhikari, S. (2020). A review on shape memory alloy reinforced polymer composite materials and structures. *Smart Materials and Structures*, *29*(7), 073001. DOI: 10.1088/1361-665X/ab8836

Bhattacharjee, J., & Roy, S. (2023). *Biogenic electrodes for low-temperature solid oxide fuel cells. ECS Meeting Abstracts, MA2023-01*(40), 2805. https://doi.org/DOI: 10.1149/MA2023-01402805mtgabs

Bhattacharjee, J., & Roy, S. (2024). Utilizing a variable material approach to combat climate change. *Materials Science Research India*, *20*(3), 141–145. DOI: 10.13005/msri/200301

Chen, W., Jia, W., Xiao, Y., Feng, Z., & Wu, G. (2021). A temperature-stable and low impedance piezoelectric MEMS resonator for drop-in replacement of quartz crystals. *IEEE Electron Device Letters*, *42*(9), 1382–1385. DOI: 10.1109/LED.2021.3094319

Chen, Y., Wang, Y., & Li, H. (2020). Smart environmental monitoring systems using IoT and intelligent materials. *Environmental Science & Technology*, *54*(18), 11569–11578. DOI: 10.1021/acs.est.0c02679

Choi, J. W.. (2024). Smart materials in next-generation energy storage. *Nature Energy*, *9*, 12–20. DOI: 10.1038/s41560-023-01357-2

Choi, N., Hwang, Y., Lee, H., Seo, M., Kim, Y., Kim, G., & Kim, K. (2022). Fabrication and simulation of a piezoelectric PIN-PMN-PT thin film for ultrahigh-frequency ultrasonic transducers. *Sensors and Actuators. A, Physical*, *347*, 113936. DOI: 10.1016/j.sna.2022.113936

Chua, M. H., Tang, T., Ong, K. H., Neo, W. T., & Xu, J. W. (2019). Chapter 1: Introduction to electrochromism. In *Electrochromic Smart Materials: Fabrication and Applications* (pp. 1–26). Royal Society of Chemistry. ISBN 978-1-78801-143-3

Collado, E., Calderón, S., Cedeño, B., De León, O., Centella, M., García, A., & Sáez, Y. (2024). Open-source Internet of Things (IoT)-based air pollution monitoring system with protective case for tropical environments. *HardwareX*, *19*, e00560. DOI: 10.1016/j.ohx.2024.e00560 PMID: 39687307

Correia, R., James, S., Lee, S. W., Morgan, S. P., & Korposh, S. (2018). Biomedical application of optical fibre sensors. *Journal of Optics*, *20*(7), 073003. DOI: 10.1088/2040-8986/aac68d

De Kanter, A. J., Jongsma, K. R., Bouten, C. V. C., & Bredenoord, A. L. (2023). How smart are smart materials? A conceptual and ethical analysis of smart lifelike materials for the design of regenerative valve implants. *Science and Engineering Ethics*, *29*(5), 33. Advance online publication. DOI: 10.1007/s11948-023-00453-1 PMID: 37668955

Dionne, J. A., Dagli, S., & Shalaev, V. M. (2023). Nanophotonics for a sustainable future. *Physics Today*, *76*(6), 24–31. DOI: 10.1063/PT.3.5254

Dwivedi, S., Jaiswal, S., & Kushwaha, A. (2022). Renewable Biofuel Resources: Introduction, Production Technologies, Challenges, and Applications. In: Chowdhary, P., Pandit, S., Khanna, N. (eds) Bio-Clean Energy Technologies Volume 2. Clean Energy Production Technologies. Springer, Singapore. https://doi.org/DOI: 10.1007/978-981-16-8094-6_3

Fakhri, P., Amini, B., Bagherzadeh, R., Kashfi, M., Latifi, M., Yavari, N., Kani, S. A., & Kong, L. (2019). Flexible hybrid structure piezoelectric nanogenerator based on ZnO nanorod/PVDF nanofibers with improved output. *RSC Advances*, *9*(18), 10117–10123. DOI: 10.1039/C8RA10315A PMID: 35520929

Fernandez, J., & Pallis, G. (2021). Privacy issues in smart environments. *IEEE Communications Surveys and Tutorials*, *23*(4), 2692–2718.

Garg, R., & Anjum, A. (2024, February 23). Smart and sustainable applications of nanocomposites. *Advances in Chemical and Materials Engineering*. Advance online publication. DOI: 10.4018/979-8-3693-3136-1.ch002

Garner, J. (2023). The transformative power of material recycling: Creating a sustainable future. *Journal of Materials Science and Nanomaterials*, *7*, 98.

Ghaffarianhoseini, A.. (2019). The application of smart materials in sustainable architecture. *Renewable & Sustainable Energy Reviews*, *101*, 778–791. DOI: 10.1016/j.rser.2018.10.042

Gong, H., Zhou, K., Zhang, Q., Liu, J., Wang, H., & Yan, H. (2020). A self-patterning multicolour electrochromic device driven by horizontal redistribution of ions. *Solar Energy Materials and Solar Cells*, *215*, 110642. DOI: 10.1016/j.solmat.2020.110642

Green, M. A., Dunlop, E. D., Yoshita, M., Kopidakis, N., Bothe, K., Siefer, G., & Hao, X. (2024). Solar cell efficiency tables (version 63). *Progress in Photovoltaics: Research and Applications*, *32*(1), 3–19. DOI: 10.1002/pip.3750

Guo, S., Duan, X., Xie, M., Aw, K. C., & Xue, Q. (2020). Composites, fabrication and application of polyvinylidene fluoride for flexible electromechanical devices: A review. *Micromachines*, *11*(11), 1076. DOI: 10.3390/mi11121076 PMID: 33287450

Gupta, A., Verma, R., & Singh, D. (2022). Smart materials and their applications: A review. *Materials Today: Proceedings*, *57*, 2250–2260. DOI: 10.1016/j.matpr.2022.04.015

Habib, M., Lantgios, I., & Hornbostel, K. (2022). A review of ceramic, polymer and composite piezoelectric materials. *Journal of Physics. D, Applied Physics*, *55*(42), 55. DOI: 10.1088/1361-6463/ac8687

Hajalilou, A., Amri Mazlan, S., Lavvafi, H., & Shameli, K. (2016). Field responsive fluids as smart materials (1st ed.). Engineering Materials, Springer. DOI: 10.1007/978-981-10-2495-5

Hård, D., Wallin, M., & Ristinmaa, M. (2024). Electrode and electroactive polymer layout design using topology optimization [Preprint]. arXiv. https://arxiv.org/abs/2401.12345

Ibn-Mohammed, T., Koh, S. C. L., Reaney, I. M., Acquaye, A., Wang, D., Taylor, S., Genovese, A., & Sinclair, D. C. (2017). Are lead-free piezoelectrics more environmentally friendly? *MRS Communications*, *7*(1), 1–7. DOI: 10.1557/mrc.2017.10

Jaiswal, S., & Khanam, S. Z. (2025). Techno-economic analysis of green hydrogen production for industrial use. In Mabrouki, J. (Ed.), *Obstacles Facing Hydrogen Green Systems and Green Energy* (pp. 479–506). IGI Global Scientific Publishing., DOI: 10.4018/979-8-3693-8980-5.ch028

Jaiswal, S., Tiwari, S., & Bahadur, P. S. (2021). Inter-stellar space discovery, spectroscopic structure identification and application of C_{60} Buckminsterfullerene. *International Journal of Engineering Research in General Science*, *9*, 12–19. https://pnrsolution.org/Datacenter/Vol9/Issue2/2.pdf

Jensen, J., Hösel, M., Dyer, A. L., & Krebs, F. C. (2015). Development and manufacture of polymer-based electrochromic devices. *Advanced Functional Materials*, *25*(14), 2073–2090. DOI: 10.1002/adfm.201403765

Johnson, T., & Woodward, K. (2025). *Enviro-IoT: Calibrating low-cost environmental sensors in urban settings* [Preprint]. arXiv. https://doi.org//arXiv.2502.07596DOI: 10.48550

Kim, J. (2017). Multifunctional smart biopolymer composites as actuators. In Thomas, S., Joseph, K., Kuruvilla, A. K., & Malhotra, S. K. (Eds.), *Biopolymer composites in electronics* (pp. 311–331). Elsevier., DOI: 10.1016/B978-0-12-809261-3.00011-5

Le, A. T., Ahmadipour, M., & Pung, S.-Y. (2022). A review on ZnO-based piezoelectric nanogenerators: Synthesis, characterization techniques, performance enhancement and applications. *Journal of Alloys and Compounds*, *844*, 156172. DOI: 10.1016/j.jallcom.2020.156172

Li, G., Zhang, B., Wang, J., Zhao, H., Ma, W., Xu, L., Zhang, W., Zhou, K., Du, Y., & He, G. (2019b). Electrochromic poly(chalcogenoviologen)s as anode materials for high-performance organic radical lithium-ion batteries. *Angewandte Chemie International Edition*, *58*(25), 8468–8473. DOI: 10.1002/anie.201903152 PMID: 30951238

Li, J., Levitt, A., Kurra, N., Juan, K., Noriega, N., Xiao, X., Wang, X., Wang, H., Alshareef, H. N., & Gogotsi, Y. (2019a). MXene-conducting polymer electrochromic microsupercapacitors. *Energy Storage Materials*, *20*, 455–461. DOI: 10.1016/j.ensm.2019.04.028

Li, R., Shi, Y., Alsaedi, M., Wu, M., Shi, L., & Yu, G. (2020). Hybrid hydrogel with high water-vapor harvesting capacity for self-sustained solar-driven clean water production. *ACS Materials Letters*, *2*(7), 854–860. DOI: 10.1021/acsmaterialslett.0c00124

Li, Z., Zhao, J., Hou, C., Fei, C., Zheng, C., Lou, L., Chen, D., Li, D., & Yang, Y. (2022). High-frequency self-focusing ultrasonic transducer with piezoelectric metamaterial. *IEEE Electron Device Letters*, *43*(6), 946–949. DOI: 10.1109/LED.2022.3170613

Lu, C., Jiang, X., Li, L., Zhou, H., Yang, A., Xin, M., Fu, G., & Wang, X. (2022). Wind energy harvester using piezoelectric materials. *The Review of Scientific Instruments*, *93*(3), 031502. DOI: 10.1063/5.0065462 PMID: 35364975

Mahapatra, S. D., Mohapatra, P. C., Aria, A. I., Christie, G., Mishra, Y. K., Hofmann, S., & Thakur, V. K. (2021). Piezoelectric materials for energy harvesting and sensing applications: Roadmap for future smart materials. *Advancement of Science*, *8*(17), 2100864. DOI: 10.1002/advs.202100864 PMID: 34254467

Maksimović, M., Vujović, V., & Omanović-Mikličanin, E. (2023). Smart materials for sustainable development. *Journal of Cleaner Production*, *405*, 136313. DOI: 10.1016/j.jclepro.2023.136313

Maskari, M. A., Aravind, N., Dhandapani, R., & AlDeen, E. E. K. M. (2025). Application of artificial intelligence in prediction of mechanical properties of sustainable plastic pavement blocks and validation through experimental analysis. [SpringerLink]. *Discover Concrete & Cement*, *1*(1), 11. DOI: 10.1007/s44416-025-00011-x

Mir, S. H., Nagahara, L. A., Thundat, T., Mokarian-Tabari, P., Furukawa, H., & Khosla, A. (2018). Review—Organic–inorganic hybrid functional materials: An integrated platform for applied technologies. *Journal of the Electrochemical Society*, *165*(8), B3137–B3156. DOI: 10.1149/2.0191808jes

Mishra, J. (2017). Smart materials: Types and their application—A review. *International Journal of Mechanical and Production Engineering*, *5*(12), 20–24.

Monk, P., Mortimer, R., & Rosseinsky, D. (2007). *Electrochromism and electrochromic devices* (1st ed.). Cambridge University Press., DOI: 10.1017/CBO9780511550959

Murat, B., & Marinella, F. (2018). *Materials that move: Smart materials, intelligent design*. Springer International Publishing., DOI: 10.1007/978-3-319-76890-8

Nguyen, T. M., Lee, J., & Kim, D. (2020). Advanced insulation materials for buildings: A review. *Construction & Building Materials*, *260*, 119400. DOI: 10.1016/j.conbuildmat.2020.119400

Niu, G., Guo, X., & Wang, L. (2015). Review of recent progress in chemical stability of perovskite solar cells. *Journal of Materials Chemistry. A, Materials for Energy and Sustainability*, *3*(17), 8970–8980. DOI: 10.1039/C4TA04994B

Olabi, A.-G. (Ed.). (2021). *Encyclopedia of smart materials* (1st ed.). Elsevier., DOI: 10.1016/C2018-0-01700-3

Park, B. R., Hong, J., Choi, E. J., Choi, Y. J., Lee, C., & Moon, J. W. (2019). Improvement in energy performance of building envelope incorporating electrochromic windows (ECWs). *Energies*, *12*(6), 1181. DOI: 10.3390/en12061181

Pop, F., Herrera, B., & Rinaldi, M. (2022). Lithium niobate piezoelectric micromachined ultrasonic transducers for high data-rate intrabody communication. *Nature Communications*, *13*(1), 17. DOI: 10.1038/s41467-022-29355-9 PMID: 35379794

Popov, A., Brasiunas, B., Mikoliunaite, L., Bagdziunas, G., Ramanavicius, A., & Ramanaviciene, A. (2019). Comparative study of PANI, PEDOT, and PANI–PEDOT films electrochemically deposited on transparent indium tin oxide electrodes. *Polymer*, *172*, 133–141. DOI: 10.1016/j.polymer.2019.03.059

Rozman, M., Alif, M., Bren, U., & Luksic, M. (2022). Electrochromic device demonstrator from household materials. *Journal of Chemical Education*, *99*(9), 3595–3600. DOI: 10.1021/acs.jchemed.2c00176 PMID: 36246424

Sakimoto, K. K., Wong, A. B., & Yang, P. (2016). Self-photosensitization of nonphotosynthetic bacteria for solar-to-chemical production. *Science*, *351*(6268), 74–77. DOI: 10.1126/science.aad3317 PMID: 26721997

Sarker, M. R., Julai, S., Sabri, M. F. M., Said, S. M., Islam, M. M., & Tahir, M. (2019). Review of piezoelectric energy harvesting systems and optimization techniques to enhance performance. *Sensors and Actuators. A, Physical*, *300*, 111634. DOI: 10.1016/j.sna.2019.111634

Seo, J., Kim, Y., & Hu, J. (2015). Pilot study for investigating the cyclic behavior of slit damper systems with recentering SMA bending bars used for seismic restrainers. *Applied Sciences (Basel, Switzerland)*, *5*(3), 187–208. DOI: 10.3390/app5030187

Singh, A., Monga, S., Sharma, N., Sreenivas, K., & Katiyar, R. S. (2022). Ferroelectric and piezoelectric mechanisms and applications. *Journal of Asian Ceramic Societies*, *10*(2), 275–291. DOI: 10.1080/21870764.2022.2075618

Smith, M., & Kar-Narayan, S. (2022). Piezoelectric polymers: Theory, challenges and opportunities. *International Materials Reviews*, *67*(1), 65–88. DOI: 10.1080/09506608.2021.1915935

Soares, N., Costa, J. J., Gaspar, A. R., & Santos, P. (2013). Review of passive PCM technology for buildings. *Renewable & Sustainable Energy Reviews*, *18*, 23–36. DOI: 10.1016/j.rser.2012.10.021

Soin, N., Zhao, P., Prashanthi, K., Chen, J., Ding, P., Zhou, E., Shah, T., Ray, S. C., Tsonos, C., Thundat, T., Siores, E., & Luo, J. (2016). High performance triboelectric nanogenerators based on PVDF–$ZnSnO_3$ membranes. *Nano Energy*, *30*, 470–480. DOI: 10.1016/j.nanoen.2016.10.040

Thakur, H., Kaur, S., Singh, A., Singh, G. B., & Mudgal, G. (2025). Smart sustainable materials: A blueprint for a better tomorrow. In Kesari, K. K., Prakash, C., Khalid, M., & Negi, A. (Eds.), *Functionalized cellulose materials* (pp. 355–382). Springer., DOI: 10.1007/978-3-031-76953-5_11

Tian, F., Liu, Y., Ma, R., Li, F., Xu, Z., & Yang, Y. (2021). Properties of PMN-PT single crystal piezoelectric material and its application in underwater acoustic transducers. *Applied Acoustics*, *175*, 107827. DOI: 10.1016/j.apacoust.2020.107827

UN Environment Programme. (2019). *2019 global status report for buildings and construction*. United Nations Environment Programme. https://www.unep.org/resources/publication/2019-global-status-report-buildings-and-construction

Wang, L., Zhao, Y., & Liu, H. (2022). Green synthesis of smart materials: Toward sustainable environmental applications. *Materials Today*, *51*, 45–57. DOI: 10.1016/j.mattod.2021.08.003

Wang, P., Tang, Q., Zhang, L., Xu, M., Sun, L., Sun, S., Zhang, J., Wang, S., & Liang, X. (2021). Ultrasmall barium titanate nanoparticles for highly efficient hypoxic tumor therapy via ultrasound-triggered piezocatalysis and water splitting. *ACS Nano*, *15*(7), 11326–11340. DOI: 10.1021/acsnano.1c00616 PMID: 34180675

Wang, X., Huan, Y., Ji, S., Zhu, Y., Wei, T., & Cheng, Z. (2022). Ultra-high piezoelectric performance by rational tuning of heterovalent-ion doping in lead-free piezoelectric ceramics. *Nano Energy*, *101*, 107580. DOI: 10.1016/j.nanoen.2022.107580

Wibowo, Y. G., Ramadan, B. S., Taher, T., & Khairurrijal, K. (2023). Advancements of nanotechnology and nanomaterials in environmental and human protection for combating COVID-19 during and post-pandemic era: A comprehensive scientific review. In *Biomedical materials & devices* (pp. 34–57). Springer., DOI: 10.1007/978-3-031-16891-1_2

World Health Organization. (2022). *Global report on equity in health technology access*. World Health Organization. https://www.who.int/publications/i/item/9789240065097

Wu, T., Li, X., Zhang, J., Wang, J., Chen, S., Zhou, Y., & Huang, S. (2023). Thermochromic perovskites for smart solar windows. *Nature Materials*, *22*, 110–118. DOI: 10.1038/s41563-022-01415-1

Xie, L. S., Skorupskii, G., & Dincă, M. (2020). Electrically conductive metal–organic frameworks. *Chemical Reviews*, *120*(16), 8536–8580. DOI: 10.1021/acs.chemrev.9b00766 PMID: 32275412

Yamashita, Y., Karaki, T., Lee, H.-Y., Wan, H., Kim, H. P., & Jiang, X. (2022). A review of lead perovskite piezoelectric single crystals and their medical transducer applications. *IEEE Transactions on Ultrasonics, Ferroelectrics, and Frequency Control*, *69*(6), 3048–3056. DOI: 10.1109/TUFFC.2022.3160526 PMID: 35298378

Yang, Z., Zhou, S., Zu, J., & Inman, D. (2018). High-performance piezoelectric energy harvesters and their applications. *Joule*, *2*(4), 642–697. DOI: 10.1016/j.joule.2018.03.011

Yu, S., & Jain, P. K. (2019). The chemical potential of plasmonic excitations. *Angewandte Chemie International Edition*, *59*(5), 2085–2088. DOI: 10.1002/anie.201914118 PMID: 31765516

Yuan, Z. Y., Banks, C. E., & Li, W. (2025). Introducing smart materials and devices. *Smart Materials and Devices*, *1*, 202511. DOI: 10.1016/j.smad.2025.202511

Yun, T. G., Park, M., Kim, D. H., Kim, D., Cheong, J. Y., Bae, J. G., Han, S. M., & Kim, I. D. (2019). All-transparent stretchable electrochromic supercapacitor wearable patch device. *ACS Nano*, *13*(3), 3141–3150. DOI: 10.1021/acsnano.8b08560 PMID: 30779547

Zhang, Y., & Lu, L. W. (2008). Introducing smart structures technology into civil engineering curriculum: Education development at Lehigh University. *Journal of Professional Issues in Engineering Education and Practice*, *134*(1), 41–48. DOI: 10.1061/(ASCE)1052-3928(2008)134:1(41)

Zhang, Y., Wang, Q., Yi, S., Lin, Z., Wang, C., Chen, Z., & Jiang, L. (2024). 4D printing of magnetoactive soft materials for on-demand magnetic actuation transformation. *ACS Applied Materials & Interfaces*, *13*(3), 4174–4184. DOI: 10.1021/acsami.0c19280 PMID: 33398983

Zhao, Y., Liu, H., Wang, L., & Li, X. (2021). Multifunctional intelligent materials for environmental remediation: Trends and prospects. *Advanced Functional Materials*, *31*(9), 2007257. DOI: 10.1002/adfm.202007257

Zhu, M., Li, H., Zhang, Y., Chen, J., Wang, X., & Zhou, Z. (2023). Flexible piezoelectric nanogenerators for energy harvesting. *Nano Energy*, *108*, 108082. DOI: 10.1016/j.nanoen.2023.108082

Chapter 9
Comparative Study of Coastal Sedimentary Systems:
Environmental Insights From El Jazira Barrier Island and the Echraf Dune Cordon, Tunisia

Hajer Haj Ali
https://orcid.org/0000-0002-6254-3778
Water, Energy and Environment Laboratory, Faculty of Sciences of Sfax, University of Sfax, Sfax, Tunisia

Jamel Touir
Water, Energy and Environment Laboratory, Faculty of Sciences of Sfax, University of Sfax, Sfax, Tunisia

Jamila Ben Ali
Water, Energy and Environment Laboratory, Faculty of Sciences of Sfax, University of Sfax, Sfax, Tunisia

ABSTRACT

This study presents a comparative analysis of two contemporary coastal sedimentary systems located in the Bekalta region esat of Tunisia: El Jazira barrier island and the Echraf dune cordon. The objective is to better understand their geomorphological development, sedimentary dynamics, and environmental evolution. Using an integrated approach that combines geomorphological mapping, sedimentological analysis, and interpretation of hydrodynamic data (waves, winds, and climatic variables), the study identifies the main morphological units of each system while detailing the spatial

DOI: 10.4018/979-8-3373-3506-3.ch009

patterns and ecological roles of vegetation cover. Special attention is given to the interaction between plant colonization and sediment stabilization. Temporal changes in the morphology and extent of these systems are assessed through a diachronic analysis of aerial photographs from the years 1962 and 1996. The results highlight the critical role of vegetation in enhancing sediment stability and buffering against erosional processes, particularly in response to seasonal climatic variations. This interaction creates a feedback mechanism that contributes to the resilience of these coastal systems. The findings provide valuable insights for integrated coastal zone management and conservation planning, especially in regions facing increasing pressure from climate change and human activities.

1- INTRODUCTION

Coastal sedimentary environments, such as barrier islands and dune cordons, are dynamic interfaces between marine and terrestrial realms. Sandy coastal environments are complex dynamic systems where multidimensional interactions between physical, biological and anthropogenic processes occur, representing nearly 31% of coastlines globally (Luijendijk et al., 2018). These systems play a crucial role in shaping coastlines, protecting inland areas, and supporting specific ecosystems. Coastal sedimentary systems play a crucial role in shaping marine and terrestrial environments, influencing biodiversity, coastal stability, and human activities. While previous studies have examined general coastal processes in Tunisia (Paskoff and Sanlaville, 1983 ; Mahmoudi, 1986 ; Jedoui et al., 2002 ; Oueslati, 1995), detailed comparative analyses of the different active sedimentary systems are limited. This paper presents a comparative study of the El Jazira barrier island and the Echraf dune cordon, both located in the Bekalta region east Tunisia. Although these two systems are geographically close (15 km apart), they have contrasting morpho-sedimentary characteristics and dynamics that have never been the subject of an in-depth comparative study, despite the pioneering work of Paskoff and Sanlaville (1983) on Tunisian dune systems and more recent studies by Oueslati (2008) and Amrouni et al. (2019), which quantified coastal erosion rates in the region (localised retreats reaching 1.5 m/year), but which suffer from several limitations, such as an often monodisciplinary approach (pure geomorphology or ecology). The study of two such systems provides valuable environmental insights into their formation, evolution, and ecological significance.

This research aims to address this gap by investigating the geomorphology, sedimentology, vegetation cover, and hydrodynamic influences on both El Jazira and Echraf coastal sedimentary systems. By analyzing aerial photographs from different periods, the study also seeks to reconstruct the spatio-temporal evolution of these systems, providing insights into their current state and potential future changes. The ultimate goal is to inform sustainable coastal management practices and conservation efforts in the face of ongoing environmental pressures.

2- STUDY FRAMEWORK

2.1. Regional Geographical and Geological Context

The study area is part of the specific geomorphological and geological setting of the Tunisian Sahel sedimentary basin, which forms a vast coastal plain stretching nearly 150 km between Sousse city at the North and Sfax city at the South (Figure 1). From a geological point of view, this area is characterised by a well-defined Miocene substrate, consisting mainly of bioclastic limestones and grey marls belonging to the Mahmoud Formation, on which rests a diverse Quaternary cover comprising marine deposits (notably shell sands), aeolian formations (dunes) and lagoon-marsh accumulations (organic mud). Geomorphologically, the region has a landscape typical of low-lying Mediterranean sandy coasts, marked by the presence of sebkhas and coastal lagoons that bear witness to recent palaeoenvironmental fluctuations. The climate is semi-arid Mediterranean, characterised by moderate average annual rainfall (250 mm) with a maximum in autumn, and a marked annual temperature range, with temperatures varying between 12°C in January and 28°C in August, which directly influences the processes of weathering and sediment transport in the region. The dominant winds are blowing NW-SE and NE-SW.

Figure 1. Morphological framework of two coastal systems: photographs highlighting the main geomorphological features of El Jazira Barrier Island and Echraf Dune Cordon.

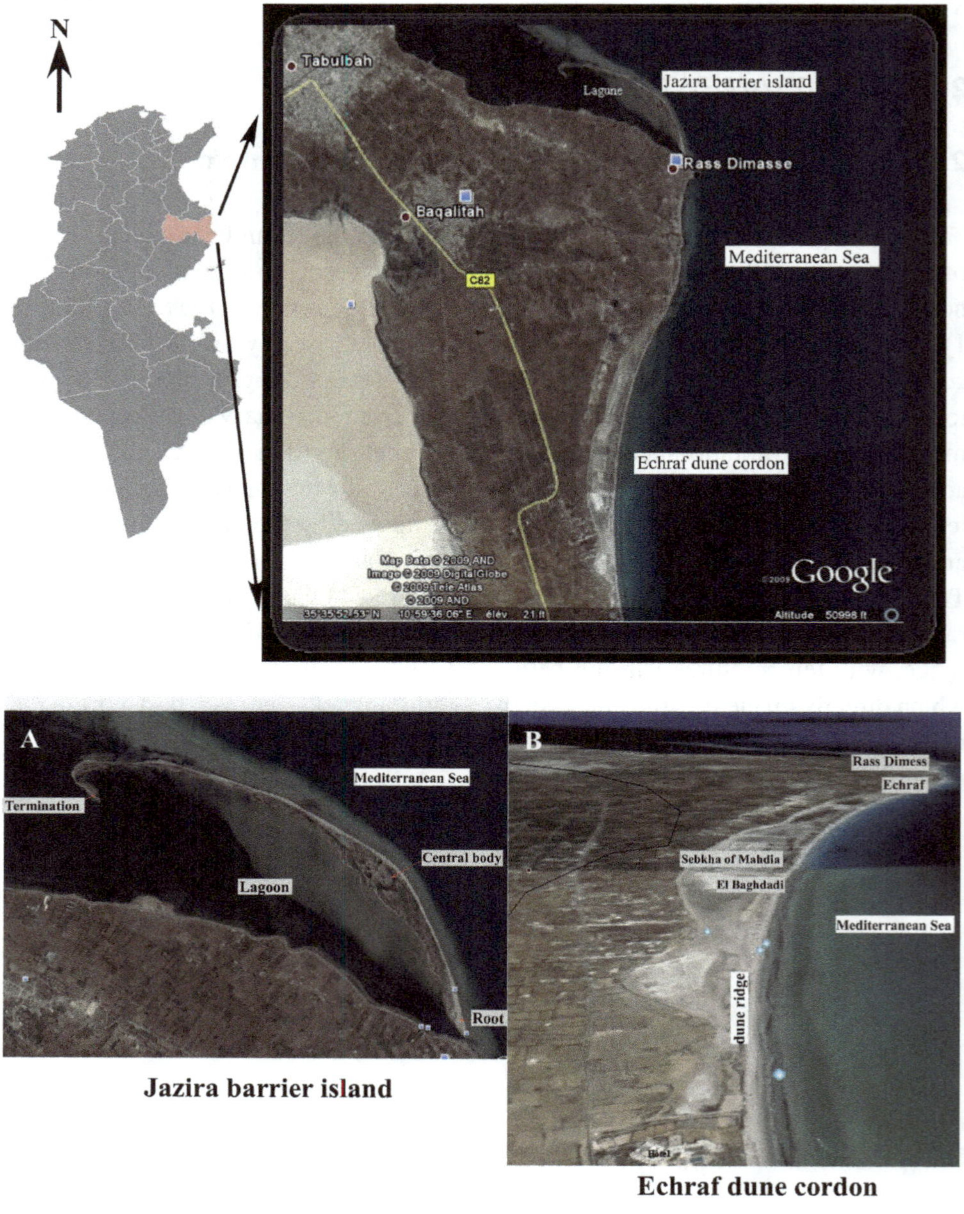

2.2. Detailed Presentation of Study Sites

2.2.1. El Jazira Barrier Island

The El Jazira barrier island (35°37'N, 10°59'E) is a typical barrier-lagoon system (Figure 1.A) stretching 2.8 km in length in a general NW-SE direction, with a distinct tripartite morphology comprising: (1) a root in the SE sector, a 180 m wide anchorage area currently marked by active erosion; (2) a central body representing the most stable part of the system and characterised by successive berms; and (3) a termination in the NW sector showing active progradation towards the NW. This geomorphological complex is associated with a 45 hectare lagoon that maintains a hydrological connection with the marine environment through a dynamic tidal channel, thus playing a crucial role in the sedimentary and hydrological exchanges of this coastal system.

2.2.2. Echraf Dune Cordon

The Echraf dune cordon (35°30'N, 11°03'E) forms a remarkable, straight coastal structure extending over 4.3 km parallel to the shoreline, acting as a natural barrier between the Mediterranean Sea and a wide coastal sebkha (Figure 1, B). This dune system exhibits a characteristic cross-sectional profile marked by pronounced asymmetry: the windward slope (facing the sea) displays a relatively gentle inclination (8–10°), whereas the leeward slope (facing the sebkha) is steeper (15–20°). The entire structure is dominated by a well-defined dune crest reaching 6.5 meters above the general ground level (GGL). The vegetation cover, mainly composed of *Tamarix africana* and *Elymus farctus*, plays a key stabilizing role by colonizing approximately 65% of the ridge surface, there by contributing to the persistence of this sedimentary formation against erosional processes

3- METHODOLOGY

This study combines geomorphological, sedimentological, and environmental analyses to characterise and compare the El Jazira barrier island and the Echraf dune cordon. Field surveys and aerial imagery were used to map geomorphological units. Surface sediment samples were analyzed for grain size and composition to interpret depositional processes using the Afnor sieving method. Local hydrodynamic data (waves, wind, and precipitation) were recovered from the Tunisian Agency of Protection and Managment of littoral (APAL) and examined to assess their impact on sediment dynamics. Vegetation cover was surveyed and different vegetal species are identified to understand its role in system stability. Finally, a diachronic analysis of aerial photographs (1962–1996) was conducted to track morphological changes over time of the studied sedimentary systems.

4- RESULTS

4.1. Morphology and Vegetation Cover

El Jazira's barrier island and Echraf's dune cordon exhibit distinct morphological units, including roots, central bodies, and termination. Both systems support diverse vegetation covers, whose density and spatial distribution vary according to their position within the sedimentary body. This vegetation is vital for stabilizing sediments, reducing erosion, and maintaining ecosystem equilibrium.

4.1.1. El Jazira Barrier Island

The El Jazira barrier island exhibits a complex geomorphological framework resulting from the dynamic interaction between marine, aeolian, and continental processes (Anthony et al., 2014; Masmoudi et al., 2018). It is characterized by an elongated sandy barrier system extending parallel to the present shoreline, separating the Mediterranean Sea from an extensive lagoonal depression (Figure 2). Besides, such barriers islands are largely occuring along the tunisian-esat coasts (Lakhdar and Soussi, 2009 ; Ben Ismail et al., 2012). The root of the barrier, located at the landward attachment point, marks the transition between the continental substrate and the littoral sandy deposits, whereas the distal sector is truncated by a tidal channel that periodically connects the lagoon with the open sea, facilitating tidal flushing and sediment exchange. Similar settings were already reported in several works in the world (Psuty, 2008; Lakhdar and Soussi, 2009 ; Short and Jackson, 2013). The back-barrier lagoon is shallow and semi-enclosed, dominated by fine-grained sandy to silty sediments showing local evaporitic crusts indicative of intermittent subaerial exposure under a semi-aride climatic conditions (Masmoudi et al., 2018; Oueslati, 1993).

Morphologically, the Jazira barrier displays a low-relief dune ridge whose seaward-facing, windward slope is gently inclined and affected by marine spray and occasional storm overwash, while the leeward slope, oriented toward the lagoon, is markedly steeper and often stabilized by vegetation (Psuty and Silveira, 2010; Hesp, 2013). The general dune crest reaches up to approximately 4.5 m above the general ground level (GGL), forming the topographic high of the coastal barrier and acting as a natural protection against marine transgression.

Figure 2. Field photograph of the vegetation cover on El Jazira Barrier Island and its distribution across its three sections.

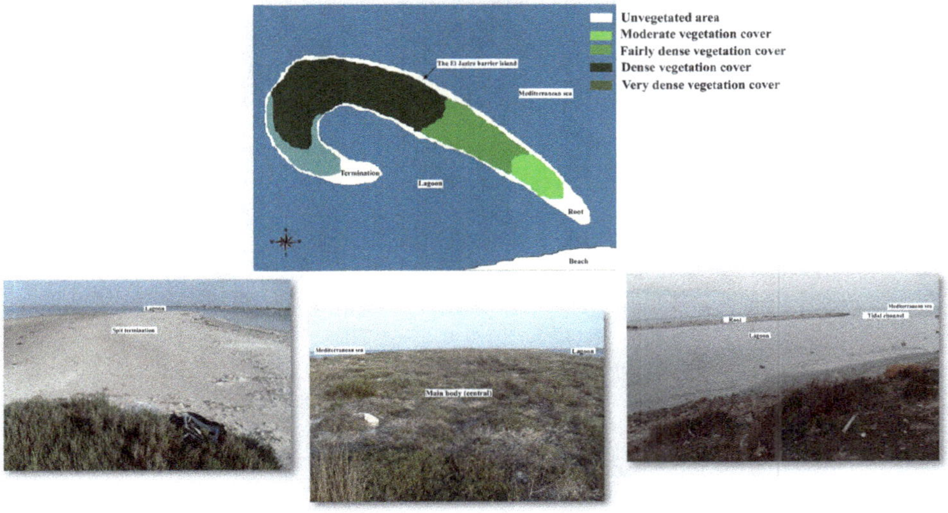

The vegetation cover plays a fundamental ecological and sedimentary role in stabilizing this geomorphic structure (Le Houérou, 1996 ; Le Floc'h et al., 2010 ; El-Gharbi et al., 2019). In the study area, the plant assemblage (Figure 3) is dominated by *halophytic* and *psammophilous* species well adapted to high salinity, wind stress, and periodic inundation. Representative taxa include *Suaeda maritima, Salsola soda, Eryngium maritimum, Plantago coronopus, Launea sp., Aeluropus littoralis, Halocnemum strobilaceum, Lotus creticus, Juncus maritimus, Pancratium maritimum, and Phragmites australis*. These species exhibit extensive rhizome and root systems capable of binding surface sediments, enhancing dune stability, and promoting vertical accretion by trapping wind-blown sand (Hesp, 2013; Maun, 2009). The vegetation cover, which colonizes nearly two-thirds of the dune surface(Figure 2), also contributes to the resilience of the Jazira barrier island against erosional forces and shoreline retreat, maintaining the integrity of the dune–lagoon continuum within a dynamic sedimentary and ecological equilibrium (Lakhdar and Soussi, 2009 ; Ben Ismail et al., 2012 ; Anthony et al., 2014)

Figure 3. Plate showing representative plant species of El Jazira Barrier Island.

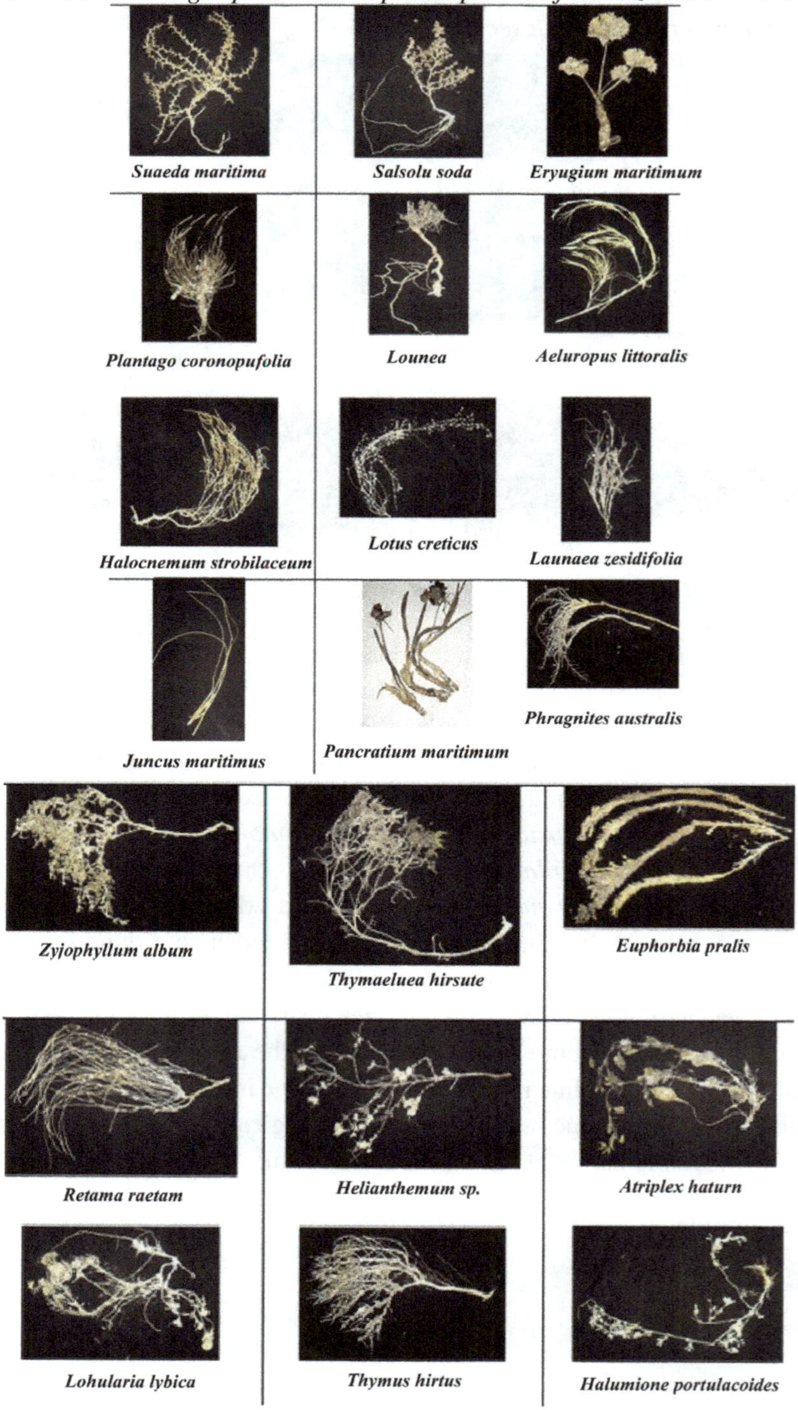

4.1.2. Echraf Dune Cordon

The Echraf dune cordon forms a remarkable, straight coastal barrier developed under the combined influence of marine, aeolian, and continental processes along the central-eastern Tunisian coast (Oueslati, 1993; Oueslati, 1995 ; Ben Ismail et al., 2012). This sandy ridge runs parallel to the Mediterranean shoreline and separates the open marine environment from a broad coastal sebkha located landward (Figure 4). Morphologically, the system consists of a continuous dune belt composed of well-sorted fine to medium quartzose sands with subordinate bioclastic fragments, forming a low-relief ridge whose crest marks the highest elevation of the coastal profile (Masmoudi et al., 2018).

Figure 4. Field photograph of the vegetation cover along El Echraf Dune Cordon.

The cross-shore profile exhibits a distinct asymmetry, with a gently inclined windward slope (8–10°) facing the sea and periodically affected by wave run-up, salt spray, and occasional storm overwash, contrasting with a steeper leeward slope (15–20°) oriented toward the sebkha, where sediments are more stabilized. This geomorphological description closely resembles observations reported for other coastal dunes in the world by several authors (Psuty, 2008; Short and Jackson, 2013). The dune crest, rising up to about 6.5 m above the general ground level (GGL), constitutes the morphological high of the system and provides an effective buffer against marine transgression and coastal flooding. Locally, the dune surface is affected by aeolian deflation hollows and small blowouts, reflecting the active wind transport regime that periodically reworks the upper ridge. The leeward slope displays localized sediment accumulation zones, occasionally affected by limited surface runoff during seasonal rains, favoring the development of thin crusts or compacted sand veneers (Masmoudi et al., 2018).

The vegetation cover plays a crucial ecological and sedimentary role in stabilizing the Echraf dune cordon. The plant assemblage is dominated by *Tamarix africana* and *Elymus farctus*, which colonize nearly 65% of the dune surface, particularly along the crest and upper backshore areas (Figure 4). Similar descriptions have been reported in previous studies of Tunisian coastal dunes, highlighting the comparable ecological and sedimentary functions of vegetation cover (Le Floc'h et al., 2010; El-Gharbi et al., 2019). These species, together with scattered *halophytic* and *psammophilous* herbs such as *Ammophila arenaria, Lotus creticus, and Pancratium maritimum* (Figure 5), form dense root and rhizome networks that anchor surface sediments, enhancing dune cohesion and reducing wind erosion (Maun, 2009; Hesp, 2013). Vegetation thus promotes vertical accretion by trapping wind-blown sand and ensures the progressive stabilization of the ridge over time. The lower backshore is characterized by *pioneer halophytes* adapted to saline, intermittently flooded substrates, marking the transition between the marine and continental environments.

Overall, the Echraf dune cordon represents a dynamic equilibrium system, where the interaction between sedimentary, hydrodynamic, and biological processes maintains the morphological and ecological stability of this coastal barrier despite the continuous action of erosional forces and climatic variability (Anthony et al., 2014; Masmoudi et al., 2018)

Figure 5. Plate illustrating representative plant species of El Echraf Dune Cordon, their depositional environment, and ecological role.

The species	Depositional environment	Role	Photos
Agropyrum junceum	Sandy coastal environments, including berms and dunes ridges, influenced by marine erosion but developing during phases of reduced hydrodynamic activity.	Stabilization of the sandy substrate, acting as a wind barrier that promotes the formation of nebkhas.	
Euphorbiapralis	Present along the margins and within the sebkha, occupying areas where no other vegetation can tolerate high salinity conditions.	Indicator of dominance phenomena within the depositional environment.	
Pancratium maritimum	Occurs in moderately sand-covered areas, characteristic of mobile dune systems, and occasionally found on the upper beach in association with *Agropyrum*.	Stabilization of drifting sand.	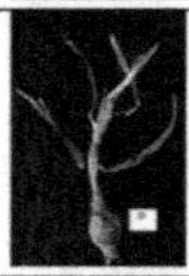
Erygium maritimum	Characteristic of mobile white dunes, occurring in association with Ammophila and Euphorbia.	Perennial sand stabilization through an extensive root system; survives winter on stored reserves, ensuring year-round substrate fixation even under marine erosion.	
Aeluropis littoralis	Characteristic of interdune areas, grey dunes, and often white dunes.	Root system promotes stabilization of sand masses.	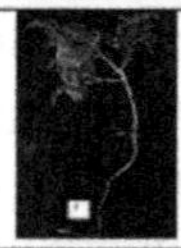
Lygeum spartum	Occupies a broad area between the dune and the sebkha.	Facilitates stabilization of grey dunes and fixed dunes.	
Asparajus stipularis	Xerophilous vegetation colonizing grey herbaceous dunes.	Stabilization and consolidation of sand.	
Salsolu soda	It is infrequent, occurring on the seaward side of the dune.	Stabilization of sand in transit.	
Ammophilia arenaria	Mobile white dunes undergoing stabilization, where it forms dense, elevated tufts that trap sand, giving the dune a hummocky appearance.	Key species in dune stabilization and consolidation; its rhizomatous growth enables resistance to sand burial and supports dune formation.	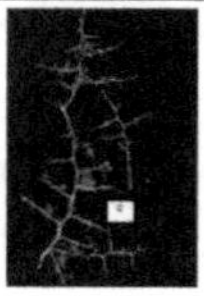
Lycium xliawii	Frequently occurs on white dunes and along the leeward slope or toe of the dune.	Occurs in colonies, trapping sand between its stems.	

4.2. Sedimentology and Depositional Facies

4.2.1. El Jazira Barrier Island

Based on the granulometric data presented in radials No. 1 to 6 (Figure 6), the El Jazira barrier island exhibits a complex sedimentary distribution, indicative of the multiple dynamic processes governing its evolution. The particle size distribution curves, which cover a wide range of grain diameters - from fine fractions (0.01 mm) to coarse sands (100–1000 mm) - highlight a mixture of sediment populations of marine, aeolian, and lagoonal origins. These multimodal distributions (Figure 6) reflect the interaction between marine sand supply, aeolian reworking on the dune ridge, and the finer deposits characteristic of the back-barrier lagoon. Such granulometric heterogeneity (Figure 7) is consistent with models of barrier islands subject to marine overwash, aeolian transport, and quiet-water sedimentation (Anthony et al., 2014; Psuty, 2008). The presence of fine fractions, particularly in samples E3 to E6, can be attributed to settling deposits in the lagoonal context, while medium to coarse sands (E1, E2) likely dominate the dune crest and seaward slope, in accordance with morphological observations describing a topographic high reaching up to 4.5 m above the general ground level. These sedimentological characteristics, combined with the stabilizing vegetation cover as already described by Hesp (2013) and Maun (2009), contribute to the resilience of the barrier island against erosional forces and saline intrusions, maintaining a fragile sedimentary equilibrium between the Mediterranean Sea and the continental lagoon

Figure 6. Grain-size curves showing sediment characteristics in El Jazira Barrier Island.

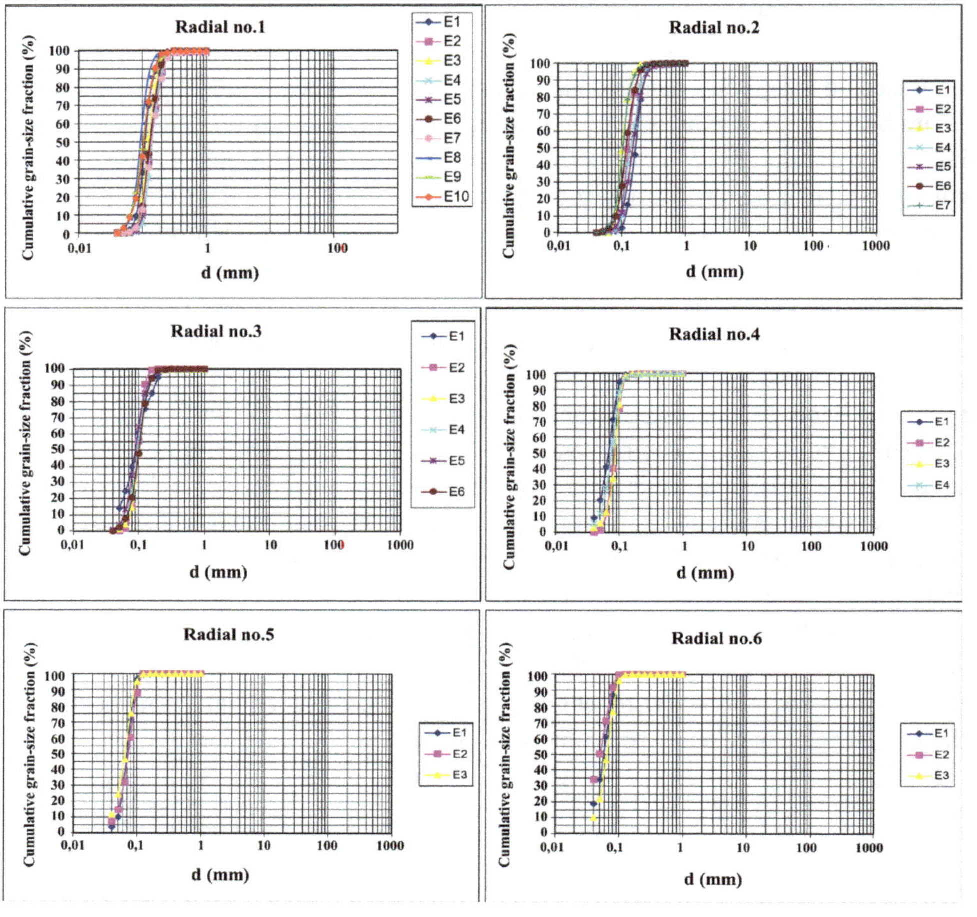

Figure 7. Spatial distribution of sedimentary facies along El Jazira Barrier Island, showing classification of sediment sorting.

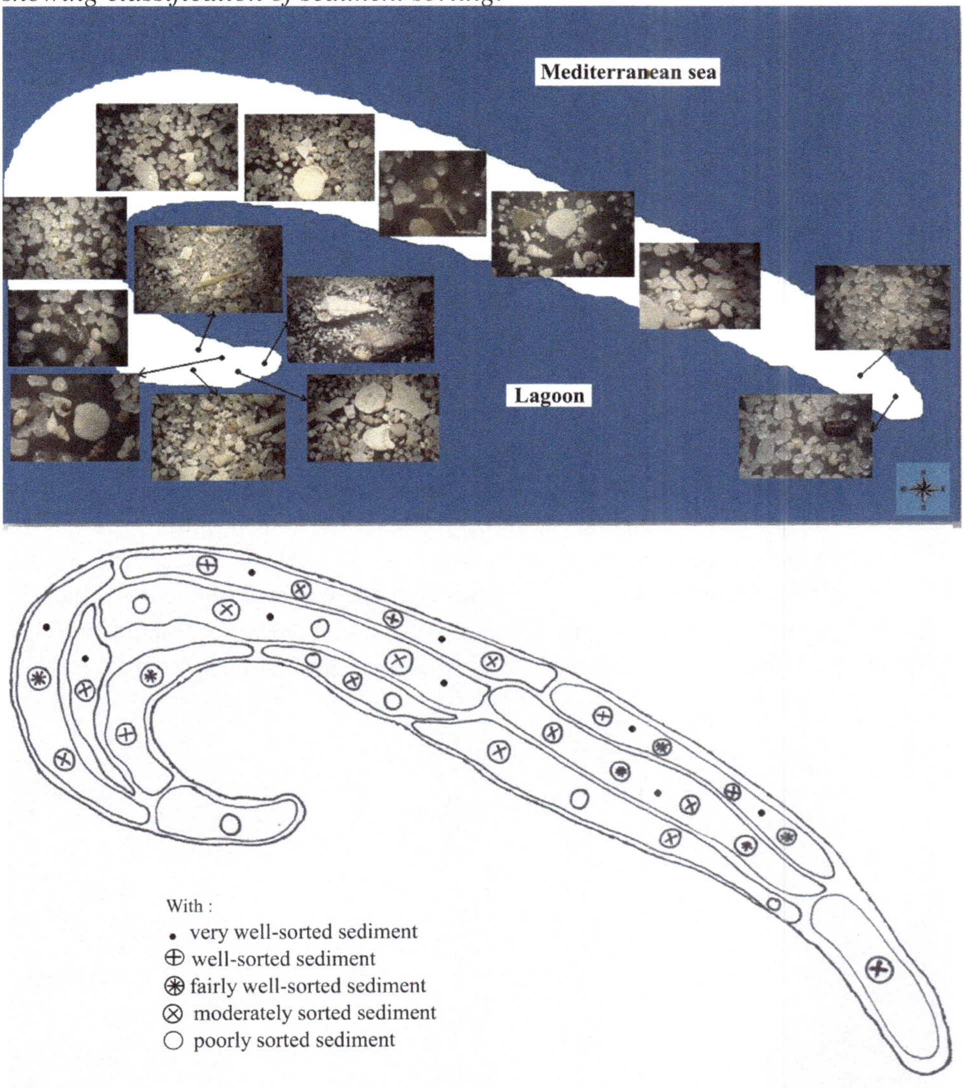

4.2.2. Echraf Dune Cordon

Based on the granulometric data from Profiles No. 1 to 7 (Figure 8), the Echraf dune cordon exhibits a well-sorted to moderately sorted sedimentary structure, characteristic of a coastal system dominated by aeolian and wave processes (Anthony et al., 2014; Psuty, 2008). The particle size distribution curves consistently show a primary mode in the fine to medium sand range (0.1–0.5 mm), with some profiles (e.g., Profile 2–E2, Profile 5–E3) (Figure 8) displaying a secondary mode in the coarse sand fraction (~1 mm), indicating periodic inputs of coarser marine-derived sediments, possibly during storm events (Short and Jackson, 2013). The generally well-sorted sediments, particularly in the upper dune crest and seaward slope samples (e.g., Profile 1–E1, Profile 7–E1), reflects efficient aeolian sorting and transport, which is responsible for building and maintaining the dune ridge (Masmoudi et al., 2018). In contrast, samples from the landward base of the dune (e.g., Profile 4–E4, E5) (Figure 8) show a wider grain size distribution, including a significant silt and clay fraction (<0.063 mm), suggesting the influence of the adjacent sebkha environment where finer particles settle under low-energy conditions (Oueslati, 1993; Ben Ismail et al., 2012). This textural contrast across the dune profile highlights the gradient from the high-energy, marine-influenced windward face to the more stable, continent-influenced leeward slope. The good sediment sorting in the dune core (Figure 9), combined with the stabilizing effect of the dominant psammophilous vegetation (e.g.,*Tamarix africana, Elymus farctus*), contributes to the geomorphic resilience of the Echraf cordon as already reported by Hesp (2013) and Maun, (2009) in different coastal sedimentary systems in the world, allowing it to function as an effective natural barrier against marine incursion into the continental sebkha, maintaining a dynamic equilibrium as described by Anthony et al. (2014).

Figure 8. Grain-size curves showing sediment characteristics in Echraf Dune Cordon.

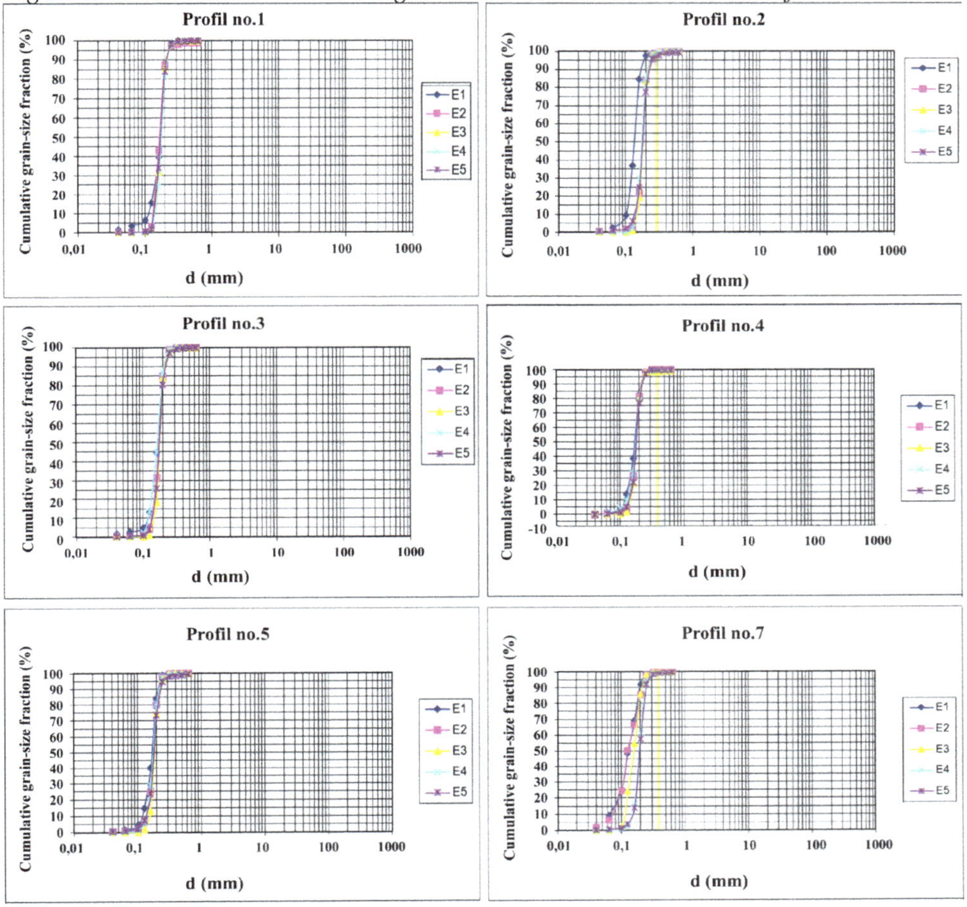

Figure 9. Spatial distribution of sedimentary facies along Echraf Dune Cordon, showing classification of sediment sorting.

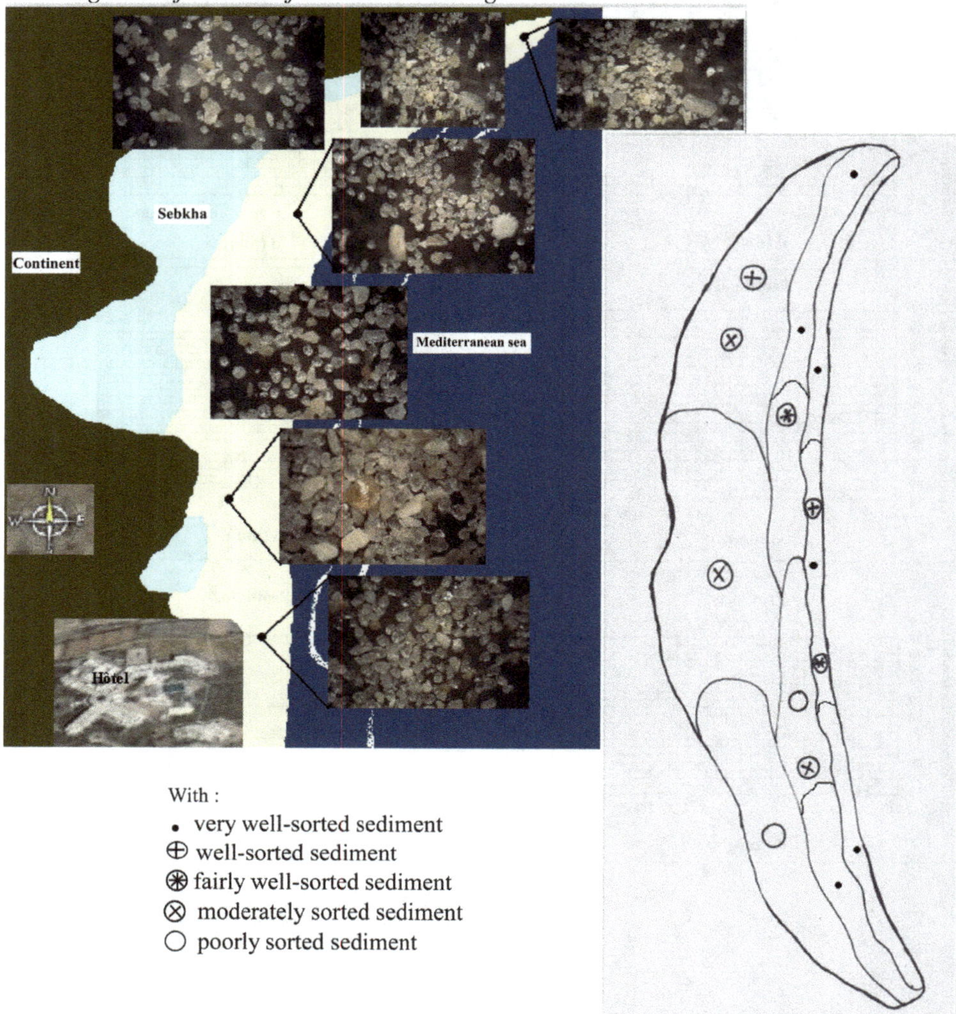

4.3. Climate and Hydrodynamic Features

The sedimentary facies of these two coastal systems reflect the dynamic interplay of hydrodynamic forces such as wave action, wind-driven processes, and seasonal climatic variations. Wave data (Table 1) indicate significant wave heights and periods that influence sediment transport and deposition patterns. Rainfall and temperature regimes contribute to sediment humidification and limit wind erosion, fostering sediment stability.

Table 1. climatic and hydrodynamic data for the study area.

Locality	Wave directions	Offshore wave height H_0 (mm)	Offshore wave period T_0 (s)	Offshore wavelength $L_0=gT_0^2/2p$ (m) (g = gravitational acceleration [m/s²])	Breaker depth $Pd=1,5H_0$ (m)	Breaking wave height $Hd=1,3H_0$ (m)
Mahdia	NW	5,7	8,6	116	8,6	7,41
	N90	4,5	7,9	98	6,8	5,85
		4	8	100	6	5,2
		4	9	128	6	5,2
		4	10	157	6	5,2
	N135	3,5	5	39	5,3	4,55
		3,5	6	57	5,3	4,55
		3,5	7	77	5,3	4,55

The climatic and hydrodynamic regime controlling the El Jazira barrier island and the Echraf dune cordon exerts a decisive influence on their morphosedimentary organization and evolution. The region is governed by a semi-arid Mediterranean climate, characterized by hot, dry summers and mild, wetter winters, with mean annual precipitation (Table 2) ranging from 250 to 350 mm and mean annual temperatures varying between 18 and 28 °C (Table 3) (Institut National de la Météorologie, 2020; Bargaoui et al., 2014). Rainfall occurs mainly between October and March, occasionally associated with convective storm events that contribute to short-lived surface runoff and temporary humidification of the sandy deposits. These conditions alternate with extended dry periods promoting aeolian deflation and sediment mobility, especially on the upper backshore and foredune sectors.

Table 2. Monthly average precipitation at mahdia meteorological station (1995–2005).

Month	Sep	Oct	Nov	Dec	Jan	Feb	Mar	Apr	May	Jun	Jul	Aug	Total
Precipitation (mm)	75,82	44,3	35,57	31,38	38,07	29,85	22,83	24,64	21,95	3,78	1,35	17,36	346,91

Table 3. Monthly mean temperatures in the study area (1968–1997, A.P.A.L. data).

	Jan	Feb	Mar	Apr	May	Jun	Jul	Aug	Sep	Oct	Nov	Dec	Mean
Mean temperature (C°)	12,1	13,2	14,4	16,2	30,2	24	26,4	27,5	25	21,5	16,5	13,7	**20**
Mean minimum (C°)	4,9	6,9	7,8	9,5	13	17,6	20	22	19,2	14,6	10	6,7	**12,68**
Mean maximum (C°)	30,2	21,7	22,2	24,8	30	32,7	34,2	33,8	31,7	28,8	28	20,5	**28,22**

Hydrodynamically, the central-eastern Tunisian coast is a wave-dominated environment influenced by swell systems from the northeast to east sectors, generating a net longshore drift directed northward (Halouani et al., 2013; Imen and Souissi, 2018). Wave height data indicate an average significant wave height between 0.6 and 1.2 m, occasionally exceeding 1.5 m during storm events (Frihy et al., 2012). These energetic conditions induce longshore sediment transport responsible for the progradation of the El Jazira barrier and the maintenance of the Echraf dune ridge as an active coastal barrier. The tidal range remains microtidal (<0.3 m), but tidal currents locally intensify near the tidal inlet connecting the El Jazira lagoon to the open sea, ensuring regular water exchange and sediment renewal (Oueslati, 1994; Yahyawi et al., 2014). Prevailing northwestern and northern winds enhance aeolian reworking during the dry season, forming low-relief foredunes and localized blowouts. Conversely, during the wet season, increased humidity and vegetation cover reduce sand mobility, enhancing dune stabilization.

Overall, the interaction between the semi-arid climatic regime, wind dynamics, and marine hydrodynamics establishes a fragile equilibrium between erosion, deposition, and stabilization that governs the morphological persistence of both the El Jazira and Echraf coastal systems under current environmental conditions.

5- DISCUSSION

5.1. The genesis of Coastal Systems

The formation and evolution of the El Jazira barrier island and the Echraf dune cordon are the result of complex interactions between Quaternary sea-level fluctuations, sediment supply, and coastal hydrodynamics along the central-eastern Tunisian margin (Figure 10). These systems developed within a morphostructural context influenced by the subsidence of the Sahel coastal plain and the gradual infilling of shallow marine depressions since the late Holocene (Oueslati, 1994; Jedoui et al.,

2003). The El Jazira barrier originated from the progradation of littoral sediments reworked by longshore drift under a microtidal regime, forming an elongated sandy ridge enclosing a back-barrier lagoon (Figure 11-A). Its evolution reflects successive phases of marine transgression and regression that alternated during the mid- to late Holocene, leading to the stabilization of the current barrier-lagoon complex (Frihy et al., 2012; Ben Ismail et al., 2016). The persistence of a tidal inlet at its distal termination demonstrates the continued hydrological exchange with the open sea, maintaining sedimentary and ecological balance within the lagoonal basin.

Figure 10. Formation and Spatio-Temporal Evolution of the study area, controlled by hydrodynamic, aeolian, and sediment supply factors.

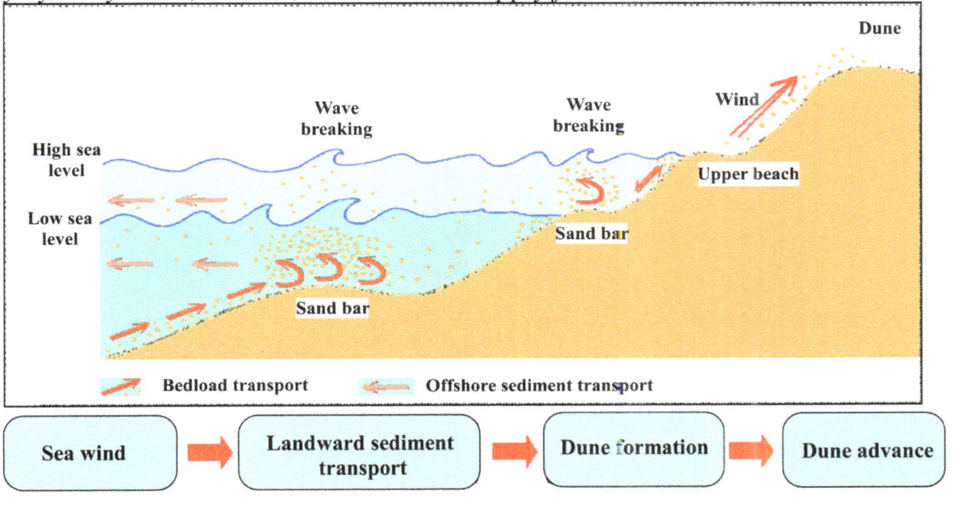

Figure 11. Mature geomorphology of the two coastal systems.

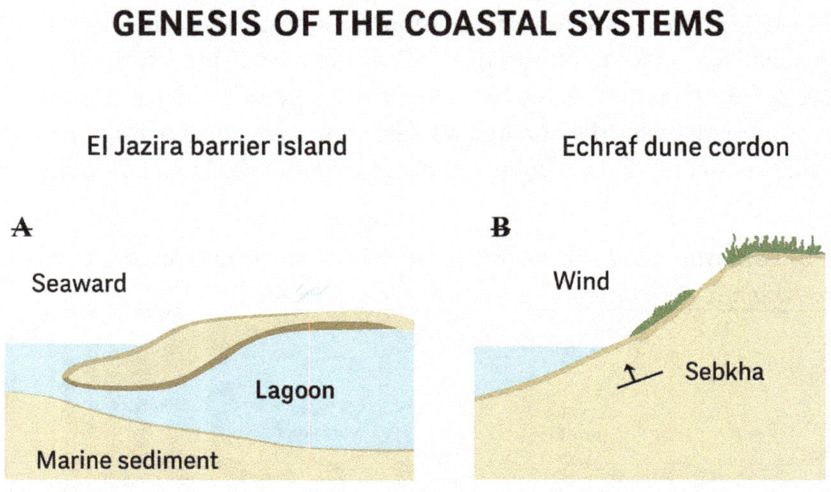

In contrast, the Echraf dune cordon represents a continental barrier generated primarily by aeolian reworking of marine and fluvial sands during the late Holocene regression. Its straight morphology and well-developed dune ridge (Figure 11-B) indicate dominant wind-driven sediment transport from the northeast, acting in conjunction with residual wave energy to build a stabilized foredune system. The alternation of humid and arid climatic phases promoted episodic dune mobilization and subsequent stabilization through vegetation colonization, as also observed along other Tunisian dune systems such as Korba and Sidi Mansour (Yahyawi et al., 2014; Halouani et al., 2013 ; Amrouni et al., 2019). Both systems thus record the imprint of postglacial sea-level oscillations and regional climatic variability that governed sediment availability, shoreline position, and dune dynamics across the Tunisian Sahel coast (Jedoui et al., 2003; Sammari et al., 2006).

The comparative analysis suggests that while the El Jazira barrier island developed under predominantly marine-dominated processes, controlled by longshore drift and tidal reworking, the Echraf dune cordon formed in a more aeolian-dominated environment, where continental and climatic factors prevailed. Their current stability reflects a dynamic equilibrium between hydrodynamic forcing, sediment supply, and biological stabilization, illustrating the complex interplay between marine and terrestrial processes in Holocene coastal evolution along the central Mediterranean margin

5.2. Spatio-Temporal Evolution of Coastal Systems

5.2.1. El Jazira Barrier Island

The comparative analysis of aerial photographs taking during 1962 and 1996 reveals a complex spatio-temporal evolution of the El Jazira barrier island, segmented into three distinct morphological units (Figure 12). The root of the spit shows clear progradation and stabilization between the two dates. The well-defined tidal channel in 1996 indicates a mature, likely flood-dominated, tidal inlet, facilitating sediment exchange between the lagoon and the open sea. This evolution aligns with the model of "inlet-driven spit growth" where sediment bypassing the inlet contributes to downdrift accretion (FitzGerald et al., 2008). The central body exhibits significant widening and appears more robust in 1996, suggesting a period of substantial sediment supply, potentially from alongshore drift or riverine inputs, leading to what is known as "barrier rollover" or "overwash" processes during high-energy events, a common auto-regulatory mechanism in barrier island systems (Carter, 1988). Finally, the tip or termination of the spit demonstrates elongation and a possible slight reorientation from the NW towards the W. Taken together, the morphological evolution of the El Jazira spit follows a south–north to northwest progradational trend initiated from the root, gradually shifting westward at the terminal end. This progressive reorientation suggests an ongoing lateral accretion that may ultimately lead to the closure of the lagoon against the mainland, promoting its transformation into a sebkha environment.This dynamic behavior is characteristic of a "free end" responding to dominant wave patterns and the relative reorientation of the longshore currents ; its evolution is highly sensitive to changes in sediment budget and can exhibit rapid accretion or erosion, as described in the "spit growth and retreat" cycles by Ashton and Giosan (2011). The overall regressive trend observed at El Jazira between 1962 and 1996 suggests a positive sediment budget during this period, likely influenced by anthropogenic factors or natural fluctuations in river sediment discharge. However, this very dynamism implies a high vulnerability to future changes in sea-level rise and reductions in sediment supply, a critical concern for Mediterranean coastal systems highlighted by many authors (e.g., Anthony et al., 2014)

Figure 12. Changes in the Morphology of El Jazira Barrier Island revealed by aerial imagery (1962 - 1996).

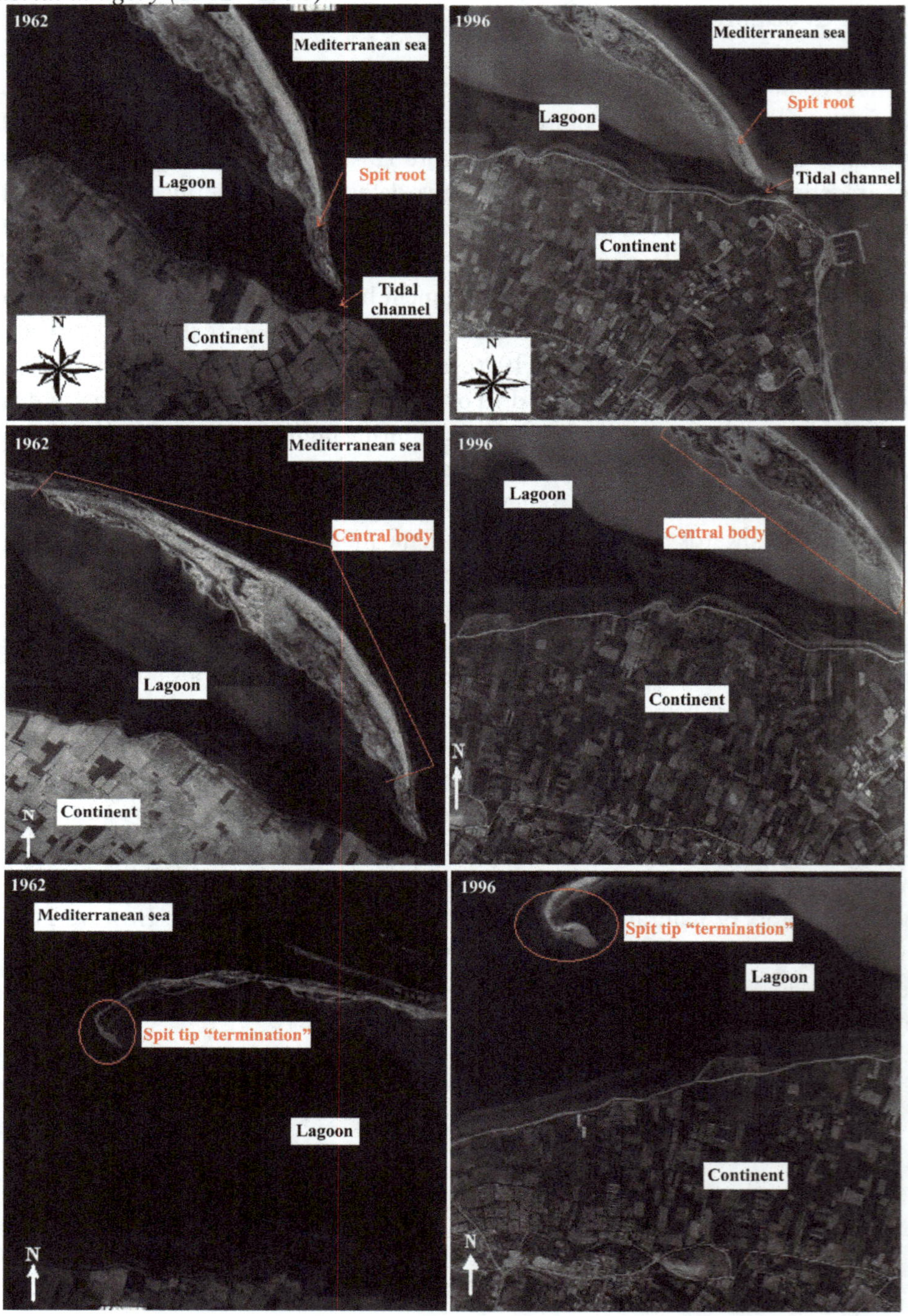

5.2.2. Echraf Dune Cordon

The spatio-temporal evolution of the Echraf barrier, as interpreted from the comparative aerial imagery taken during 1962 and 1996, reveals a complex morphodynamic behavior characteristic of wave-dominated, sediment-starved coasts. The analysis can be segmented into three primary morphological units (Figure 13): the root, the central body, and the distal tip. The root of the barrier, where it attaches to the mainland, shows evidence of both stability and minor landward retreat over the 34-year period. This suggests the influence of rising sea levels or a reduction in sediment supply from the adjacent headland, leading to "rollover" processes through overwash and aeolian transport, a fundamental transgressive response of barrier islands (Carter, 1988; Ashton and Giosan, 2011). The central body of the Echraf cordon exhibits the most significant changes, including thinning and probable fragmentation. This morphological degradation is a classic indicator of sediment deficit, where the rate of lateral sediment input from longshore drift is insufficient to maintain the barrier's volume against erosional forces, a phenomenon extensively documented in micro-tidal Mediterranean settings (Anthony et al., 2014). The incipient breaching in this section signals a potential for future segmentation into a smaller barrier island, a critical threshold in barrier disintegration (FitzGerald et al., 2008). Finally, the distal tip demonstrates pronounced elongation and a slight curvature, reflecting its dynamic nature as a "free end" actively receiving and reworking sediment from longshore currents. However, its narrow and attenuated form in later imagery suggests it is becoming increasingly vulnerable to breaching during storm events. The overall transgressive trajectory of the Echraf cordon, moving from a progradational system in the mid-20th century towards a state of destabilization and potential disintegration, aligns with the regional paradigm of coastal squeeze affecting the southern Mediterranean. This evolution can be largely driven by not only anthropogenic constraints on sediment fluxes, such as damming of rivers, coupled with the impacts of accelerated sea-level rise (Stanley and Clemente, 2014; Besset et al., 2017), but also by the absence of feeding weder stream like in our study area, due to the semi-aride climatic conditions.

Figure 13. Changes in the morphology of Echraf Dune Cordon revealed by aerial imagery from (1962 - 1996).

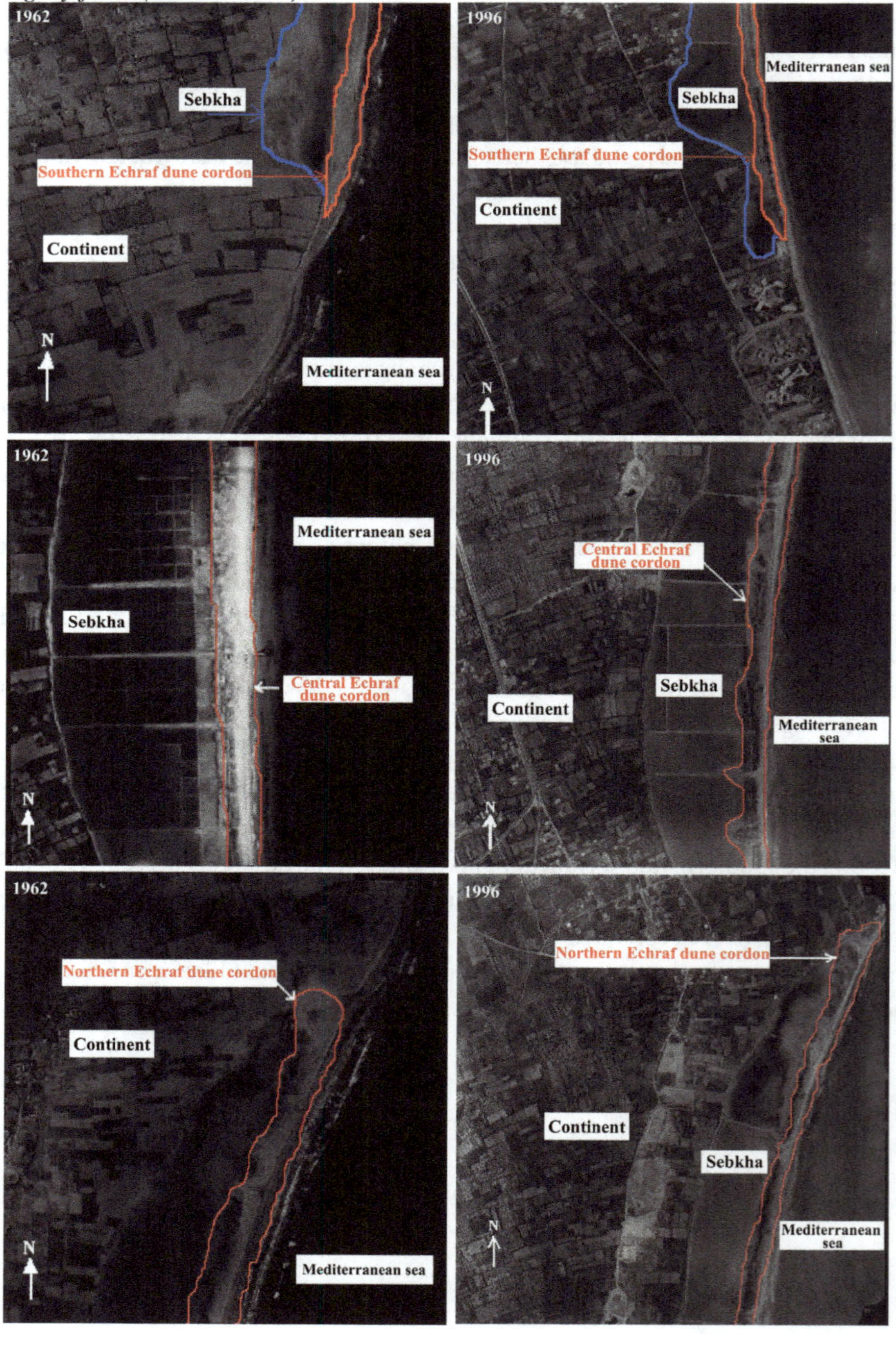

5.3. Environmental Implications

The contrasting environmental implications of the coastal evolution at El Jazira and Echraf littoral structures reveal critical insights into coastal vulnerability under climate change scenarios. At El Jazira, the spit's hydrodynamic dominance and high wave energy render it acutely vulnerable to accelerated sea-level rise, as documented in Mediterranean deltaic systems by Anthony et al. (2014). The prolonged post-storm recovery period (2-3 years) reflects a system where natural sediment replenishment is outpaced by erosional forces, consistent with observations of transgressive barrier systems along sediment-starved coasts (Besset et al., 2017). This implies an increasing risk of permanent land loss and habitat degradation, particularly for shallow marine ecosystems that depend on stable sedimentary environments.

For the Echraf cordon, the environmental implications highlight the crucial role of biogeomorphic feedbacks in coastal resilience. The system's dependence on psammophilic vegetation for stability aligns with models of dune evolution under climatic stress proposed by Durán and Moore (2013). However, the observed sediment deficit and dune fragmentation indicate the system is approaching a geomorphic threshold, where the loss of vegetative cover could trigger irreversible transgression, as documented in similar Mediterranean dune systems by Slimani et al. (2021). This vulnerability is particularly significant given the projected intensification of aeolian processes under climate change scenarios in North Africa (Meddi et al., 2022).

The divergence in the evolutionary pathways of the two studied littoral structures underscores the complex interplay between sediment supply, hydrodynamic forces, and ecological factors in determining coastal resilience. As noted in national studies of Moroccan coastal dynamics (Snoussi et al., 2020), these systems require distinct management approaches: conserving the vegetative cover at Echraf while addressing sediment starvation at El Jazira. Their continuing evolution will provide crucial insights into the adaptive capacity of Mediterranean coastal landforms under accelerating environmental change.

6- CONCLUSION

- This comparative study of the El Jazira and Echraf coastal systems reveals fundamentally different morphodynamic behaviors despite their geographical proximity and exposure to similar regional climatic forcing. The El Jazira spit demonstrates a predominantly hydrodynamic-controlled evolution, characterized by initial progradation followed by increasing vulnerability to transgressive processes. In contrast, the Echraf dune cordon exhibits aeolian-dominated dynamics where vegetation plays a crucial stabilizing role, though its resilience appears threatened by sediment deficit and climatic pressures.

- The divergent evolution of these systems highlights the critical importance of local environmental controls in determining coastal response to regional forcing. At El Jazira, the high wave/current ratio and prolonged recovery periods reflect a system with limited sediment replenishment capacity, while at Echraf, the dense psammophilic vegetation provides significant but potentially fragile protection against erosion. These findings underscore that coastal sedimentary systems, even within the same regional context, follow distinct evolutionary pathways governed by their specific sediment supply, hydrodynamic conditions, and ecological characteristics.
- The identification of these differential controls provides essential insights for understanding coastal vulnerability in Mediterranean environments. The contrasting behaviors of El Jazira and Echraf systems illustrate how similar climatic pressures can produce markedly different morphological responses, emphasizing the need for site-specific understanding of coastal dynamics in both research and management contexts. This study establishes that comprehensive analysis of these systems must account for their unique environmental settings and the complex interactions between physical and biological processes shaping their evolution.

CONFLICT OF INTEREST

On behalf of all authors, the corresponding author states that there is no conflict of interest.

ACKNOWLEDGEMENT

I wish to express my heartfelt gratitude to the Doctoral School of Fundamental Sciences, Faculty of Sciences of Sfax, University of Sfax for providing the framework and resources that made this research possible.

I also wish to thank the Laboratory of Spectroscopy and Optical Characterisation of Materials (LaSCOM), Faculty of Sciences of Sfax, University of Sfax, for the opportunity to pursue my postdoctoral research within its team.

REFERENCES

Amrouni, O., Sánchez, A., Khélifi, N., Benmoussa, T., Chiarella, D., Mahé, G., & McLaren, P. (2019). Sensitivity assessment of the deltaic coast of Medjerda based on fine-grained sediment dynamics, Gulf of Tunis, *Western Mediterranean. Journal of Coastal Conservation*, *23*(3), 571–587. DOI: 10.1007/s11852-019-00687-x

Anthony, E. J., Marriner, N., & Morhange, C. (2014). Human influence and the changing geomorphology of Mediterranean deltas and coasts over the last 6000 years: From progradation to destruction phase. *Earth-Science Reviews*, *139*, 336–361. DOI: 10.1016/j.earscirev.2014.10.003

Ashton, A. D., & Giosan, L. (2011). Wave-angle control of delta evolution. *Geophysical Research Letters*, *38*(13), n/a. DOI: 10.1029/2011GL047630

Bargaoui, Z., Tramblay, Y., & Servat, E. (2014). Seasonal precipitation variability in regional climate simulations over Northern basins of Tunisia. *International Journal of Climatology*, *34*(1), 235–248. DOI: 10.1002/joc.3683

Ben Ismail, M., Masmoudi, H., & Oueslati, A. (2012). Morphodynamics and sedimentary evolution of the central-eastern Tunisian coastal barriers. *Journal of Coastal Research, SI*, *65*, 1115–1120.

Ben Ismail, M. H., Jedoui, Y., & Oueslati, A. (2016). Holocene sea-level changes and coastal evolution in eastern Tunisia. *Quaternary International*, *401*, 45–60.

Besset, M., Anthony, E. J., & Sabatier, F. (2017). River delta shoreline reworking and erosion in the Mediterranean and Black Seas: A meta-analysis. *Marine Geology*, *391*, 10–25.

Carter, R. W. G. (1988). *Coastal Environments: An Introduction to the Physical, Ecological, and Cultural Systems of Coastlines*. Academic Press.

Durán, O., & Moore, L. J. (2013). Vegetation controls on the maximum size of coastal dunes. *Proceedings of the National Academy of Sciences of the United States of America, 110*(43), 17217–17222. DOI: 10.1073/pnas.1307580110 PMID: 24101481

El-Gharbi, S., Masmoudi, H., & Zargouni, F. (2019). Coastal dune vegetation and its role in dune stabilization along the eastern coast of Tunisia. *Environmental Earth Sciences*, *78*, 512.

FitzGerald, D. M., Buynevich, I. V., & Argow, B. A. (2008). Model of tidal inlet and barrier island dynamics in a regime of accelerated sea-level rise. *Journal of Coastal Research, SI*, (55), 1071–1078.

Frihy, O. E., Gueddari, M., & Halouani, N. (2012). Coastal processes and sediment dynamics along southern Mediterranean coasts. *Journal of Coastal Research, 28*(4), 859–870.

Halouani, N., Gueddari, M., & Frihy, O. (2013). The Northwestern Mediterranean Coast of Tunisia: Wave processes and shoreline dynamics. *Arabian Journal for Science and Engineering*, *38*, 3219–3230. DOI: 10.1007/s13369-012-0401-4

Hesp, P. A. (2013). Concepts in coastal dune geomorphology. In *Treatise on Geomorphology* (Vol. 10, pp. 328–355). Academic Press. DOI: 10.1016/B978-0-12-374739-6.00310-9

Imen, T., & Souissi, R. (2018). Coastal erosion in the south-eastern Mediterranean: Case of beaches in North Tunisia. *Arabian Journal of Geosciences, 11*(14), 386. DOI: 10.1007/s12517-018-3716-y

Institut National de la Météorologie (INM) (2020). Données climatiques de la région du Sahel et du Centre-Est tunisien.

Jedoui, Y., Davaud, E., Ben Ismaïl, H., & Reyss, J. L. (2002). Sedimentological analysis of marine Pleistocene deposits of southeastern Tunisia: Evidence of two positive eustatic pulsations during the marine isotopic substage 5e (Eemian, Tyrrhenian). *Bulletin de la Société Géologique de France, 173*(3), 255–264. DOI: 10.2113/173.3.255

Jedoui, Y., Oueslati, A., & Ben Ismail, H. (2003). Les variations du niveau marin holocène et la morphogenèse littorale en Tunisie orientale. Geomorphologie: Relief, Processus. *Environnement*, *9*(4), 277–286.

Lakhdar, R., & Soussi, M. (2009). Les dépôts lagunaires holocènes du Sud-Est de la Tunisie: extension et organisation spatio-temporelle. *In Conférence Méditerranéenne Côtiere et Maritime*, pp. 115e118.

Le Floc'h, E., Boulos, L., Véla, E., & Domina, G. (2010). Orobanchaceae. In *Catalogue synonymique commenté de la Flore de Tunisie* (Vol. 2010, pp. 272–275). Ministère de l'Environnement et du Dévelopment Durable.

Le Houérou, H. N. (1996). Climate change, drought and desertification. *Journal of Arid Environments*, *34*(2), 133–185. DOI: 10.1006/jare.1996.0099

Luijendijk, A., Hagenaars, G., Ranasinghe, R., Baart, F., Donchyts, G., & Aarninkhof, S. (2018). The state of the world's beaches. *Scientific Reports*, *8*(1), 6641. DOI: 10.1038/s41598-018-24630-6 PMID: 29703960

Mahmoudi, M. (1986). Stratigraphie, sédimentologie et diagenèse des dépôts tyrrhéniens du Sahel tunisien (Tunisie orientale) (*Thèse de doctorat*, Paris 11).

Masmoudi, H., El-Gharbi, S., & Zargouni, F. (2018). Morphological dynamics of Tunisian barrier–lagoon systems. *Quaternary International*, *475*, 48–60.

Maun, M. A. (2009). *The biology of coastal sand dunes*. Oxford University Press. DOI: 10.1093/oso/9780198570356.001.0001

Meddi, M., Toumi, S., & Assani, A. A. (2022). Analysis of meteorological drought sequences in various climatic zones of Algeria. Revue. *Journal of African Earth Sciences*, *185*, 104397.

Oueslati, A. (1993). *Les côtes de la Tunisie: Géomorphologie et environnement et aptitudes à l'aménagement*. Faculté des Sciences Humaines et Sociales de Tunisie.

Oueslati, A. (1994). Dynamique littorale et évolution des côtes de la Tunisie orientale. *Bulletin de la Société Géologique de France*.

Oueslati, A. (1995). The evolution of low Tunisian coasts in historical times: From progradation to erosion and salinization. *Quaternary International*, *29*, 41–47. DOI: 10.1016/1040-6182(95)00006-5

Oueslati, A. (2008). Le littoral de Oued Laou: de l'apport de l'étude géomorphologique à la connaissance de ses aptitudes à l'aménagement et à la prévention des risques naturels et de la dégradation des paysages. *Travaux de l'Institut Scientifique, Rabat, série générale*, (5), 1-16.

Paskoff, R., & Sanlaville, P. (1983). Les côtes de la Tunisie. Variations du niveau marin depuis le Tyrrhénien (Vol. 14, No. 1, pp. 0-0). *Maison de l'Orient et de la Méditerranée Jean Pouilloux*.

Psuty, N. P. (2008). The coastal foredune: A morphological basis for regional coastal dune development. *Ecological Studies*, *171*, 11–27. DOI: 10.1007/978-3-540-74002-5_2

Psuty, N. P., & Silveira, T. M. (2010). Global morphodynamics of coastal foredunes: The role of vegetation. *Geomorphology*, *123*, 201–211.

Sammari, C., Koutitonsky, V., & Moussa, M. (2006). Hydrodynamic modeling of the central Tunisian shelf: Implications for sediment transport. *Continental Shelf Research*, *26*, 1763–1783.

Short, A. D., & Jackson, D. W. T. (2013). Geomorphology of coastal dunes. In Shroder, J. (Ed.), *Treatise on Geomorphology* (Vol. 10, pp. 166–182). Academic Press.

Slimani, H., Aydda, A., & Irigaray, C. (2021). Assessment of the impact of climate change on the dune systems of the Mediterranean coast of Morocco. *Catena*, *207*, 105627.

Snoussi, M., Haida, S., & Imassi, S. (2020). Vulnerability assessment of the Moroccan coastal zone to sea-level rise: The case of the Mediterranean coast. *Journal of Coastal Conservation*, *24*(2), 20.

Stanley, D. J., & Clemente, P. L. (2014). Clay distributions, grain sizes, sediment thicknesses, and compaction rates to interpret subsidence in Egypt's northern Nile Delta. *Journal of Coastal Research*, *30*(1), 88–101.

Yahyawi, Z., Rbaib, N., & Saadi, A. (2014). Étude descriptive de l'évolution saisonnière des dunes bordières de la plage de Korba (façade orientale, Cap Bon, Tunisie) par la methode de des Modeles Numeriques de Terrain Residuel. *INSTM Bulletin: Marine and Freshwater Sciences*, *41*, 143–158.

Compilation of References

Abbass, K., Qasim, M. Z., Song, H., Murshed, M., Mahmood, H., & Younis, I. (2022). A review of the global climate change impacts, adaptation, and sustainable mitigation measures. *Environmental Science and Pollution Research International*, *29*(28), 42539–42559. DOI: 10.1007/s11356-022-19718-6 PMID: 35378646

Abdallah, H., Sassi, S., Meister, C., & Souissi, R. (2000). Stratigraphie séquentielle et paléogéographie à la limite Cénomanien-Turonien dans la région de Gafsa-Chotts (Tunisie centrale). *Cretaceous Research*, *2000*(21), 35106.

Abu-Allaban, M., & Abu-Qudais, H. (2011). Impact assessment by ambient air quality by Cement Industry: A case study in Jordan. *Aerosol and Air Quality Research*, *11*(7), 802–810. DOI: 10.4209/aaqr.2011.07.0090

Abulude, F. O., Ndamitso, M. M., Iyaka, Y. A., & Abdulkadir, A. (2018a). Study of particulate matter of Akure, Nigeria using a sharp-cut inertial filter combined with an impactor-a preliminary study. *Journal of Atmospheric and Solar-Terrestrial Physics*, *179*, 396–404. DOI: 10.1016/j.jastp.2018.09.005

Abulude, F. O., Ndamitso, M. M., Iyaka, Y. A., & Abdulkadir, A. (2018b). An assessment of meteorological conditions in Akure, Ondo State, Nigeria. *Bartin University International Journal of Natural and Applied Sciences*, *1*(2), 73–78.

Adebiyi, F. M., Ezeh, G. C., Adeyemi, F. A., & Obioh, I. B. (2015). Trace elements in particulate matter of ambient air at petroleum filling stations. *Toxicological and Environmental Chemistry*, *97*(7), 847–856. DOI: 10.1080/02772248.2015.1071377

Ahmed, A., Ali, A., Alzahrani, B., & Sanada, K. (2023). Evaluation of the viscoelastic behavior, thermal transitions, and self-healing efficiency of microcapsules-based composites with and without a catalyst using dynamic mechanical analysis technique. *Journal of Applied Polymer Science*, *140*(34), e54323. DOI: 10.1002/app.54323

Ahn, C., Hong, P. H., Lee, J., Kim, J., Moon, G., Lee, S., Park, I., Han, H., & Hong, S. W. (2022). Highly self-healable polymeric coating materials with enhanced mechanical properties based on the charge transfer complex. *Polymers*, *14*(23), 5181. DOI: 10.3390/polym14235181 PMID: 36501576

Ahuja, S., & Kumar, J. (2023). Ethical aspects of health sensing applications in e-textiles. *Engineering Proceedings*, *30*(1), 4. DOI: 10.3390/engproc2023030004

Alahmr, F. O. M., Othman, M., Abd Wahid, N. B., Abdul Halim, A. A., & Latif, M. T. (2012). Composition of dust falls around semi-urban areas in Malaysia. *Aerosol and Air Quality Research*, *12*(4), 629–642. DOI: 10.4209/aaqr.2012.02.0027

Al-Duais, I. N., Ahmad, S., Al-Osta, M. A., Maslehuddin, M., & Saleh, T. A. (2024). Properties of alkali-activated concrete made using the optimum combinations of precursor materials and activation parameters. *Journal of Materials in Civil Engineering*, *36*(4), 04024042. DOI: 10.1061/JMCEE7.MTENG-16944

Alkahtani, M. E., Elbadawi, M., Chapman, C. A. R., Green, R. A., Gaisford, S., Orlu, M., & Basit, A. W. (2023). Electroactive polymers for on-demand drug release. *Advanced Healthcare Materials*, *13*(3), 2301759. Advance online publication. DOI: 10.1002/adhm.202301759 PMID: 37861058

Alkhatib, A. J., Boran, A. M., Al-Hourani, Z., Al-Balwi, A., Muhammad, A., Bello, A. A. M., Qasem, A., Alkhatatbeh, M. A., & Adamu, A. (2014). Occupational Exposure To Nickel, Cadmium And Copper Among Workers In Jewelry Manufacturing. *European Scientific Journal*, *10*(15), 159–169.

Aller-Pellitero, M., Fremeau, J., Villa, R., Guirado, G., Lakard, B., Hihn, J. Y., & del Campo, F. J. (2019). Electrochromic biosensors based on screen-printed Prussian Blue electrodes. *Sensors and Actuators. B, Chemical*, *290*, 591–597. DOI: 10.1016/j.snb.2019.03.100

Alonso-Zarza, A. M., & Martin-Perez, A. (2008). Dolomite in caves: Recent dolomite formation in oxic, non-sulfate environments. Castañar Cave, Spain. *Sedimentary Geology*, *205*(3-4), 160–164.

Alpdogan, G., San, N., & Zor, S. D. (2019). Analysis of Some Trace Metals in Fish Species after Preconcentration with Congo Red on Amberlite XAD-7 Resin by Flame Atomic Absorption Spectrometry. *Journal of Chemistry*. Advance online publication. DOI: 10.1155/2016/1365171

Al-Thani, H., Koc, M., & Isaifan, R. J. (2018). Investigations on Deposited Dust Fallout in Urban Doha: Characterization, Source Apportionment and Mitigation. *Environment and Ecology Research*, *6*(5), 493–506. DOI: 10.13189/eer.2018.060510

Amodio, M., Catino, S., Dambruoso, P. R., deGennaro, G., Di Gilio, A., Guingato, P., Laiola, E., Marzocca, A., Mazzone, A., Sardaro, A., & Tutino, M. (2014). Atmospheric deposition: Sampling procedures, analytical methods, and main recent findings from the scientific literature. *Advances in Meteorology*, *2014*, 1–28. DOI: 10.1155/2014/161730

Amrouni, O., Sánchez, A., Khélifi, N., Benmoussa, T., Chiarella, D., Mahé, G., & McLaren, P. (2019). Sensitivity assessment of the deltaic coast of Medjerda based on fine-grained sediment dynamics, Gulf of Tunis, *Western Mediterranean. Journal of Coastal Conservation*, *23*(3), 571–587. DOI: 10.1007/s11852-019-00687-x

Anthony, E. J., Marriner, N., & Morhange, C. (2014). Human influence and the changing geomorphology of Mediterranean deltas and coasts over the last 6000 years: From progradation to destruction phase. *Earth-Science Reviews*, *139*, 336–361. DOI: 10.1016/j.earscirev.2014.10.003

Aromolo, R., Savi, F., Salvati, L., Ilardi, F., Moretti, V., Fares, S. (2015). Particulate matter and meteorological conditions in Castelporziano forest: a brief commentary. 26: 269-273.

Ashton, A. D., & Giosan, L. (2011). Wave-angle control of delta evolution. *Geophysical Research Letters*, *38*(13), n/a. DOI: 10.1029/2011GL047630

Ayala-Ruiz, D., Atoche, A. C., Ruiz-Ibarra, E., Osorio-de-la-Rosa, E., & Vázquez-Castillo, J. (2019). A self-powered pmfc-based wireless sensor node for smart city applications. *Wireless Communications and Mobile Computing*, *2019*, 1–10. DOI: 10.1155/2019/8986302

Ayari, J., Barbieri, M., Barhoumi, A., Boschetti, T., Braham, A., Dhaha, F., & Charef, A. (2023). Trace metal element pollution in media from the abandoned Pb and Zn mine of Lakhouat, Northern Tunisia. *Journal of Geochemical Exploration*, *247*, 107180. DOI: 10.1016/j.gexplo.2023.107180

Ayari, J., Barbieri, M., Boschetti, T., Sellami, A., Ballirano, P., & Charef, A. (2025). Geological, Mineralogical, and Alteration Insights of the Intermediate-Sulfidation Epithermal Mineralization in the Sidi Aissa District, Northern Tunisia. *Geosciences*, *15*(7), 269. DOI: 10.3390/geosciences15070269

Bachari, M., Negra, M. H., Geraud, Y., Grosheny, D., & Soltani, A. (2022). Petrophysics and sedimentology of the Turonian-Coniacian Bireno and Douleb reservoir analogues in north-central Tunisia. *Journal of African Earth Sciences*, *196*, 104728.

Badiozamani, K. (1973). The dorag dolomitization model-application to the middle Ordovician of Wisconsin. *Journal of Sedimentary Petrology*, *4*(43), 965–984.

Bahloul, M., Chabbi, I., Dammak, R., Amdouni, R., Medhioub, K., & Azri, C. (2015). Geochemical behavior of PM_{10} aerosol constituents under the influence of succeeding anticyclonic/cyclonic situations: Case of Sfax City, Southern Tunisia. *Environmental Monitoring and Assessment*, *187*(12), 757. DOI: 10.1007/s10661-015-4980-x PMID: 26577217

Bahl, S., Nagar, H., Singh, I., & Sehgal, S. (2020). Smart material types, properties and applications: A review. *Materials Today: Proceedings*, *28*, 1302–1306. DOI: 10.1016/j.matpr.2020.04.505

Bai, H., Kim, Y., & Shepherd, R. (2022). Autonomous self-healing optical sensors for damage intelligent soft-bodied systems. *Science Advances*, *8*(49), eabq2104. Advance online publication. DOI: 10.1126/sciadv.abq2104 PMID: 36475793

Ballová, Z., Janiga, M., & Hančinský, R. (2019). Comparison of Element Concentrations (Ba, Mn, Pb, Sr, Zn) in the Bones and Teeth of Wild Ruminants from the West Carpathians and the Tian-Shan Mountains as Indicators of Air Pollution. *Atmosphere (Basel)*, *10*(2), 64. DOI: 10.3390/atmos10020064

Bargaoui, Z., Tramblay, Y., & Servat, E. (2014). Seasonal precipitation variability in regional climate simulations over Northern basins of Tunisia. *International Journal of Climatology*, *34*(1), 235–248. DOI: 10.1002/joc.3683

Barrias, A., Casas, J. R., & Villalba, S. (2016). A review of distributed optical fibre sensors for civil engineering applications. *Sensors (Basel)*, *16*(5), 748. DOI: 10.3390/s16050748 PMID: 27223289

Barros, L. B., Knockaert, M., & Tenório, R. (2023). An analysis of the commercialisation barriers of self-healing concrete. MATEC Web of Conferences, 378, 10001. https://doi.org/DOI: 10.1051/matecconf/202337810001

Basahi, J. M., Ismail, I. M., Hassan, I. A., Almeelbi, T., Haiba, N. S., & Hammam, E. (2017). Total suspended particulate matter (TSP) and its associated heavy metals in atmosphere on the Western Coast of Saudi Arabia. *Polish Journal of Environmental Studies*, *26*(5), 2419–2424. DOI: 10.15244/pjoes/69102

Batchelor, C., Martin-Martinez, F. J., Griffiths, C., Mabbett, I., Smith, J., & Jewell, E. (2024). Computationally accelerated prediction of polyester-melamine coatings degradation to design sustainable organically coated steels for outdoor applications. *RSC Advances*, *14*(26), 18343–18354. DOI: 10.1039/D3RA06744K PMID: 38915881

Bathurst, R. G. C. (1971). *Carbonate Sediments and their diagenesis*. Elsevier.

Bathurst, R. G. C. (1975). *Carbonate Sediments and their diagenesis, Developments in gfsedimentology n° 12*. Elsevier.

Belie, N. D., Gruyaert, E., Al-Tabbaa, A., Antonaci, P., Baeră, C., Bajāre, D., & Jonkers, H. (2018). A review of self-healing concrete for damage management of structures. *Advanced Materials Interfaces*, *5*(17), 1800074. Advance online publication. DOI: 10.1002/admi.201800074

Ben Ayed N., Ghali A. et Touir J. (1991). Sur les orientations des paléocontraintes tectoniques durant la sédimentation du CénomanienTuronien de Tunisie. Association Tunisienne des Etudes Internationales de Géologie, Tunis, 31 Mai et 1er Juin 1991.

Ben Ismail, M. H., Jedoui, Y., & Oueslati, A. (2016). Holocene sea-level changes and coastal evolution in eastern Tunisia. *Quaternary International*, *401*, 45–60.

Ben Ismail, M., Masmoudi, H., & Oueslati, A. (2012). Morphodynamics and sedimentary evolution of the central-eastern Tunisian coastal barriers. *Journal of Coastal Research, SI*, *65*, 1115–1120.

Benedict, B., Bell, B., & Miller, M. (2017). Electrochromic windows: Technology and energy performance. *Building and Environment*, *125*, 172–181. DOI: 10.1016/j.buildenv.2017.08.037

Berkeley Earth. (n.d.). https://berkeleyearth.org/data/

Besset, M., Anthony, E. J., & Sabatier, F. (2017). River delta shoreline reworking and erosion in the Mediterranean and Black Seas: A meta-analysis. *Marine Geology*, *391*, 10–25.

Bhaskar, J., Sharma, A. K., Bhattacharya, B., & Adhikari, S. (2020). A review on shape memory alloy reinforced polymer composite materials and structures. *Smart Materials and Structures*, *29*(7), 073001. DOI: 10.1088/1361-665X/ab8836

Bhattacharjee, J., & Roy, S. (2023). *Biogenic electrodes for low-temperature solid oxide fuel cells. ECS Meeting Abstracts, MA2023-01*(40), 2805. https://doi.org/DOI: 10.1149/MA2023-01402805mtgabs

Bhattacharjee, J., & Roy, S. (2024). Utilizing a variable material approach to combat climate change. *Materials Science Research India*, *20*(3), 141–145. DOI: 10.13005/msri/200301

Bhuiyan, M. A. H., Parvez, L., Islam, M. A., Dampare, S. B., & Suzuki, S. (2010). Heavy metal pollution of coal mine-affected agricultural soils in the northern part of Bangladesh. *Journal of Hazardous Materials*, *173*(1-3), 384–392. DOI: 10.1016/j.jhazmat.2009.08.085 PMID: 19744789

Bjørlykke, K. (1998). Compaction, porosity reduction and fluid flow in sedimentary basins. *Marine and Petroleum Geology*, *15*, 321–340.

Boahen, E., Pan, B., Kweon, H., Kim, J., Choi, H., Kong, Z., & Kim, D. (2022). Ultrafast, autonomous self-healable iontronic skin exhibiting piezo-ionic dynamics. https://doi.org/DOI: 10.21203/rs.3.rs-2135447/v1

Boopathy, G., Srinivasan, V., Kamalakar, V., & Bose, G. C. (2025). Sustainability through digital twins: Optimizing energy usage and reducing waste. In *Accelerating product development cycles with digital twins and IoT integration* (pp. 13–48). IGI Global Scientific Publishing., DOI: 10.4018/979-8-3373-2028-1.ch002

Bouaziz, S., Barrier, E., Soussi, M., Turki, M. M., & Zouari, H. (2002). Tectonic evolution of the northern African margin in Tunisia from paleostress data and sedimentary record. *Tectonophysics*, *357*(1-4), 227–253.

Bourrouilh F. (1973). Dolomitisation actuelle dans le monde. une revue-sc. Terre, Nancy, Annales, t. XVIII, N°3, p. 279-298, 4 fig.

Boussen, S. (2010). Evolution de haldes plombo-zincifères dans le nord de la Tunisie: l'exemple d'un contexte carbonaté. Ph.D. Thesis, Faculté des Sciences de Tunis, Université de Tunis El Manar, Tunis, Tunisia

Bruker (2019). X-ray Fluorescence (XRF). Elemental Analysis, Process and Quality Control. www.bruker.com

Buat-Menard, P., & Chesselet, R. (1979). Variable influence of the atmospheric flux on the trace metal chemistry of oceanic suspended matter. *Earth and Planetary Science Letters*, *42*(3), 399–411. DOI: 10.1016/0012-821X(79)90049-9

Butle, V. V., Srinivasan, V., Reddy, D. S., & Gouda, J. (2025). Comparative assessment of limestone-calcined clay blend and fly ash as binders in roller-compacted concrete pavement: Evaluation of fresh and hardened properties. *Road Materials and Pavement Design*, •••, 1–36. DOI: 10.1080/14680629.2025.2486526

Cadamuro, F., Ardenti, V., Nicotra, F., & Russo, L. (2023). Alginate–gelatin self-healing hydrogel produced via static–dynamic crosslinking. *Molecules (Basel, Switzerland)*, *28*(6), 2851. DOI: 10.3390/molecules28062851 PMID: 36985823

Camacho-Alcantar, M., González-Méndez, B., Loredo-Portales, R., Romo-Leon, J. R., & Molina-Freaner, F. (2025). Evaluating potentially toxic element accumulation in crops near abandoned mine tailings in northwestern Mexico: A One Health perspective. *Environmental Monitoring and Assessment*, *197*(10), 1–18. DOI: 10.1007/s10661-025-14587-1 PMID: 40973812

Cardoso, J., Almeida, S. M., Nunes, T., Almida-Silva, M., Cerqueira, M., Alves, C., Rocha, F., Chaves, P., Reis, M., Salvador, P., Artinano, B., & Pio, C. (2018). Source apportionment of atmospheric aerosol in a marine dusty environment by ionic/composition mass balance (IMB). *Atmospheric Chemistry and Physics*, *18*(17), 13215–13230. DOI: 10.5194/acp-18-13215-2018

Carrillo Beltran, R., Martinez, A., Lopez, J., & Gonzalez, P. (2025a). Geopolymers manufactured by the alkali activation of mining, ceramics, and olive industry by-products. *Materials (Basel)*, *18*(3), 688. DOI: 10.3390/ma18030688 PMID: 39942354

Carrillo Beltran, R., Picazo Camilo, E., Perea Toledo, G., & Corpas Iglesias, F. A. (2025b). Study of Novel Geopolymer Concrete Prepared with Slate Stone Cutting Sludge, Chamotte, Steel Slag and Activated with Olive Stone Bottom Ash. *Materials (Basel)*, *18*(9), 1974. DOI: 10.3390/ma18091974 PMID: 40363481

Carter, R. W. G. (1988). *Coastal Environments: An Introduction to the Physical, Ecological, and Cultural Systems of Coastlines*. Academic Press.

Chaouachi, M. C. H. (1988). Etude sédimentologique des séries du Permien supérieur du Jebel Tebaga de Medenine Sud-Est de la Tunisie. Genèse, Diagenèse et Potentiel Réservoir des corps récifaux, Thèse 3ème cycle, Tunis, 285 p.

Chen, P., Bi, X., Zhang, J., Wu, J., & Feng, Y. (2015). Assessment of heavy metal pollution characteristics and human health risk of exposure to ambient $PM_{2.5}$ in Tianjin, China. *Particuology*, *20*, 104–109. DOI: 10.1016/j.partic.2014.04.020

Chen, W., Jia, W., Xiao, Y., Feng, Z., & Wu, G. (2021). A temperature-stable and low impedance piezoelectric MEMS resonator for drop-in replacement of quartz crystals. *IEEE Electron Device Letters*, *42*(9), 1382–1385. DOI: 10.1109/LED.2021.3094319

Chen, Y. C., Chiu, H. W., Su, Y. F., Wu, Y. C., & Cheng, K. S. (2017). Does urbanization increase diurnal land surface temperature variation? Evidence and implications. *Landscape and Urban Planning*, *157*, 247–258. DOI: 10.1016/j.landurbplan.2016.06.014

Chen, Y., Wang, Y., & Li, H. (2020). Smart environmental monitoring systems using IoT and intelligent materials. *Environmental Science & Technology*, *54*(18), 11569–11578. DOI: 10.1021/acs.est.0c02679

Chihi L. et Ben Ayed N. (1991). Le rôle de la fracturation précoce sur la distribution des structures récentes le long du décrochement de Kasserine. Annales tectoniques, vol. V, N°1, pp. 6473.

Chihi L., Dlala M. et Ben Ayed N. (1984). Manifestations tectoniques synsédimentaires et polyphasées d'age Crétacé moyen dans l'Atlas Tunisie centrale (région de Kasserine).C.R.A.S., Paris, t. 298, N°4, p. 141146.

Choi, J. W.. (2024). Smart materials in next-generation energy storage. *Nature Energy*, *9*, 12–20. DOI: 10.1038/s41560-023-01357-2

Choi, N., Hwang, Y., Lee, H., Seo, M., Kim, Y., Kim, G., & Kim, K. (2022). Fabrication and simulation of a piezoelectric PIN-PMN-PT thin film for ultrahigh-frequency ultrasonic transducers. *Sensors and Actuators. A, Physical*, *347*, 113936. DOI: 10.1016/j.sna.2022.113936

Choquette P. W. et Pray L. C. (1970). Geological nomenclature and classification of porosity in sedimentary carbonates. An. Assoc. petrol. Geol. Bull., 54, 2, 207250.

Chua, M. H., Tang, T., Ong, K. H., Neo, W. T., & Xu, J. W. (2019). Chapter 1: Introduction to electrochromism. In *Electrochromic Smart Materials: Fabrication and Applications* (pp. 1–26). Royal Society of Chemistry. ISBN 978-1-78801-143-3

Clarke, D. R., Oechsner, M., & Padture, N. P. (2012). Thermal-Barrier Coatings for More Efficient Gas-Turbine Engines. *MRS Bulletin*, *37*(10), 891–898. Advance online publication. DOI: 10.1557/mrs.2012.232

Co, H. X., Dung, N. T., Oanh, N. T. K., Hang, N. T., Phuc, N. H., & Lee, H. A. (2014). Level, and composition of ambient particulate matter at a mountainous rural site in Northern Vietnam. *Aerosol and Air Quality Research*, *14*(7), 1917–1928. DOI: 10.4209/aaqr.2013.09.0300

Collado, E., Calderón, S., Cedeño, B., De León, O., Centella, M., García, A., & Sáez, Y. (2024). Open-source Internet of Things (IoT)-based air pollution monitoring system with protective case for tropical environments. *HardwareX*, *19*, e00560. DOI: 10.1016/j.ohx.2024.e00560 PMID: 39687307

Contini, D., Genga, A., Cesari, D., Siciliano, M., Donateo, A., Bove, M. C., & Cuascito, M. R. (2010). Characterization and source apportionment of PM_{10} in an urban background site in Lecce. *Atmospheric Research*, *95*(1), 40–54. DOI: 10.1016/j.atmosres.2009.07.010

Correia, R., James, S., Lee, S. W., Morgan, S. P., & Korposh, S. (2018). Biomedical application of optical fibre sensors. *Journal of Optics*, *20*(7), 073003. DOI: 10.1088/2040-8986/aac68d

CPCB. (2008). Stationary Sources Emission Profiles. Central Pollution Control Board, New Delhi, India. https://cpcb.nic.in/Stationary_Sources_Emission_Profiles .xls, Last Access: 09.10.2014.

Croize, D., Bjørlykke, K., & Jahren, J. (2010). Experimental mechanical compaction of clay mineral aggregates—Changes in physical properties of mudstones during burial. *Marine and Petroleum Geology*, *27*, 1765–1774.

da Silva Oliveira, H., & de Brito Nunes, H. K. (2025). Utilização de geotecnologia no estudo das "Terras Caídas" na Terra Indígena barreira da Missão, Tefé/Amazonas/Brasil: Use of geotechnology in the study of the "Fallen Lands" in the Indigenous Land of the Mission barrier, Tefé/Amazonas/Brazil. *Revista de Geociências do Nordeste*, *11*(1), 342–357. DOI: 10.21680/2447-3359.2025v11n1ID34396

Datta, D., Colaco, V., Bandi, S., Dhas, N., Janardhanam, L., Singh, S., & Vora, L. (2025). Stimuli-responsive self-healing ionic gels: A promising approach for dermal and tissue engineering applications. *ACS Biomaterials Science & Engineering*, *11*(3), 1338–1372. DOI: 10.1021/acsbiomaterials.4c02264 PMID: 39999055

Davydovich, D., & Urban, M. W. (2020). Water accelerated self-healing of hydrophobic copolymers. *Nature Communications*, *11*(1), 5743. Advance online publication. DOI: 10.1038/s41467-020-19405-5 PMID: 33184268

de Carvalho Nasser, F., de Mello, D. C., Francelino, M. R., Krause, M. B., de Moura Soares, H., & Demattê, J. A. (2024). Mapping deactivated mine areas in the amazon forest impacted by seasonal flooding: Assessing soil-hydrological processes and quality dynamics by remote sensing and geophysical techniques. *Remote Sensing Applications: Society and Environment*, *34*, 101148. DOI: 10.1016/j.rsase.2024.101148

De Kanter, A. J., Jongsma, K. R., Bouten, C. V. C., & Bredenoord, A. L. (2023). How smart are smart materials? A conceptual and ethical analysis of smart lifelike materials for the design of regenerative valve implants. *Science and Engineering Ethics*, *29*(5), 33. Advance online publication. DOI: 10.1007/s11948-023-00453-1 PMID: 37668955

De La Cruz, A. H., Roca, Y. B., Suarez-Salas, L., Pomalaya, J., Tolentino, D. A., & Gioda, A. (2019). Chemical Characterization of $PM_{2.5}$ at Rural and Urban Sites around the Metropolitan Area of Huancayo (Central Andes of Peru). *Atmosphere (Basel)*, *10*(1), 21. DOI: 10.3390/atmos10010021

de Oliveira, H. F. E., de Castro, L. E. V., Sousa, C. M., Alves Júnior, L. R., Mesquita, M., Silva, J. A. O. S., Faria, L. C., da Silva, M. V., Giongo, P. R., de Oliveira Júnior, J. F., de Siqueira, V. S., & da Silva, J. L. B. (2024). Geotechnologies in Biophysical Analysis through the Applicability of the UAV and Sentinel-2A/MSI in Irrigated Area of Common Beans: Accuracy and Spatial Dynamics. *Remote Sensing (Basel)*, *16*(7), 1254. DOI: 10.3390/rs16071254

Dhirani, L. L., Mukhtiar, N., Chowdhry, B. S., & Newe, T. (2023). Ethical dilemmas and privacy issues in emerging technologies: A review. *Sensors (Basel)*, *23*(3), 1151. DOI: 10.3390/s23031151 PMID: 36772190

Diffo, B. K., Elimbi, A., Cyr, M., Manga, J. D., & Kouamo, H. T. (2015). Effect of the rate of calcination of kaolin on the properties of metakaolin-based geopolymers. *Journal ofAsian Ceramic Societies*, *3*(1), 130–138. DOI: 10.1016/j.jascer.2014.12.003

Ding, C., Tian, M., Ren, F., Dang, Y., & Zhang, M. (2020). Novel self-healing hydrogel with injectable, ph-responsive, strain-sensitive, promoting wound-healing, and hemostatic properties based on collagen and chitosan. ACS Biomaterials Science &Amp. *ACS Biomaterials Science & Engineering*, *6*(7), 3855–3867. DOI: 10.1021/acsbiomaterials.0c00588 PMID: 33463340

Ding, S., Zhang, J., Zhou, L., & Luo, Y. (2020). Promoting healing progress in polymer composites based on diels-alder reaction by constructing silver bridges. *Polymers for Advanced Technologies*, *32*(3), 1239–1250. DOI: 10.1002/pat.5173

Dionne, J. A., Dagli, S., & Shalaev, V. M. (2023). Nanophotonics for a sustainable future. *Physics Today*, *76*(6), 24–31. DOI: 10.1063/PT.3.5254

Du, J., Li, Y., Jiu-chun, W., Wang, C., Liu, D., Wang, G., & Liu, S. (2020). Mechanically robust, self-healing, polymer blends and polymer/small molecule blend materials with high antibacterial activity. ACS Applied Materials &Amp. *Interfaces*, *12*(24), 26966–26972. DOI: 10.1021/acsami.0c06591 PMID: 32466641

Durán, O., & Moore, L. J. (2013). Vegetation controls on the maximum size of coastal dunes. *Proceedings of the National Academy of Sciences of the United States of America*, *110*(43), 17217–17222. DOI: 10.1073/pnas.1307580110 PMID: 24101481

Dušeková, L., Herich, P., Pukanská, K., Bartoš, K., Kseňak, Ľ., Šveda, J., & Fehér, J. (2024). Comparison of non-contact measurement technologies applied on the underground glacier—The choice for long-term monitoring of ice changes in Dobšiná ice cave. *Remote Sensing (Basel)*, *16*(20), 3870. DOI: 10.3390/rs16203870

Dwivedi, S., Jaiswal, S., & Kushwaha, A. (2022). Renewable Biofuel Resources: Introduction, Production Technologies, Challenges, and Applications. In: Chowdhary, P., Pandit, S., Khanna, N. (eds) Bio-Clean Energy Technologies Volume 2. Clean Energy Production Technologies. Springer, Singapore. https://doi.org/DOI: 10.1007/978-981-16-8094-6_3

EEA (European Environment Agency). (2014). Air quality in Europe — 2014 report. Report No. 5/2014, Copenhagen (84 pages).

Ekeocha, J., Ellingford, C., Pan, M., Wemyss, A. M., Bowen, C., & Wan, C. (2021). Challenges and opportunities of self-healing polymers and devices for extreme and hostile environments. *Advanced Materials*, *33*(33), 2008052. Advance online publication. DOI: 10.1002/adma.202008052 PMID: 34165832

El-Gharbi, S., Masmoudi, H., & Zargouni, F. (2019). Coastal dune vegetation and its role in dune stabilization along the eastern coast of Tunisia. *Environmental Earth Sciences*, *78*, 512.

Enamorado-Báez, S. M., Gómez-Guzmán, J. M., Chamizo, E., & Abril, J. M. (2015). Levels of 25 trace elements in high-volume air filter samples from Seville (2001–2002): Sources, enrichment factors and temporal variations. *Atmospheric Research*, *155*, 118–129. DOI: 10.1016/j.atmosres.2014.12.005

Ezeh, G. C., Abiye, O. E., & Obioh, I. B. (2017). Elemental analyses and source apportionment of $PM_{2.5}$ and $PM_{2.5-10}$ aerosols from Nigerian urban cities. *Cogent Environmental Science*, *3*(1), 1323376. DOI: 10.1080/23311843.2017.1323376

Ezeh, G. C., Obioh, I. B., Asubiojo, O. I., & Abiye, O. E. (2012). PIXE characterization of PM_{10} and $PM_{2.5}$ particulates sizes collected in Ikoyi, Lagos, Nigeria. *Toxicological and Environmental Chemistry*, *94*(5), 884–894. DOI: 10.1080/02772248.2012.674133

Fakhri, P., Amini, B., Bagherzadeh, R., Kashfi, M., Latifi, M., Yavari, N., Kani, S. A., & Kong, L. (2019). Flexible hybrid structure piezoelectric nanogenerator based on ZnO nanorod/PVDF nanofibers with improved output. *RSC Advances*, *9*(18), 10117–10123. DOI: 10.1039/C8RA10315A PMID: 35520929

Fernandez, J., & Pallis, G. (2021). Privacy issues in smart environments. *IEEE Communications Surveys and Tutorials*, *23*(4), 2692–2718.

Ferrazzo, S. T., de Araújo, M. T., Consoli, N. C., & da Rocha, C. G. (2024). Geotech social impacts: Development, application, and comparative analysis. *Environmental Impact Assessment Review*, *108*, 107577. DOI: 10.1016/j.eiar.2024.107577

FitzGerald, D. M., Buynevich, I. V., & Argow, B. A. (2008). Model of tidal inlet and barrier island dynamics in a regime of accelerated sea-level rise. *Journal of Coastal Research, SI*, (55), 1071–1078.

Flügel, E., & Munnecke, A. (2010). Microfacies of carbonate rocks: analysis, interpretation and application (Vol. 976, p. 2004). Berlin: springer.

Flügel, E. (2004). Diagenesis, porosity, and dolomitization. In *Microfacies of carbonate rocks: Analysis, interpretation and application* (pp. 267–338). Springer Berlin Heidelberg.

Folk, R. L. (1965). Some aspects of recrystallization in ancient limestones, Soc, Econ. Paleont. *Mineral. Spec. Pub.*, *13*, 1448.

Folk, R. L., & Land, L. S. (1975). Mg/Ca Ratio and salinity; two controls over crystallization of dolomite. *Bulletin of the American Association of Petroleum Geologists*, *59*, 6068.

Folk, R. L., & Pittman, J. S. (1971). Length-slow chalcedony: A new testament for vanished evaporites. *Journal of Sedimentary Petrology*, *41*, 1045–1058.

Folk, R. L., & Siedelcka, A. (1974). The Bschizohaline^ environment: Its sedi mentary and diagenetic fabrics as exemplified by Late Paleozoic rocks of Bear Island, Svalbard. *Sedimentary Geology*, *11*, 1–15.

Fortoul, T. I., Rodriguez-Lara, V., Gonzalez-Villalva, A., Rojas-Lemus, M., Colin-Barenque, L., Bizarro-Nevares, P., García-Peláez, I., Ustarroz-Cano, M., López-Zepeda, S., Cervantes-Yépez, S., López-Valdez, N., Meléndez-García, N., Espinosa-Zurutuza, M., Cano-Gutierrez, G., & Cano-Rodríguez, M. C. (2015). Health Effects of Metals in Particulate Matter. *Current Air Quality Issues. Ch, 25*, 572–607. DOI: 10.5772/59749

Frei, R., McWilliam, R., Derrick, B., Purvis, A., Tiwari, A., & Serugendo, G. D. M. (2013). Self-healing and self-repairing technologies. *International Journal of Advanced Manufacturing Technology*, *69*(5-8), 1033–1061. DOI: 10.1007/s00170-013-5070-2

Frihy, O. E., Gueddari, M., & Halouani, N. (2012). Coastal processes and sediment dynamics along southern Mediterranean coasts. *Journal of Coastal Research*, *28*(4), 859–870.

Fruehauf, A., Pellegrino, P., & Lombardo, M. (2025). The importance of urban greenery in the construction of a smart landscape to reduce negative environmental and climate impacts. *Environmental Science & Sustainable Development*, *10*(1), 53–60. DOI: 10.21625/essd.v10i1.1111

Fujiwara, F. G., Gómez, D. R., Dawidowski, L., Perelman, P., & Faggi, A. (2011). Metals associated with airborne particulate matter in road dust and tree bark collected in a megacity (Buenos Aires, Argentina). *Ecological Indicators*, *11*(2), 240–247. DOI: 10.1016/j.ecolind.2010.04.007

Fulan, M. H., Vendruscolo, J., Fulan, J. A., da Silva Souza, R. F., dos Santos, M. A., dos Santos Hara, F. A., & Cavalheiro, W. C. S. (2024). Geotechnologies in the assessment of eutrophication in the barra bonita reservoir, São Paulo, Brazil. *Revista de Gestão Social e Ambiental*, *18*(2), 1–14. DOI: 10.24857/rgsa.v18n2-063

Furuuchi, M. M., Hata, M., Nuttapon, K., & Hongtieab, S. (2017). Nano sampler (NS) Standard Operate Procedure. Kanazawa University, 1-6.

Galindo, A. L., Ruiz, J. T., & Lopez, J. G. (1996). Mineral quantification in sepiolite-palygorskite deposits using X-ray diffraction and chemical data. *Clay Minerals*, *31*(2), 217–224. DOI: 10.1180/claymin.1996.031.2.07

Gao, L., Yang, Y., Xie, J., Zhang, S., Hu, J., Zeng, R., He, J., Li, Q., & Wang, Q. (2020). Autonomous self-healing of electrical degradation in dielectric polymers using in situ electroluminescence. *Matter*, *2*(2), 451–463. DOI: 10.1016/j.matt.2019.11.012

Garg, R., & Anjum, A. (2024, February 23). Smart and sustainable applications of nanocomposites. *Advances in Chemical and Materials Engineering*. Advance online publication. DOI: 10.4018/979-8-3693-3136-1.ch002

Garner, J. (2023). The transformative power of material recycling: Creating a sustainable future. *Journal of Materials Science and Nanomaterials*, *7*, 98.

Gell, M., Wang, J., Kumar, R., Roth, J., Jiang, C., & Jordan, E. H. (2018). Higher Temperature Thermal Barrier Coatings with the Combined Use of Yttrium Aluminum Garnet and the Solution Precursor Plasma Spray Process. *Journal of Thermal Spray Technology*, *27*(4), 543–555. DOI: 10.1007/s11666-018-0701-7

Genua, F., Lancellotti, I., & Leonelli, C. (2025). Geopolymer-based stabilization of heavy metals, the role of chemical agents in encapsulation and adsorption. *Polymers*, *17*(5), 670. DOI: 10.3390/polym17050670 PMID: 40076162

GeoNames Geographical Database. **(2012). Population of Akure, Nigeria.** https://population.mongabay.com/population/nigeria/2350841/akure (Retrieved 9th January 2016).

Georgopoulou, A., Bosman, A. W., Brancart, J., Vanderborght, B., & Clemens, F. (2021). Supramolecular self-healing sensor fiber composites for damage detection in piezoresistive electronic skin for soft robots. *Polymers*, *13*(17), 2983. DOI: 10.3390/polym13172983 PMID: 34503023

Ge, Y., Della Porta, G., Pederson, C. L., Lokier, S. W., Hoffmann, R., & Immenhauser, A. (2021). Botryoidal and spherulitic aragonite in carbonates associated with microbial mats : Precipitation or diagenetic replacement product? *Frontiers of Earth Science*, *9*, 698952.

Ghaffarianhoseini, A.. (2019). The application of smart materials in sustainable architecture. *Renewable & Sustainable Energy Reviews*, *101*, 778–791. DOI: 10.1016/j.rser.2018.10.042

Giampaoli, P., Wannaz, E. D., Tavares, A. R., & Domingos, M. (2016). Suitability of *Tillandsia usneoides* and *Aechmea fasciata* for biomonitoring toxic elements under tropical seasonal climate. *Chemosphere*, *149*, 14–23. DOI: 10.1016/j.chemosphere.2016.01.080 PMID: 26844661

Gili, E., Masse, J. P., & Skelton, P. (1995, November). Rudists as gregarious sediment-dwellers, not reef-builders, on Cretaceous carbonate platforms. *Palaeogeography, Palaeoclimatology, Palaeoecology*, *118*(3-4), 245–267.

Gistemp, T. (2024). *GISS Surface Temperature Analysis (GISTEMP), version 4. NASA Goddard Institute for Space Studies, 2023, NASA [data set].*

Golik, V. I., Razorenov, Y. I., Valiev, N. G., & Gavrina, O. A. (2024). Environmentally sound geotechnologies for leaching metals from polymetallic ore processing wastes and wastewater. *Mining Science and Technology (Russia)*, *9*(3), 271–282. DOI: 10.17073/2500-0632-2023-11-184

Gong, H., Zhou, K., Zhang, Q., Liu, J., Wang, H., & Yan, H. (2020). A self-patterning multicolour electrochromic device driven by horizontal redistribution of ions. *Solar Energy Materials and Solar Cells*, *215*, 110642. DOI: 10.1016/j.solmat.2020.110642

Gowd, S. S., Reddy, M. R., & Govil, P. K. (2010). Assessment of heavy metal contamination in soils at Jajmau (Kanpur) and Unnao industrial areas of the Ganga Plain, Uttar Pradesh, India. *Journal of Hazardous Materials*, *174*(1-3), 113–121. DOI: 10.1016/j.jhazmat.2009.09.024 PMID: 19837511

Green, M. A., Dunlop, E. D., Yoshita, M., Kopidakis, N., Bothe, K., Siefer, G., & Hao, X. (2024). Solar cell efficiency tables (version 63). *Progress in Photovoltaics: Research and Applications*, *32*(1), 3–19. DOI: 10.1002/pip.3750

Gregg J. M. and Sibley D. F. (1984). Epigenetic dolomitization and the origin of xenotopic dolomite texture. Jour. sedim. Petrology, 54, 3, 908931.

Grimm, N. B., Chapin, F. S.III, Bierwagen, B., Gonzalez, P., Groffman, P. M., Luo, Y., Melton, F., Nadelhoffer, K., Pairis, A., Raymond, P. A., Schimel, J., & Williamson, C. E. (2013). The impacts of climate change on ecosystem structure and function. *Frontiers in Ecology and the Environment*, *11*(9), 474–482. DOI: 10.1890/120282

Guan, Q., Cai, A., Wang, F., Yang, L., Xu, C., & Liu, Z. (2017). Spatio-temporal variability of particulate matter in the key part of Gansu Province, Western China. *Environmental Pollution*, *230*, 189–198. DOI: 10.1016/j.envpol.2017.06.045 PMID: 28651090

Guo, S., Duan, X., Xie, M., Aw, K. C., & Xue, Q. (2020). Composites, fabrication and application of polyvinylidene fluoride for flexible electromechanical devices: A review. *Micromachines*, *11*(11), 1076. DOI: 10.3390/mi11121076 PMID: 33287450

Gupta, A., Verma, R., & Singh, D. (2022). Smart materials and their applications: A review. *Materials Today: Proceedings*, *57*, 2250–2260. DOI: 10.1016/j.matpr.2022.04.015

Habib, M., Lantgios, I., & Hornbostel, K. (2022). A review of ceramic, polymer and composite piezoelectric materials. *Journal of Physics. D, Applied Physics*, *55*(42), 55. DOI: 10.1088/1361-6463/ac8687

Haj Ali, H., Belghithi, H., Ouali, J. A., & Touir, J. (2016b). Relationship between fracturing and porogenesis in a carbonate reservoir: Example from the Middle Turonian Bireno Member in Jebel M'rhila, Central Tunisia. *Journal of African Earth Sciences*, *121*, 342–357.

Haj Ali, H., Essefi, E., Belghithi, H., Yaich, C., & Touir, J. (2016a). Repercussions of the dolomitization of the fractured Middle Turonian succession of Bireno platform central part of Tunisia. *Arabian Journal of Geosciences*, *9*(3), 213.

Hajalilou, A., Amri Mazlan, S., Lavvafi, H., & Shameli, K. (2016). Field responsive fluids as smart materials (1st ed.). Engineering Materials, Springer. DOI: 10.1007/978-981-10-2495-5

Hakanson, L. (1980). An ecological risk index for aquatic pollution control. A sediment logical approach. *Water Research*, *14*(8), 975–1001. DOI: 10.1016/0043-1354(80)90143-8

Halouani, N., Gueddari, M., & Frihy, O. (2013). The Northwestern Mediterranean Coast of Tunisia: Wave processes and shoreline dynamics. *Arabian Journal for Science and Engineering*, *38*, 3219–3230. DOI: 10.1007/s13369-012-0401-4

Han, S., Liu, C., Lin, X., Zheng, J., Wu, J., & Liu, C. (2020). Dual conductive network hydrogel for a highly conductive, self-healing, anti-freezing, and non-drying strain sensor. *ACS Applied Polymer Materials*, *2*(2), 996–1005. DOI: 10.1021/acsapm.9b01198

Hanshaw, B. B., Back, W., & Deike, R. G. (1971). A geochemical hypothesis for dolomitization by groundwater. *Economic Geology and the Bulletin of the Society of Economic Geologists*, *66*, 710–724.

Hao, W., Hao, H., Kanwal, H., & Jiang, S. (2023). Evaluation of self-healing efficiency of microcapsule-based self-healing cementitious composites based on acoustic emission. *Journal of Renewable Materials*, *11*(4), 1687–1697. DOI: 10.32604/jrm.2022.023795

Haq, B. U., Hardenbol, J. A. N., & Vail, P. R. (1987). Chronology of fluctuating sea levels since the Triassic. *Science*, *235*(4793), 1156–1167.

Hård, D., Wallin, M., & Ristinmaa, M. (2024). Electrode and electroactive polymer layout design using topology optimization [Preprint]. arXiv. https://arxiv.org/abs/2401.12345

Haverkos, B. M., Pan, Z., Gru, A. A., Freud, A. G., Rabinovitch, R., Xu-Welliver, M., Otto, B., Barrionuevo, C., Baiocchi, R. A., Rochford, R., & Porcu, P. (2016). Extranodal NK/T cell lymphoma, nasal type (ENKTL-NT): An update on epidemiology, clinical presentation, and natural history in North American and European cases. *Current Hematologic Malignancy Reports*, *11*(6), 514–527. DOI: 10.1007/s11899-016-0355-9 PMID: 27778143

Hesp, P. A. (2013). Concepts in coastal dune geomorphology. In *Treatise on Geomorphology* (Vol. 10, pp. 328–355). Academic Press. DOI: 10.1016/B978-0-12-374739-6.00310-9

Hood, S. D., Nelson, C. S., & Kamp, P. J. J. (2004). Burial dolomitisation in a nontropical carbonate petroleum reservoir : The Oligocène Tikorangi Formation, Taranaki Basin, New Zealand. *Sedimentary Geology*, *172*, 117138.

Hsu, C. Y., Chiang, H. C., Lin, S. L., Chen, M. J., Lin, T. Y., & Chen, Y. C. (2016). Elemental characterization and source apportionment of PM_{10} and $PM_{2.5}$ in the western coastal area of central Taiwan. *The Science of the Total Environment*, *15*(541), 1139–1150. DOI: 10.1016/j.scitotenv.2015.09.122 PMID: 26473714

HU, Y., Cai, C., Pederson, C. L., Liu, D., Jiang, L., He, X., & Immenhauser, A. (2020). Dolomitization history and porosity evolution of a giant, deeply buried Ediacaran gas field (Sichuan Basin, China). *Precambrian Research*, *338*, 105595.

Huang, B., Menne, M. J., Boyer, T., Freeman, E., Gleason, B. E., Lawrimore, J. H., Liu, C., Rennie, J. J., Schreck, C. J., Sun, F., Vose, R., Williams, C. N., Yin, X., & Zhang, H.-M. (2020). Uncertainty estimates for sea surface temperature and land surface air temperature in NOAAGlobalTemp version 5. *Journal of Climate*, *33*(4), 1351–1379. DOI: 10.1175/JCLI-D-19-0395.1

Huang, J. L., Li, Z. Y., Mao, J. Y., Chen, Z. M., Liu, H. L., Liang, G. Y., Zhang, D.-B., Wen, P.-J., Mo, Z.-Y., & Jiang, Y. M. (2024). Contamination and health risks brought by arsenic, lead and cadmium in a water-soil-plant system nearby a non-ferrous metal mining area. *Ecotoxicology and Environmental Safety*, *270*, 115873. DOI: 10.1016/j.ecoenv.2023.115873 PMID: 38150749

Huang, L., & Wang, G. (2014). Chemical characteristics and source apportionment of atmospheric particles during the heating period in Harbin, China. *Journal of Environmental Sciences (China)*, *26*(12), 2475–2483. DOI: 10.1016/j.jes.2014.04.008 PMID: 25499495

Huang, W., Cheng, S., Wang, X., Zhang, Y., Chen, L., & Zhang, L. (2021). Noncompressible hemostasis and bone regeneration induced by an absorbable bioadhesive self-healing hydrogel. *Advanced Functional Materials*, *31*(22), 2009189. Advance online publication. DOI: 10.1002/adfm.202009189

Huang, X., Zhang, A., Tan, Q., Gou, K., Chen, Y., Nie, Y., & Weng, G. (2024). Highly stretchable and self-healable tough polymer elastomer through dy3+ and cu2+ coordination cross-linking. *Macromolecules*, *57*(3), 963–975. DOI: 10.1021/acs.macromol.3c02347

Hu, S., Zhong, L., Yang, X., Bai, H., Ren, B., Zhao, Y., Zhang, W., Ju, X., Wen, H., Mao, S., Tao, R., & Li, C. (2020). Synthesis of rare earth tailing-based geopolymer for efficiently immobilizing heavy metals. *Construction & Building Materials*, *254*, 119273. DOI: 10.1016/j.conbuildmat.2020.119273

Ibn-Mohammed, T., Koh, S. C. L., Reaney, I. M., Acquaye, A., Wang, D., Taylor, S., Genovese, A., & Sinclair, D. C. (2017). Are lead-free piezoelectrics more environmentally friendly? *MRS Communications*, *7*(1), 1–7. DOI: 10.1557/mrc.2017.10

Ihenketu, C. (2024). Reduction in dilapidation and high cost of maintenance on buildings using self-healing concrete for building construction. *International Research Journal of Modernization in Engineering Technology and Science*. Advance online publication. DOI: 10.56726/IRJMETS48855

Imen, T., & Souissi, R. (2018). Coastal erosion in the south-eastern Mediterranean: Case of beaches in North Tunisia. *Arabian Journal of Geosciences*, *11*(14), 386. DOI: 10.1007/s12517-018-3716-y

Institut National de la Météorologie (INM) (2020). Données climatiques de la région du Sahel et du Centre-Est tunisien.

Ismail, N. A., Khan, A., Fayyad, E. M., Kahraman, R., Abdullah, A. M., & Shakoor, A. (2021). Self-healing performance of smart polymeric coatings modified with tung oil and linalyl acetate. *Polymers*, *13*(10), 1609. DOI: 10.3390/polym13101609 PMID: 34067528

Jain, S., Sharma, S. K., Mandal, T. K., & Saxena, M. (2018). Source apportionment of PM_{10} in Delhi, India using PCA/APCS, UNMIX, and PMF. *Particuology*, *37*, 107–118. DOI: 10.1016/j.partic.2017.05.009

Jaiswal, S., & Khanam, S. Z. (2025). Techno-economic analysis of green hydrogen production for industrial use. In Mabrouki, J. (Ed.), *Obstacles Facing Hydrogen Green Systems and Green Energy* (pp. 479–506). IGI Global Scientific Publishing., DOI: 10.4018/979-8-3693-8980-5.ch028

Jaiswal, S., Tiwari, S., & Bahadur, P. S. (2021). Inter-stellar space discovery, spectroscopic structure identification and application of C_{60} Buckminsterfullerene. *International Journal of Engineering Research in General Science*, *9*, 12–19. https://pnrsolution.org/Datacenter/Vol9/Issue2/2.pdf

James, G., Witten, D., Hastie, T., & Tibshirani, R. (2015). *An Introduction to Statistical Learning with Applications in R* (Olkin, G. C. S. F. I., Ed.). Vol. 6). Springer.

Janta, R., & Chantara, S. (2017). Tree bark as a bio indicator of a metal accumulation from road traffic and air quality map: A case study of Chiang Mai, Thailand. *Atmospheric Pollution Research*, *8*(5), 956–967. DOI: 10.1016/j.apr.2017.03.010

Jedoui, Y., Davaud, E., Ben Ismaïl, H., & Reyss, J. L. (2002). Sedimentological analysis of marine Pleistocene deposits of southeastern Tunisia: Evidence of two positive eustatic pulsations during the marine isotopic substage 5e (Eemian, Tyrrhenian). *Bulletin de la Société Géologique de France*, *173*(3), 255–264. DOI: 10.2113/173.3.255

Jedoui, Y., Oueslati, A., & Ben Ismail, H. (2003). Les variations du niveau marin holocène et la morphogenèse littorale en Tunisie orientale. Geomorphologie: Relief, Processus. *Environnement*, *9*(4), 277–286.

Jemmali, N., Souissi, F., Carranza, E. J. M., Vennemann, T. W., & Bogdanov, K. (2014). Geochemical constraints on the genesis of the Pb–Zn deposit of Jalta (northern Tunisia): Implications for timing of mineralization, sources of metals and relationship to the Neogene volcanism. *Chemie der Erde*, *74*(4), 601–613. DOI: 10.1016/j.chemer.2014.01.002

Jensen, J., Hösel, M., Dyer, A. L., & Krebs, F. C. (2015). Development and manufacture of polymer-based electrochromic devices. *Advanced Functional Materials*, *25*(14), 2073–2090. DOI: 10.1002/adfm.201403765

Jha, S., Ramesh, B., & Macrin, D. (2023). Study on the flexural strength of polypropylene fiber reinforced m20 grade self-compacting bacterial concrete. Cardiometry, (25), 1763-1770. https://doi.org/DOI: 10.18137/cardiometry.2022.25.17631770

Jia, C., Kong, K., Yao, Y., Yang, X., Wang, D., & Shao, S. (2024). Combining multi-source data to identify the paleochannel system in the saltwater intrusion area. *Marine Georesources and Geotechnology*, *42*(5), 562–576. DOI: 10.1080/1064119X.2023.2207562

Jia, H., Yang, D., Deng, W., Wei, Q., & Jiang, W. (2021). Predicting land surface temperature with geographically weighed regression and deep learning. *Wiley Interdisciplinary Reviews. Data Mining and Knowledge Discovery*, *11*(1), e1396. DOI: 10.1002/widm.1396

Jindal, B. B., Alomayri, T., Hasan, A., & Kaze, C. R. (2023). Geopolymer concrete with metakaolin for sustainability: A comprehensive review on raw material's properties, synthesis, performance, and potential application. *Environmental Science and Pollution Research International*, *30*(10), 25299–25324. DOI: 10.1007/s11356-021-17849-w PMID: 35000173

Jin, Q., Du, R., Tang, H., Zhao, Y., Peng, W., Li, Y., Zhang, J., Zhu, T., Huang, X., Kong, D., He, Y., Bao, T., Kong, D., Wang, X., Wang, R., Zhang, Q., & Jia, X. (2023). Quadruple h-bonding and polyrotaxanes dual cross-linking supramolecular elastomer for high toughness and self-healing conductors. *Angewandte Chemie International Edition*, *62*(26), e202305282. Advance online publication. DOI: 10.1002/anie.202305282 PMID: 37186156

Johnson, T., & Woodward, K. (2025). *Enviro-IoT: Calibrating low-cost environmental sensors in urban settings* [Preprint]. arXiv. https://doi.org//arXiv.2502.07596DOI: 10.48550

Jolliffe, I. T., Cadima, J., & Cadima, J. (2016). Principal component analysis: A review and recent developments. *Philosophical Transactions. Series A, Mathematical, Physical, and Engineering Sciences*, *A374*(2065), 20150202. DOI: 10.1098/rsta.2015.0202 PMID: 26953178

Jonnalagadda, P. (2019). *Thermal Barrier Coatings : Failure Mechanisms and Life Prediction*. DOI: 10.3384/diss.diva-154777

Juvanhol, R. S., da Silva, E. F., da Paschoa Manhães, L., Santos, J. S., Silva, J. P. M., Vieira, G. C., Viana, J. C. C., & da Silva, M. L. M. (2024). Mathematical programming and geotechnologies applied to allocation of forest fire detection towers. *Stochastic Environmental Research and Risk Assessment*, *38*(12), 4795–4808. DOI: 10.1007/s00477-024-02831-y

Kafy, A. A., Shuvo, R. M., Naim, M. N. H., Sikdar, M. S., Chowdhury, R. R., Islam, M. A., Sarker, M. H. S., Khan, M. H. H., & Kona, M. A. (2021). Remote sensing approach to simulate the land use/land cover and seasonal land surface temperature change using machine learning algorithms in a fastest-growing megacity of Bangladesh. *Remote Sensing Applications: Society and Environment*, *21*, 100463. DOI: 10.1016/j.rsase.2020.100463

Kassa, A. E., Shibeshi, N. T., & Tizazu, B. Z. (2022). Kinetic analysis of dehydroxylation of Ethiopian kaolinite during calcination. *Journal of Thermal Analysis and Calorimetry*, *147*(22), 12837–12853. DOI: 10.1007/s10973-022-11452-y

Kazemi Garajeh, M., Salmani, B., Zare Naghadehi, S., Valipoori Goodarzi, H., & Khasraei, A. (2023). An integrated approach of remote sensing and geospatial analysis for modeling and predicting the impacts of climate change on food security. *Scientific Reports*, *13*(1), 1057. DOI: 10.1038/s41598-023-28244-5 PMID: 36658205

Kchih, H., Perrino, C., & Cherif, S. (2015). Investigation of desert dust contribution to source apportionment of PM_{10} and $PM_{2.5}$ from Southern Mediterranean Coast. *Aerosol and Air Quality Research*, *15*(2), 454–464. DOI: 10.4209/aaqr.2014.10.0255

Kekez, S., Novakova, I., Lach, M., Setlak, K., Fiala, L., Přikryl, J., Furtos, G., Petean, I., Alexandersson, K. F., Baronins, J., & Hager, I. (2025). Waste valorization for fabrication of geopolymers. *Case Studies in Construction Materials*, *22*, e04388. DOI: 10.1016/j.cscm.2025.e04388

Khademi, H., Gabarron, M., Abbaspour, A., Martínez-Martínez, S., Faz, A., & Acosta, J. A. (2019). Environmental impact assessment of industrial activities on heavy metals distribution in street dust and soil. *Chemosphere*, *217*, 695–705. DOI: 10.1016/j.chemosphere.2018.11.045 PMID: 30448749

Khalil, A., Ait-Khouia, Y., Beniddar, H., El Ghorfi, M., Hakkou, R., Taha, Y., & Benzaazoua, M. (2025). Sustainable reprocessing of Pb–Zn mine tailings through froth flotation for resource recovery and environmental remediation in abandoned mining regions. *Minerals Engineering*, *222*, 109132. DOI: 10.1016/j.mineng.2024.109132

Khan, A., & Rabnawaz, M. (2021). Base-layer-driven self-healing materials. *ACS Applied Polymer Materials*, *3*(8), 3922–3928. DOI: 10.1021/acsapm.1c00476

Khillare, P. S., & Sarkar, S. (2012). Airborne inhalable metals in residential areas of Delhi, India: Distribution, source apportionment, and health risks. *Atmospheric Pollution Research*, *3*(1), 46–54. DOI: 10.5094/APR.2012.004

Kılıçarslan, B., Bozyel, I., Gökcen, D., & Bayram, C. (2022). Sustainable macromolecular materials in flexible electronics. *Macromolecular Materials and Engineering*, *307*(6), 2100978. Advance online publication. DOI: 10.1002/mame.202100978

Kim, J. (2017). Multifunctional smart biopolymer composites as actuators. In Thomas, S., Joseph, K., Kuruvilla, A. K., & Malhotra, S. K. (Eds.), *Biopolymer composites in electronics* (pp. 311–331). Elsevier., DOI: 10.1016/B978-0-12-809261-3.00011-5

Kothai, P., Saradhi, I. V., Pandit, G. G., Markwitz, A., & Puranik, V. D. (2011). **Chemical** characterization and source identification of particulate matter at an urban site of Navi Mumbai, India. *Aerosol and Air Quality Research*, *11*(5), 560–569. DOI: 10.4209/aaqr.2011.02.0017

Kulkarni, A., Wang, Z., Nakamura, T., Sampath, S., Goland, A., Herman, H., Allen, J., Ilavsky, J., Long, G., Frahm, J., & Steinbrech, R. W. (2003). Comprehensive microstructural characterization and predictive property modeling of plasma-sprayed zirconia coatings. *Acta Materialia*, *51*(9), 2457–2475. https://doi.org/https://doi.org/10.1016/S1359-6454(03)00030-2. DOI: 10.1016/S1359-6454(03)00030-2

Kumar, B., Verma, K., & Kulshrestha, U. (2014). Deposition and mineral characteristics of atmospheric dust in relation to land use and land cover change in Delhi (India). *The Geographical Journal*, *2014*, 1–11. Advance online publication. DOI: 10.1155/2014/325612

Lakhdar, R., & Soussi, M. (2009). Les dépôts lagunaires holocènes du Sud-Est de la Tunisie: extension et organisation spatio-temporelle. *In Conférence Méditerranéenne Côtiere et Maritime*, pp. 115e118.

Land, L. S. (1985). The origin of massive dolomite. *Journal of Geological Education*, *33*(2), 112–125.

Lazorenko, G., Ivanov, S., Petrova, T., & Dimitrov, V. (2025). Towards coal mining waste valorization: Gangue as a resource for geopolymers and related alkali-activated materials. *Journal of Hazardous Materials*, *420*(3), 126–138. DOI: 10.1016/j.grets.2025.100205

Le Floc'h, E., Boulos, L., Véla, E., & Domina, G. (2010). Orobanchaceae. In *Catalogue synonymique commenté de la Flore de Tunisie* (Vol. 2010, pp. 272–275). Ministère de l'Environnement et du Dévelopment Durable.

Le Houérou, H. N. (1996). Climate change, drought and desertification. *Journal of Arid Environments, 34*(2), 133–185. DOI: 10.1006/jare.1996.0099

Le, A. T., Ahmadipour, M., & Pung, S.-Y. (2022). A review on ZnO-based piezoelectric nanogenerators: Synthesis, characterization techniques, performance enhancement and applications. *Journal of Alloys and Compounds, 844*, 156172. DOI: 10.1016/j.jallcom.2020.156172

Lee, J., Park, S. U., Park, C., Kwon, O. S., Chung, S., Shim, J., Lee, C.-S., & Bae, J. (2018). Effect of a surfactant in microcapsule synthesis on self-healing behavior of capsule embedded polymeric films. *Polymers, 10*(6), 675. DOI: 10.3390/polym10060675 PMID: 30966709

Lee, S., Hong, P. H., Kim, J., Choi, K., Moon, G., Kang, J., Lee, S., Ahn, J. B., Eom, W., Ko, M. J., & Hong, S. W. (2020). Highly self-healable polymeric blend synthesized using polymeric glue with outstanding mechanical properties. *Macromolecules, 53*(6), 2279–2286. DOI: 10.1021/acs.macromol.9b02359

Lee, W. K. W., & Van Deventer, J. S. J. (2003). Use of infrared spectroscopy to study geopolymerization of heterogeneous amorphous aluminosilicates. *Langmuir, 19*(21), 8726–8734. DOI: 10.1021/la026127e

Li, F., Wang, X., Zuo, J., Chen, C., Chen, J., Zhu, J., & Ying, W. B. (2024). Oxime-urethane-based self-healing polyurethane for achieving complex structures via 3d printing. *ACS Applied Polymer Materials, 6*(7), 4070–4077. DOI: 10.1021/acsapm.4c00140

Li, G., Zhang, B., Wang, J., Zhao, H., Ma, W., Xu, L., Zhang, W., Zhou, K., Du, Y., & He, G. (2019b). Electrochromic poly(chalcogenoviologen)s as anode materials for high-performance organic radical lithium-ion batteries. *Angewandte Chemie International Edition, 58*(25), 8468–8473. DOI: 10.1002/anie.201903152 PMID: 30951238

Li, J., Levitt, A., Kurra, N., Juan, K., Noriega, N., Xiao, X., Wang, X., Wang, H., Alshareef, H. N., & Gogotsi, Y. (2019a). MXene-conducting polymer electrochromic microsupercapacitors. *Energy Storage Materials, 20*, 455–461. DOI: 10.1016/j.ensm.2019.04.028

Li, R., Shi, Y., Alsaedi, M., Wu, M., Shi, L., & Yu, G. (2020). Hybrid hydrogel with high water-vapor harvesting capacity for self-sustained solar-driven clean water production. *ACS Materials Letters, 2*(7), 854–860. DOI: 10.1021/acsmaterialslett.0c00124

Li, S., Dong, Q., Peng, X., Chen, Y., Yang, H., Xu, W., Zhao, Y., Xiao, P., & Zhou, Y. (2022). Self-healing hyaluronic acid nanocomposite hydrogels with platelet-rich plasma impregnated for skin regeneration. *ACS Nano*, *16*(7), 11346–11359. DOI: 10.1021/acsnano.2c05069 PMID: 35848721

Liu, F., Wang, X., Sun, F., Wang, H., Wu, L., Zhang, X., Liu, W., & Che, H. (2022). Correction of overestimation in observed land surface temperatures based on machine learning models. *Journal of Climate*, *35*(16), 5359–5377. DOI: 10.1175/JCLI-D-21-0447.1

Liu, H., Scneider, P., Haugen, R., & Vogt, M. (2019). Performance Assessment of a Low-Cost $PM_{2.5}$ Sensor for a near Four-Month Period in Oslo, Norway. *Atmosphere (Basel)*, *10*(2), 41. DOI: 10.3390/atmos10020041

Liu, J., Wang, L., Bai, L., Wang, W., Yang, L., Chen, H., Yang, H., & Wei, D. (2024). Fabrication of nanocomposite hydrogels based on cellulose nanocrystals and multi-walled carbon nanotubes for human motion monitoring. *Macromolecular Chemistry and Physics*, *225*(21), 2400207. Advance online publication. DOI: 10.1002/macp.202400207

Liu, X., Ren, Z., Liu, F., Li, Z., Ling, Q., & Gu, H. (2021). Multifunctional self-healing dual network hydrogels constructed via host–guest interaction and dynamic covalent bond as wearable strain sensors for monitoring human and organ motions. ACS Applied Materials &Amp. *Interfaces*, *13*(12), 14612–14622. DOI: 10.1021/acsami.1c03213 PMID: 33723988

Liu, Y., Lin, S.-H., Chuang, W.-T., Dai, N.-T., & Hsu, S. (2022). Biomimetic Strain-Stiffening in Chitosan Self-Healing Hydrogels. *ACS Applied Materials & Interfaces*, *14*(14), 16032–16046. DOI: 10.1021/acsami.2c01720 PMID: 35321544

Liu, Z., Bisoyi, H. K., Huang, Y., Wang, M., Yang, H., & Li, Q. (2021). Thermo- and mechanochromic camouflage and self-healing in biomimetic soft actuators based on liquid crystal elastomers. *Angewandte Chemie International Edition*, *61*(8), e202280862. Advance online publication. DOI: 10.1002/anie.202280862 PMID: 34904346

Liu, Z., Zhong, Y., Li, S., Yu, S., Zhong, J., Yang, Y., & Shen, L. (2024). Hybrid self-repairing polymer composites based on a mixture of intrinsic and extrinsic self-healing. *Macromolecular Chemistry and Physics*, *225*(18), 2400079. Advance online publication. DOI: 10.1002/macp.202400079

Li, W., Ni, L., Li, Z. L., Duan, S. B., & Wu, H. (2019). Evaluation of machine learning algorithms in spatial downscaling of MODIS land surface temperature. *IEEE Journal of Selected Topics in Applied Earth Observations and Remote Sensing*, *12*(7), 2299–2307. DOI: 10.1109/JSTARS.2019.2896923

Li, Y. J., Lee, B. P., Su, L., Fung, J. C. H., & Chan, C. K. (2015). Seasonal characteristics of fine particulate matter (PM) based on high-resolution time-of-flight aerosol mass spectrometric (HR-ToF-AMS) measurements at the HKUST Supersite in Hong Kong. *Atmospheric Chemistry and Physics*, *15*(1), 37–53. DOI: 10.5194/acp-15-37-2015

Li, Z., Zhang, S., Zuo, Y., Chen, W., & Ye, G. (2019). Chemical deformation of metakaolin based geopolymer. *Cement and Concrete Research*, *120*, 108–118. DOI: 10.1016/j.cemconres.2019.03.017

Li, Z., Zhao, J., Hou, C., Fei, C., Zheng, C., Lou, L., Chen, D., Li, D., & Yang, Y. (2022). High-frequency self-focusing ultrasonic transducer with piezoelectric metamaterial. *IEEE Electron Device Letters*, *43*(6), 946–949. DOI: 10.1109/LED.2022.3170613

Lu, C., Jiang, X., Li, L., Zhou, H., Yang, A., Xin, M., Fu, G., & Wang, X. (2022). Wind energy harvester using piezoelectric materials. *The Review of Scientific Instruments*, *93*(3), 031502. DOI: 10.1063/5.0065462 PMID: 35364975

Lucia, F. J. (1999). Rock-Fabric, Petrophysical Parameters, and Classification. In *Carbonate Reservoir Characterization* (pp. 23–57). Springer Berlin Heidelberg.

Luijendijk, A., Hagenaars, G., Ranasinghe, R., Baart, F., Donchyts, G., & Aarninkhof, S. (2018). The state of the world's beaches. *Scientific Reports*, *8*(1), 6641. DOI: 10.1038/s41598-018-24630-6 PMID: 29703960

Lyashenko, V., & Andreev, B. (2024). Increasing The Efficiency And Environmental Safety Of Combined Geotechnologies For Leaching Metals From And Off-Balance Ores. Norwegian Journal of development of the International Science No, 144, 123. DOI: 10.2478/minrv-2025-0007

M'rabet, A., Ghenima, R., & el Euchi, H. (1996). MidCretaceous source rocks and reservoirs in central Tunisia. The fifth Tunisian petroleum exploration conference, ETAP, N°11, p2945.

Machel, H. G. (1999). Effects of groundwater flow on mineral diagenesis, with emphasis on carbonate aquifers. *Hydrogeology Journal*, *7*, 94–107.

Machel, H. G. (2004). "Concepts and models of dolomitization: A critical reappraisal." In The Geometry and Petrogenesis of Dolomite Hydrocarbon Reservoirs. *Geological Society of London, Special Publications, 235*, 7–63.

Mahapatra, S. D., Mohapatra, P. C., Aria, A. I., Christie, G., Mishra, Y. K., Hofmann, S., & Thakur, V. K. (2021). Piezoelectric materials for energy harvesting and sensing applications: Roadmap for future smart materials. *Advancement of Science, 8*(17), 2100864. DOI: 10.1002/advs.202100864 PMID: 34254467

Mahmoudi, M. (1986). Stratigraphie, sédimentologie et diagenèse des dépôts tyrrhéniens du Sahel tunisien (Tunisie orientale) (*Thèse de doctorat*, Paris 11).

Maksimović, M., Vujović, V., & Omanović-Mikličanin, E. (2023). Smart materials for sustainable development. *Journal of Cleaner Production, 405*, 136313. DOI: 10.1016/j.jclepro.2023.136313

Mandal, P., Saud, T., Sarkar, R., Mandal, A., Sharma, S. K., Mandal, T. K., & Bassin, J. K. (2014). High seasonal variation of atmospheric C and particle concentrations in Delhi, India. *Environmental Chemistry Letters, 12*(1), 225–230. DOI: 10.1007/s10311-013-0438-y

Mansourmoghaddam, M., Rousta, I., Ghafarian Malamiri, H., Sadeghnejad, M., Krzyszczak, J., & Ferreira, C. S. S. (2024). Modeling and Estimating the Land Surface Temperature (LST) Using Remote Sensing and Machine Learning (Case Study: Yazd, Iran). *Remote Sensing (Basel), 16*(3), 454. DOI: 10.3390/rs16030454

Maqbool, Q., Mobili, A., Blasi, E., Sabbatini, S., Ruello, M. L., & Tittarelli, F. (2023). Sustainable Alkali-Activated Mortars for the Immobilization of Heavy Metals from Copper Mine Tailings. *ACS Sustainable Resource Management, 1*(1), 154–164. DOI: 10.1021/acssusresmgt.3c00002

Martínez-Ibernón, A., Antonaci, P., & Anglani, G. (2023). Initial proposal of a smart cement-based material to enhance the service-life of reinforcement concrete structures. MATEC Web of Conferences, 378, 05003. https://doi.org/DOI: 10.1051/matecconf/202337805003

Maskari, M. A., Aravind, N., Dhandapani, R., & AlDeen, E. E. K. M. (2025). Application of artificial intelligence in prediction of mechanical properties of sustainable plastic pavement blocks and validation through experimental analysis. [SpringerLink]. *Discover Concrete & Cement, 1*(1), 11. DOI: 10.1007/s44416-025-00011-x

Masmoudi, H., El-Gharbi, S., & Zargouni, F. (2018). Morphological dynamics of Tunisian barrier–lagoon systems. *Quaternary International, 475*, 48–60.

Maun, M. A. (2009). *The biology of coastal sand dunes*. Oxford University Press. DOI: 10.1093/oso/9780198570356.001.0001

Meddi, M., Toumi, S., & Assani, A. A. (2022). Analysis of meteorological drought sequences in various climatic zones of Algeria. Revue. *Journal of African Earth Sciences*, *185*, 104397.

Mendelsohn, R., Morrison, W., Schlesinger, M. E., & Andronova, N. G. (2000). Country-specific market impacts of climate change. *Climatic Change*, *45*(3), 553–569. DOI: 10.1023/A:1005598717174

Merino, E. (2006). Self-accelerating dolomite-for-calcite replacement and displacive zebra veins: Dynamics of burial dolomitization. *Geochimica et Cosmochimica Acta Supplement*, *70*(18), A417–A417.

Mir, S. H., Nagahara, L. A., Thundat, T., Mokarian-Tabari, P., Furukawa, H., & Khosla, A. (2018). Review—Organic–inorganic hybrid functional materials: An integrated platform for applied technologies. *Journal of the Electrochemical Society*, *165*(8), B3137–B3156. DOI: 10.1149/2.0191808jes

Mishra, A. K., Tripathy, A. K., Obaidat, M. S., Zhi-yuan, T., Prasad, M., Sadoun, B., & Puthal, D. (2018). A chain topology for efficient monitoring of food grain storage using smart sensors. *Proceedings of the 15th International Joint Conference on E-Business and Telecommunications*. https://doi.org/DOI: 10.5220/0006850602550264

Mishra, J. (2017). Smart materials: Types and their application—A review. *International Journal of Mechanical and Production Engineering*, *5*(12), 20–24.

Mobbs, P. M. (2005). The mineral industry of Tunisia. Minerals Yearbook: Area Reports, International. *Africa and the Middle East.*, *III*, 37.

Mokhtari, P., Rahimi, H., & Hosseini, R. (2025). Valorization of mine tailings for sustainable geopolymer production. *International Journal of Ceramic Engineering & Science*, *7*(4), e70014. DOI: 10.1002/ces2.70014

Monk, P., Mortimer, R., & Rosseinsky, D. (2007). *Electrochromism and electrochromic devices* (1st ed.). Cambridge University Press., DOI: 10.1017/CBO9780511550959

Moore, C. H. (2001). Diagenetic environments of porosity modification and tools for their recognition in the geologic record. Carbonate reservoirs porosity evolution and diagenesis in a sequence stratigraphic framework. *Developments in Sedimentology*, *55*, 61–88.

Morales Aranibar, C. G., Torres, J., & Ramirez, L. (2025). Reuse of mine tailings through geopolymerization and 3D printing: A sustainable approach. *Sustainability, 17*(6), 2617. DOI: 10.3390/su17062617

Morang, S., Bandyopadhyay, A., Borah, N., Kar, A., Mandal, B. B., & Karak, N. (2024). Photoluminescent self-healable waterborne polyurethane/mo and s codoped graphitic carbon nitride nanocomposite with bioimaging and encryption capability. *ACS Applied Bio Materials*, *7*(3), 1910–1924. DOI: 10.1021/acsabm.3c01259 PMID: 38391158

Morice, C.P., Kennedy, J.J., Rayner, N.A., Winn, J.P., Hogan, E., Killick, R.E., Dunn, R.J.H., Osborn, T.J., Jones, P.D. & Simpson, I.R. (2021). An updated assessment of near-surface temperature change from 1850: The HadCRUT5 data set. *Journal of Geophysical Research: Atmospheres, 126*(3), p.e2019JD032361.

Moses, E. A., & Orok, U. B. (2015). Contamination and health risk assessment of suspended particulate matter (SPM) in Uyo, Nigeria Delta, Nigeria. *Journal of Scientific Research and Reports*, *6*(4), 276–286. DOI: 10.9734/JSRR/2015/16296

Motora, K. G., Wu, C., Chang, C., & Liao, J. (2021). Nir light stimulated self-healing reduced tungsten oxide/polyurethane nanocomposite based on the diels–alder reaction. *Macromolecular Materials and Engineering*, *306*(11), 2100438. Advance online publication. DOI: 10.1002/mame.202100438

Murat, B., & Marinella, F. (2018). *Materials that move: Smart materials, intelligent design*. Springer International Publishing., DOI: 10.1007/978-3-319-76890-8

National Population Commission (NPC). (2010). *Federal Republic of Nigeria. 2006 Population and Housing Census. Priority Table* (Vol. III). Population Distribution – By Sex, State, LGA and Senatorial District.

Ndamitso, M.M., Abdulkadir, A., Abulude, F.O. (2015). Total atmospheric deposit source apportionment: A review. *Environmental Skeptics and Critics,* 2016, 5(4): 63-78.

Ndamitso, M. M., Abdulkadir, A., & Abulude, F. O. (2016). Source Apportionment: Case Studies of Selected African Countries. *International Journal of Advancements in Research & Technology*, *5*(3), 1–25.

Neumann, A. C. (1965). Processes of recent carbonate sedimentation in Harrington sound, Bermuda. *Bulletin of Marine Science*, *15*, 987–1035.

Nguyen, T. M., Lee, J., & Kim, D. (2020). Advanced insulation materials for buildings: A review. *Construction & Building Materials*, *260*, 119400. DOI: 10.1016/j.conbuildmat.2020.119400

Niu, G., Guo, X., & Wang, L. (2015). Review of recent progress in chemical stability of perovskite solar cells. *Journal of Materials Chemistry. A, Materials for Energy and Sustainability*, *3*(17), 8970–8980. DOI: 10.1039/C4TA04994B

Oh, S., & Kim, Y. J. (2025). Evaluation of permeability recovery in precast concrete with hybrid capsules using constant-head permeability test as smart construction material. *Materials (Basel)*, *18*(2), 220. DOI: 10.3390/ma18020220 PMID: 39859692

Okuo, J. M., Chiedu, I. E., Anegbe, B., Oyibo, F. O., & Ojo, W. (2017). Elemental Characterization and Source Identification of Fine Particulate Matter (PM2.5) in an Industrial Area of Lagos State, Nigeria. *Physical Science International Journal*, *16*(2), 1–11. DOI: 10.9734/PSIJ/2017/36683

Olabi, A.-G. (Ed.). (2021). *Encyclopedia of smart materials* (1st ed.). Elsevier., DOI: 10.1016/C2018-0-01700-3

Onabowale, M. K., & Owoade, O. K. (2015). Assessment of residential indoor-outdoor airborne particulate matter in Ibadan, Southwestern Nigeria. *Donnish Journal of Physical Sciences*, *1*(1), 1–7.

Onwudiegwu, C. A., Ezeh, G. C., & Obioh, I. B. (2016). Trace Metals in Total Atmospheric Depositions (TAD) of a Nigerian Island. *Journal of Atmospheric Pollution*, *4*(1), 15–22. DOI: 10.12691/jap-4-1-2

Otgonbayar, M., Atzberger, C., Mattiuzzi, M., & Erdenedalai, A. (2019). Estimation of climatologies of average monthly air temperature over mongolia using MODIS land surface temperature (LST) time series and machine learning techniques. *Remote Sensing (Basel)*, *11*(21), 2588. DOI: 10.3390/rs11212588

Oueslati, A. (2008). Le littoral de Oued Laou: de l'apport de l'étude géomorphologique à la connaissance de ses aptitudes à l'aménagement et à la prévention des risques naturels et de la dégradation des paysages. *Travaux de l'Institut Scientifique, Rabat, série générale*, (5), 1-16.

Oueslati, A. (1993). *Les côtes de la Tunisie: Géomorphologie et environnement et aptitudes à l'aménagement*. Faculté des Sciences Humaines et Sociales de Tunisie.

Oueslati, A. (1994). Dynamique littorale et évolution des côtes de la Tunisie orientale. *Bulletin de la Société Géologique de France*.

Oueslati, A. (1995). The evolution of low Tunisian coasts in historical times: From progradation to erosion and salinization. *Quaternary International*, *29*, 41–47. DOI: 10.1016/1040-6182(95)00006-5

Padture, N., Gell, M., & Jordan, E. (2002). TBCs for Gas-Turbine Engine Applications. *Science*, *296*, 280–284. DOI: 10.1126/science.1068609 PMID: 11951028

Palma, G., Bolaños, H., Huamani, R., Clements, C., & Hedayat, A. (2024). Optimization of Geopolymers for Sustainable Management of Mine Tailings: Impact on Mechanical, Microstructural, and Toxicological Properties. *Minerals (Basel)*, *14*(10), 997. DOI: 10.3390/min14100997

Pande, C. B., Egbueri, J. C., Costache, R., Sidek, L. M., Wang, Q., Alshehri, F., Din, N. M., Gautam, V. K., & Pal, S. C. (2024). Predictive modeling of land surface temperature (LST) based on Landsat-8 satellite data and machine learning models for sustainable development. *Journal of Cleaner Production*, *444*, 141035. DOI: 10.1016/j.jclepro.2024.141035

Pandey, B., Agrawal, M., & Singh, S. (2015). Ecological risk assessment of soil contamination by trace elements around the coal mining area. *Journal of Soils and Sediments*. Advance online publication. DOI: 10.1007/s11368-015-1173-8

Paolillo, S., Bose, R. K., Santana, M. H., & Grande, A. M. (2021). Intrinsic self-healing epoxies in polymer matrix composites (pmcs) for aerospace applications. *Polymers*, *13*(2), 201. DOI: 10.3390/polym13020201 PMID: 33429922

Park, B. R., Hong, J., Choi, E. J., Choi, Y. J., Lee, C., & Moon, J. W. (2019). Improvement in energy performance of building envelope incorporating electrochromic windows (ECWs). *Energies*, *12*(6), 1181. DOI: 10.3390/en12061181

Parlak, M., Tunçay, T., & Parlak, A. Ö. (2024). Heavy metals in tailings and soils in the Pb-Zn mining areas of North-west Türkiye and health risk evaluations. *International Journal of Agriculture Environment and Food Sciences*, *8*(1), 131–148. DOI: 10.31015/jaefs.2024.1.14

Paskoff, R., & Sanlaville, P. (1983). Les côtes de la Tunisie. Variations du niveau marin depuis le Tyrrhénien (Vol. 14, No. 1, pp. 0-0). *Maison de l'Orient et de la Méditerranée Jean Pouilloux.*

Peng, Y., Gu, S., Wu, Q., Xie, Z., & Wu, J. (2023). High-performance self-healing polymers. *Accounts of Materials Research*, *4*(4), 323–333. DOI: 10.1021/accountsmr.2c00174

Perez-Aguilar, L. Y., Plata-Rocha, W., Monjardin-Armenta, S. A., & Lopez-Osorio, R. F. (2025). Implementation of a web-based system for monitoring and simulation of arid zones in northwestern Mexico. Region of Mexico. *Ecological Modelling*, *501*, 111024. DOI: 10.1016/j.ecolmodel.2025.111024

Perin, D., Dorigato, A., & Pegoretti, A. (2023). Thermoplastic self-healing polymer blends for structural composites: Development of polyamide 6 and cyclic olefinic copolymer blends. *Journal of Applied Polymer Science, 140*(16), e53751. Advance online publication. DOI: 10.1002/app.53751

Pervinquiere, L. (1903). Etude géologique de la Tunisie centrale thèse. Sc. Paris, de Rudeval édit., Vol. gr. in. 4, 360 p.

Petrakis, E. G. M., Sotiriadis, S., Soultanopoulos, T., Renta, P. T., Buyya, R., & Bessis, N. (2018). Internet of things as a service (itaas): Challenges and solutions for management of sensor data on the cloud and the fog. *Internet of Things : Engineering Cyber Physical Human Systems, 3-4*, 156–174. DOI: 10.1016/j.iot.2018.09.009

Philip, J. (1972). Les formations calcaires à rudistes du Crétacé supérieur provençal et rhodanien. Thèse Doc. Etat, Marseille, 443 p., VIII pl.

Police, S., Sahu, S. K., & Pandit, G. G. (2016). Chemical characterization of atmospheric particulate matter and their source apportionment at an emerging industrial coastal city, Visakhapatnam, India. *Atmospheric Pollution Research, 7*(4), 725–733. DOI: 10.1016/j.apr.2016.03.007

Pongpiachan, S., & Iijima, A. (2016). Assessment of selected metals in the ambient air PM_{10} in urban sites of Bangkok (Thailand). *Environmental Science and Pollution Research International, 23*(3), 2948–2961. DOI: 10.1007/s11356-015-5877-5 PMID: 26631022

Poojalakshmi, E. S., Nagarajan, P., Sudhakumar, J., & Thomas, B. S. (2025). Impact of alkaline activator concentration on mechanical properties and microstructure of a ternary blended one-part geopolymer cement. *Scientific Reports, 15*(1), 33808. DOI: 10.1038/s41598-025-01610-1 PMID: 41028007

Pop, F., Herrera, B., & Rinaldi, M. (2022). Lithium niobate piezoelectric micromachined ultrasonic transducers for high data-rate intrabody communication. *Nature Communications, 13*(1), 17. DOI: 10.1038/s41467-022-29355-9 PMID: 35379794

Popov, A., Brasiunas, B., Mikoliunaite, L., Bagdziunas, G., Ramanavicius, A., & Ramanaviciene, A. (2019). Comparative study of PANI, PEDOT, and PANI–PEDOT films electrochemically deposited on transparent indium tin oxide electrodes. *Polymer, 172*, 133–141. DOI: 10.1016/j.polymer.2019.03.059

Psuty, N. P. (2008). The coastal foredune: A morphological basis for regional coastal dune development. *Ecological Studies, 171*, 11–27. DOI: 10.1007/978-3-540-74002-5_2

Psuty, N. P., & Silveira, T. M. (2010). Global morphodynamics of coastal foredunes: The role of vegetation. *Geomorphology*, *123*, 201–211.

Purdy, E. G. (1968). Carbonate diagenesis : An environmental survey. *Geologica Romana*, *7*, 183–228.

Purohit, P., Amann, M., Kiesewtter, G., Chaturvedi, V., Rafj, P., Dholakia, H. H., Koti, P. N., Klimont, Z., Borken-Kleefeld, J., Gomez-Sanabria, A. G., Schopp, W., & Sander, R. (2019). *Pathways to Achieve National Ambient Air Quality Standards (NAAQS) in India. International Institute for Applied System Analysis (IIASA). The Council on Energy, Environment and Water*. CEEW.

Purser, B. H. (1978). Early diagenesis and the preservation of porosity in Jurassic limestones. Jour. *Petroleum Geology*, *1*(2), 8394.

Querol, X., Alastuey, A., Moreno, T., Viana, M. M., Castillo, S., Pey, J., Rodríguez, S., Artiñano, B., Salvador, P., Sánchez, M., Garcia Dos Santos, S., Herce Garraleta, M. D., Fernandez-Patier, R., Moreno-Grau, S., Negral, L., Minguillón, M. C., Monfort, E., Sanz, M. J., Palomo-Marín, R., & Sánchez de la Campa, A. (2008). Spatial and temporal variations in airborne particulate matter (PM_{10} and $PM_{2.5}$) across Spain 1999–2005. *Atmospheric Environment*, *42*(17), 3964–3979. DOI: 10.1016/j.atmosenv.2006.10.071

Rahimi, M., Wood, C. M., & Kallivokas, L. F. (2024). A comparative study of using geophysical methods for imaging subsurface voids of various sizes and at different depths. *Engineering Geology*, *341*, 107711. DOI: 10.1016/j.enggeo.2024.107711

Rajczakowska, M., Szeląg, M., Habermehl-Cwirzen, K., Hedlund, H., & Ćwirzeń, A. (2023). Interpretable machine learning for prediction of post-fire self-healing of concrete. *Materials (Basel)*, *16*(3), 1273. DOI: 10.3390/ma16031273 PMID: 36770279

Ranganathan, P., Chen, C., Tasi, M., Rwei, S., & Lee, Y. (2021). Biomass thermoplastic (co)polyamide elastomers synthesized from a fatty dimer acid: A sustainable route toward a new era of uniform and bimodal foams. Industrial &Amp. *Industrial & Engineering Chemistry Research*, *60*(33), 12139–12154. DOI: 10.1021/acs.iecr.1c01533

Rao, F., & Liu, Q. (2015). Geopolymerization and its potential application in mine tailings consolidation: A review. *Mineral Processing and Extractive Metallurgy Review*, *36*(6), 399–409. DOI: 10.1080/08827508.2015.1055625

Reeder, R. J., & Prosky, J. L. (1986). Compositional sector zoning in dolomiteJour. Sed. *Petro.*, *56*, 237–247.

Rich, D. Q., Zhang, W., Lin, S., Squizzato, S., Thurston, S. W., van Wijngaarden, E., Croft, D., Masiol, M., & Hopke, P. K. (2019). Triggering of cardiovascular hospital admissions by source-specific fine particle concentrations in urban centers of New York State. *Environment International*, *126*, 387–394. DOI: 10.1016/j.envint.2019.02.018 PMID: 30826617

Rosas, J. T. F., Demattê, J. A., Rosin, N. A., dos Anjos Bartsch, B., Poppiel, R. R., Rodriguez-Albarracin, H. S., & Alves, M. R. (2024). Geotechnologies on the phosphorus stocks determination in tropical soils: General impacts on society. *The Science of the Total Environment*, *938*, 173537. DOI: 10.1016/j.scitotenv.2024.173537 PMID: 38802008

Rozman, M., Alif, M., Bren, U., & Luksic, M. (2022). Electrochromic device demonstrator from household materials. *Journal of Chemical Education*, *99*(9), 3595–3600. DOI: 10.1021/acs.jchemed.2c00176 PMID: 36246424

Sadiku, M. N. O., Suman, G., & Musa, S. M. (2021). Artificial intelligence in materials science. *International Journal of Advances in Scientific Research and Engineering*, *07*(08), 77–81. DOI: 10.31695/IJASRE.2021.34057

Sahraoui, H., Attia, R., Hamrouni, H., & Hachicha, M. (2016). Assessment of trace elements contamination of agricultural topsoil around Lakhouat mine area, Tunisia. [WJST]. *Walailak Journal of Science and Technology*, *13*(11), 965–976.

Saidi, F., Ben Ismail, M. H., & M'rabet, A. (1997). Le Turonien de Tunisie centrooccidentale : Faciès, paléogéographie et stratigraphie séquentielle d'une plate forme carbonatée ennoyée. *Cretaceous Research*, *1997*(18), 6385.

Sakimoto, K. K., Wong, A. B., & Yang, P. (2016). Self-photosensitization of nonphotosynthetic bacteria for solar-to-chemical production. *Science*, *351*(6268), 74–77. DOI: 10.1126/science.aad3317 PMID: 26721997

Sammari, C., Koutitonsky, V., & Moussa, M. (2006). Hydrodynamic modeling of the central Tunisian shelf: Implications for sediment transport. *Continental Shelf Research*, *26*, 1763–1783.

Sampath, S., Schulz, U., Jarligo, M., & Kuroda, S. (2012). Processing Science of Advanced Thermal-Barrier Systems. *MRS Bulletin*, *37*(10), 903–910. Advance online publication. DOI: 10.1557/mrs.2012.233

Santos, A., Andrejkovičová, S., Perna, I., Almeida, F., & Rocha, F. (2024). Mechanical and thermal properties of geopolymers derived from metakaolin with iron mine waste. *Applied Clay Science*, *258*, 107452. DOI: 10.1016/j.clay.2024.107452

Sarafanov, M., Kazakov, E., Nikitin, N. O., & Kalyuzhnaya, A. V. (2020). A machine learning approach for remote sensing data gap-filling with open-source implementation: An example regarding land surface temperature, surface albedo and NDVI. *Remote Sensing (Basel)*, *12*(23), 3865. DOI: 10.3390/rs12233865

Sarker, M. R., Julai, S., Sabri, M. F. M., Said, S. M., Islam, M. M., & Tahir, M. (2019). Review of piezoelectric energy harvesting systems and optimization techniques to enhance performance. *Sensors and Actuators. A, Physical*, *300*, 111634. DOI: 10.1016/j.sna.2019.111634

Sayadi, S., Irimia, L., & Jefferson, T. (2024). Biomimetic materials in construction industry: the necessity of simulation. https://doi.org/DOI: 10.18573/conf1.n

Scherb, S., Köberl, M., Beuntner, N., Thienel, K. C., & Neubauer, J. (2020). Reactivity of metakaolin in alkaline environment: Correlation of results from dissolution experiments with XRD quantifications. *Materials (Basel)*, *13*(10), 2214. DOI: 10.3390/ma13102214 PMID: 32408524

Schmidt, J., Marques, M. R. G., Botti, S., & Marques, M. A. L. (2019). Recent advances and applications of machine learning in solid-state materials science. *npj Computational Materials*, *5*(1), 83. Advance online publication. DOI: 10.1038/s41524-019-0221-0

Schneider, S., Sarukhan, J., Adejuwon, J., Azar, C., Baethgen, W., Hope, C., Moss, R., Leary, N., Richels, R., & Van Ypersele, J. P. (2001). Overview of impacts, adaptation, and vulnerability to climate change. *Climatic Change*, 75–103.

Selvaraj, S., Prabhu Kavin, B., Kavitha, C., & Lai, W. C. (2022). A Multiclass Fault Diagnosis Framework Using Context-Based Multilayered Bayesian Method for Centrifugal Pumps. *Electronics (Basel)*, *11*(23), 4014. DOI: 10.3390/electronics11234014

Seo, J., Kim, Y., & Hu, J. (2015). Pilot study for investigating the cyclic behavior of slit damper systems with recentering SMA bending bars used for seismic restrainers. *Applied Sciences (Basel, Switzerland)*, *5*(3), 187–208. DOI: 10.3390/app5030187

Sesana, E., Gagnon, A. S., Ciantelli, C., Cassar, J., & Hughes, J. J. (2021). Climate change impacts on cultural heritage: A literature review. *Wiley Interdisciplinary Reviews: Climate Change*, *12*(4), e710. DOI: 10.1002/wcc.710

Sharanya, S., & Karthikeyan, S. (2022). Local Time Invariant Learning from Industrial Big Data for Predictive Maintenance in Smart Manufacturing. In *Big Data Analytics in Smart Manufacturing* (pp. 69–85). Chapman and Hall/CRC. DOI: 10.1201/9781003202776-4

Shariati, S., Pourbabaee, A. A., Alikhani, H. A., & Rezaei, K. A. (2019). Investigation of Heavy Metal Contamination in the Surface Sediments of Anzali Wetland in North of Iran. *Pollution*, *5*(1), 211–224. DOI: 10.22059/poll.2018.257276.438

Shields, Y., Belie, N. D., Jefferson, T., & Tittelboom, K. V. (2021). A review of vascular networks for self-healing applications. *Smart Materials and Structures*, *30*(6), 063001. DOI: 10.1088/1361-665X/abf41d

Sibley, D. F., & Gregg, J. M. (1987). Classification of dolomite rock textures. *Journal of Sedimentary Petrology*, *57*, 967–975.

Sima, W., Chen, L., Sun, P., Yang, M., Zhu, C., Yuan, T., & Deng, Q. (2021). Novel smart insulating materials achieving targeting self-healing of electrical trees: High performance, low cost, and eco-friendliness. *ACS Applied Materials & Interfaces*, *13*(28), 33485–33495. DOI: 10.1021/acsami.1c07469 PMID: 34232014

Singh, A., Monga, S., Sharma, N., Sreenivas, K., & Katiyar, R. S. (2022). Ferroelectric and piezoelectric mechanisms and applications. *Journal of Asian Ceramic Societies*, *10*(2), 275–291. DOI: 10.1080/21870764.2022.2075618

Siyal, A. A., Mohamed, R. M. S. R., Shamsuddin, R., & Ridzuan, M. B. (2024). A comprehensive review of synthesis kinetics and formation mechanism of geopolymers. *RSC Advances*, *14*(1), 446–462. DOI: 10.1039/D3RA06205H PMID: 38173565

Slimani, H., Aydda, A., & Irigaray, C. (2021). Assessment of the impact of climate change on the dune systems of the Mediterranean coast of Morocco. *Catena*, *207*, 105627.

Smith, M., & Kar-Narayan, S. (2022). Piezoelectric polymers: Theory, challenges and opportunities. *International Materials Reviews*, *67*(1), 65–88. DOI: 10.1080/09506608.2021.1915935

Snoussi, M., Haida, S., & Imassi, S. (2020). Vulnerability assessment of the Moroccan coastal zone to sea-level rise: The case of the Mediterranean coast. *Journal of Coastal Conservation*, *24*(2), 20.

Soares, N., Costa, J. J., Gaspar, A. R., & Santos, P. (2013). Review of passive PCM technology for buildings. *Renewable & Sustainable Energy Reviews*, *18*, 23–36. DOI: 10.1016/j.rser.2012.10.021

Soin, N., Zhao, P., Prashanthi, K., Chen, J., Ding, P., Zhou, E., Shah, T., Ray, S. C., Tsonos, C., Thundat, T., Siores, E., & Luo, J. (2016). High performance triboelectric nanogenerators based on PVDF–$ZnSnO_3$ membranes. *Nano Energy*, *30*, 470–480. DOI: 10.1016/j.nanoen.2016.10.040

Song, Y., Kim, H., & Chung, C. (2022). Repeatable self-healing of a protective coating based on vegetable-oil-loaded microcapsules. *Polymers, 14*(10), 2013. DOI: 10.3390/polym14102013 PMID: 35631895

Song, Y., Yang, S., Li, D., Shao, Y., Cheng, J., Shan, X., Cao, L., Li, S., Xie, C., Shan, L., Tian, D., Wang, H., Jia, L., Huang, J., Tian, Z., & Yue, J. (2025). Changes of Yancheng coastline near 20 years based on remote sensing technology. *Marine Georesources and Geotechnology, 43*(1), 27–35. DOI: 10.1080/1064119X.2024.2310736

Srivastava, P. K., Han, D., Ramirez, M. R., & Islam, T. (2013). Machine learning techniques for downscaling SMOS satellite soil moisture using MODIS land surface temperature for hydrological application. *Water Resources Management, 27*(8), 3127–3144. DOI: 10.1007/s11269-013-0337-9

Stanley, D. J., & Clemente, P. L. (2014). Clay distributions, grain sizes, sediment thicknesses, and compaction rates to interpret subsidence in Egypt's northern Nile Delta. *Journal of Coastal Research, 30*(1), 88–101.

Stern, K. H. (n.d.). *Coatings Ceramic Protective.*

Sun, F., Liu, L., Liu, T., Wang, X., Qi, Q., Hang, Z., Chen, K., Xu, J. H., & Fu, J. (2023). Vascular smooth muscle-inspired architecture enables soft yet tough self-healing materials for durable capacitive strain-sensor. *Nature Communications, 14*(1), 130. Advance online publication. DOI: 10.1038/s41467-023-35810-y PMID: 36624140

Sun, Y., Gao, C., Li, J., Wang, R., & Liu, J. (2019). Quantifying the effects of urban form on land surface temperature in subtropical high-density urban areas using machine learning. *Remote Sensing (Basel), 11*(8), 959. DOI: 10.3390/rs11080959

Surenbaatar, U., Lee, S., Kwon, J.-Y., Lim, H., Kim, J.-J., Kim, Y.-H., & Hong, Y.-S. (2023). Bioaccumulation of Lead, Cadmium, and Arsenic in a Mining Area and Its Associated Health Effects. *Toxics, 11*(6), 519. DOI: 10.3390/toxics11060519 PMID: 37368619

Sutherland, R. A. (2000). Bed sediment-associated trace metals in an urban stream, Oahu, Hawaii. *Environmental Geology (Berlin), 39*(6), 611–637. DOI: 10.1007/s002540050473

Tan, J., NourEldeen, N., Mao, K., Shi, J., Li, Z., Xu, T., & Yuan, Z. (2019). Deep learning convolutional neural network for the retrieval of land surface temperature from AMSR2 data in China. *Sensors (Basel), 19*(13), 2987. DOI: 10.3390/s19132987 PMID: 31284617

Taylor, S. R. (1964). Abundance of chemical elements in the continental crust: A new table. *Geochimica et Cosmochimica Acta*, *28*(8), 1273–1285. DOI: 10.1016/0016-7037(64)90129-2

Terryn, S., Langenbach, J., Roels, E., Brancart, J., Bakkali-Hassani, C., Poutrel, Q., Georgopoulou, A., George Thuruthel, T., Safaei, A., Ferrentino, P., Sebastian, T., Norvez, S., Iida, F., Bosman, A. W., Tournilhac, F., Clemens, F., Van Assche, G., & Vanderborght, B. (2021). A review on self-healing polymers for soft robotics. *Materials Today*, *47*, 187–205. DOI: 10.1016/j.mattod.2021.01.009

Thakur, H., Kaur, S., Singh, A., Singh, G. B., & Mudgal, G. (2025). Smart sustainable materials: A blueprint for a better tomorrow. In Kesari, K. K., Prakash, C., Khalid, M., & Negi, A. (Eds.), *Functionalized cellulose materials* (pp. 355–382). Springer., DOI: 10.1007/978-3-031-76953-5_11

Tian, F., Liu, Y., Ma, R., Li, F., Xu, Z., & Yang, Y. (2021). Properties of PMN-PT single crystal piezoelectric material and its application in underwater acoustic transducers. *Applied Acoustics*, *175*, 107827. DOI: 10.1016/j.apacoust.2020.107827

Tian, Q., Bai, Y., Pan, Y., Chen, C., Yao, S., Sasaki, K., & Zhang, H. (2022). Application of geopolymer in stabilization/solidification of hazardous pollutants: A review. *Molecules (Basel, Switzerland)*, *27*(14), 4570. DOI: 10.3390/molecules27144570 PMID: 35889449

Tian, Y., Cong, S., Su, W., Chen, H., Li, Q., Geng, F., & Zhao, Z. (2014). Synergy of w18o49 and polyaniline for smart supercapacitor electrode integrated with energy level indicating functionality. *Nano Letters*, *14*(4), 2150–2156. DOI: 10.1021/nl5004448 PMID: 24593047

Tomas, W. M., Baggio, R., Berlinck, C. N., Camilo, A. R., Cunha, C. N. D., Damasceno-Junior, G., Durigan, G., Dutra-Silva, R., Fidélis, A., Garcia, L. C., Herrera, H. M., Libonati, R., Marengo, J. A., Oliveira, M. R., Overbeck, G. E., Pereira, A. M. M., Pillar, V. D. P., Pivello, V. R., Ribeiro, D. B., & Walter, B. M. T. (2024). Challenges in the conservation and management of legal reserve areas in Brazilian grassland and savanna ecosystems in the face of global climate change. *Pesquisa Agropecuária Brasileira*, *59*, e03491. DOI: 10.1590/s1678-3921.pab2024.v59.03491

Tositti, L., Brattich, E., Maisiol, M., Baldacci, D., Ceccato, D., Parmeggiani, S., Stracquadanio, M., & Zappoli, S. (2013). Source apportionment of particulate matter in a large city of southeastern Po Valley (Bologna, Italy). *Environmental Science and Pollution Research International*, *21*(2), 872–890. DOI: 10.1007/s11356-013-1911-7 PMID: 23828727

Touir, J., M'rabet, A., & Ben Ayed, N. (1987). Sédimentation et tectonique précoce des séquences turoniennes du Jebel M'rhila, Tunisie centrale. 8th IAS Regional Meeting of SedimentologyTunis 87.

Touir, J., Ben Haj Ali, N., Donze, P., Maamouri, A. L., Memmi, L., M'rabet, A., & Zaghbib-Turki, D. (1989). Biostratigraphie et sédimentologie des séquences du Crétacé supérieur du Jebel Mrhila-Tunisie centrale. *Géologie Méditerranéenne*, *16*(1), 55–66.

Touir, J., & Soussi, M. (2003). The Growth and Migration of two Turonian Rudist-Bearing Carbonate Platforms in central TunisiaEustatic and tectonic controls. North African Cretaceous Carbonate platform. *System*, •••, 5381.

Touir, J., Soussi, M., & Troudi, H. (2009). Polyphased dolomitization of a shoal rimmed carbonate platform: Example from the middle Turonian Bireno dolomites of central Tunisia. *Cretaceous Research*, *30*, 785–804.

Tran, D. X., Pla, F., Latorre-Carmona, P., Myint, S. W., Caetano, M., & Kieu, H. V. (2017). Characterizing the relationship between land use land cover change and land surface temperature. *ISPRS Journal of Photogrammetry and Remote Sensing*, *124*, 119–132. DOI: 10.1016/j.isprsjprs.2017.01.001

Trost, C. O. W., Lassnig, A., Kreiml, P., Jörg, T., Terziyska, V. L., Mitterer, C., & Cordill, M. J. (2024). Enthalpy-Driven Self-Healing in Thin Metallic Films on Flexible Substrates. *Advanced Materials*, *36*(29), 2401007. DOI: 10.1002/adma.202401007 PMID: 38695220

Troudi, H. (1997). Les réservoirs et faciès associés du Crétacé supérieur en Tunisie centrale : Sédimentologie, Stratigraphie séquentielle et Diagenèse. Thèse, Sc., Fac. Sc. Tunis, 287 p.

Troudi, H. (1990). *Les faciès cénomanoturoniens du Jebel Essif (Tunisie centrale). Stratigraphie séquentielle, sédimentologie et aspect réservoir*. D. E. A., Fac. Sc.

Troudi, H., Negra, M. H., & M'rabet, A. (2000). Predictive evaluation of lowermiddle Turonian carbonate reservoir in central Tunisia through integration of deposition environment and diagenesis. *Annales des Mines et de la Géologie*, *2000*(40), 81–98.

Troudi, H., Saidi, M., Akrout, N., & Kilani, F. (2002). Jurassic and cretaceous petroleum systems in central Tunisia. The eigfth tunisian petroleum exploration and production conference. *ETAP*, (18), 1926.

Tucker, M. E., & Wright, V. P. (2009). *Carbonate sedimentology*. John Wiley & Sons.

Tuyan, M., Andiç-Çakir, Ö., & Ramyar, K. (2018). Effect of alkali activator concentration and curing condition on strength and microstructure of waste clay brick powder-based geopolymer. *Composites. Part B, Engineering*, *135*, 242–252. DOI: 10.1016/j.compositesb.2017.10.013

Ullah, S., Qiao, X., & Abbas, M. (2024). Addressing the impact of land use land cover changes on land surface temperature using machine learning algorithms. *Scientific Reports*, *14*(1), 18746. DOI: 10.1038/s41598-024-68492-7 PMID: 39138301

UN Environment Programme. (2019). *2019 global status report for buildings and construction*. United Nations Environment Programme. https://www.unep.org/resources/publication/2019-global-status-report-buildings-and-construction

USEPA (2002). Methods for Measuring the Acute Toxicity of Effluents and Receiving Waters to Freshwater and Marine Organisms. EPA-821-R-02-012.

Utrera-Barrios, S., Santana, M., Verdejo, R., & López-Manchado, M. (2020). Design of rubber composites with autonomous self-healing capability. *ACS Omega*, *5*(4), 1902–1910. DOI: 10.1021/acsomega.9b03516 PMID: 32039326

Varol, M. (2011). Assessment of heavy metal contamination in sediments of the Tigris River (Turkey) using pollution indices and multivariate statistical techniques. *Journal of Hazardous Materials*, *195*, 355–364. DOI: 10.1016/j.jhazmat.2011.08.051 PMID: 21890271

Vaßen, R., Giesen, S., & Stöver, D. (2009). Lifetime of Plasma-Sprayed Thermal Barrier Coatings: Comparison of Numerical and Experimental Results. *Journal of Thermal Spray Technology*, *18*(5-6), 835–845. DOI: 10.1007/s11666-009-9389-z

Vaßen, R., & Stöver, D. (2001). *Conventional and New Materials for Thermal Barrier Coatings*., DOI: 10.1007/978-94-010-0702-3_21

Viao, P., Magli, E., Caliendo, G., Corvino, A., Fiorino, F., Frecentese, F., Saccone, I., Santagada, V., Severino, B., Onorati, G., Freda, G., Manzo, C., & Perissutti, E. (2018). Heavy Metals Size Distribution in PM_{10} and Environmental-Sanitary Risk Analysis in Acerra (Italy). *Atmosphere (Basel)*, *9*(2), 5. DOI: 10.3390/atmos9020058

Wan, D., Jiang, Q., Song, Y., Pan, J., Qi, T., & Li, G. L. (2020). Biomimetic tough self-healing polymers enhanced by crystallization nanostructures. *ACS Applied Polymer Materials*, *2*(2), 879–886. DOI: 10.1021/acsapm.9b01094

Wang, H., Yao, W., Wang, R., Yang, J., Xu, W., An, N., & Zhou, Y. (2024, June). Analysis of Changes in Mining Geological Environment in a Certain Area of Northeast China Based on Domestic Satellite Monitoring Technology. In International Conference on Mineral Resources, Geotechnology and Geological Exploration (pp. 441-451). Cham: Springer Nature Switzerland. DOI: 10.1007/978-3-031-78690-7_50

Wang, A., Silva, K. D., Jones, M. I., & Gao, W. (2024). Cr-free anticorrosive primers for marine propeller applications. *Polymers*, *16*(3), 408. DOI: 10.3390/polym16030408 PMID: 38337297

Wang, B., Liu, J., Niu, D., Wu, N., Yun, W., Wang, W., Zhang, K., Li, G., Yan, S., Xu, G., & Yin, J. (2021). Mussel-inspired bisphosphonated injectable nanocomposite hydrogels with adhesive, self-healing, and osteogenic properties for bone regeneration. ACS Applied Materials &Amp. *Interfaces*, *13*(28), 32673–32689. DOI: 10.1021/acsami.1c06058 PMID: 34227792

Wang, L., Li, W., Sun, Y., Tao, M., Xin, J., Song, T., Li, X., Zhang, N., Ying, K., & Wang, Y. (2018). $PM_{2.5}$ Characteristics and Regional Transport Contribution in Five Cities in Southern North China Plain, During 2013–2015. *Atmosphere (Basel)*, *9*(4), 157. DOI: 10.3390/atmos9040157

Wang, L., Zhao, Y., & Liu, H. (2022). Green synthesis of smart materials: Toward sustainable environmental applications. *Materials Today*, *51*, 45–57. DOI: 10.1016/j.mattod.2021.08.003

Wang, P., Hu, M., Wang, H., Chen, Z., Feng, Y., Wang, J., Ling, W., & Huang, Y. (2020). The evolution of flexible electronics: From nature, beyond nature, and to nature. *Advancement of Science*, *7*(20), 2001116. Advance online publication. DOI: 10.1002/advs.202001116 PMID: 33101851

Wang, P., Tang, Q., Zhang, L., Xu, M., Sun, L., Sun, S., Zhang, J., Wang, S., & Liang, X. (2021). Ultrasmall barium titanate nanoparticles for highly efficient hypoxic tumor therapy via ultrasound-triggered piezocatalysis and water splitting. *ACS Nano*, *15*(7), 11326–11340. DOI: 10.1021/acsnano.1c00616 PMID: 34180675

Wang, Q., Kwan, M., Zhou, K., Fan, J., Wang, Y., & Zhang, D. (2019). The impacts of urbanization on fine particulate matter ($PM_{2.5}$) concentrations: Empirical evidence from 135 countries worldwide. *Environmental Pollution*, *247*, 989–998. DOI: 10.1016/j.envpol.2019.01.086 PMID: 30823354

Wang, X., Huan, Y., Ji, S., Zhu, Y., Wei, T., & Cheng, Z. (2022). Ultra-high piezoelectric performance by rational tuning of heterovalent-ion doping in lead-free piezoelectric ceramics. *Nano Energy*, *101*, 107580. DOI: 10.1016/j.nanoen.2022.107580

Wang, X., Sun, Q., Wang, S., Shao, R., & Su, J. (2023). Multiscale mathematical analysis of influencing factors and experimental verification of microcrack self-healing efficiency of bitumen composites using microcapsules. *Materials (Basel), 16*(14), 5073. DOI: 10.3390/ma16145073 PMID: 37512348

Wang, Y. N., Peng, J. R., Bogireddy, C., Rattan, B., & Tan, M. X. (2024). Artificial islands in modern development: Construction, applications, and environmental challenges. *Marine Georesources and Geotechnology, 42*(12), 1896–1906. DOI: 10.1080/1064119X.2023.2295034

Wang, Y., Wang, Y., Li, T., Wang, X., Ren, H., Lou, C., & Lin, J. (2023). Preparation and properties of polyaniline self-healing conductive materials based on hydrogen bonding and electrostatic interaction forces. *ACS Applied Polymer Materials, 5*(11), 9585–9593. DOI: 10.1021/acsapm.3c01239

Wang, Z., Wu, T., Shi, G., Fu, X., Tian, Y., Feng, Y., Wu, X., Wu, G., Bai, Z., & Zhang, W. (2012). Potential source analysis for PM_{10} and $PM_{2.5}$ in autumn in a Northern city in China. *Aerosol and Air Quality Research, 12*(1), 39–48. DOI: 10.4209/aaqr.2011.04.0045

Ward, W. C., & Halley, R. B. (1985). Dolomitization in a mixing zone of near seawater composition, Late Pleistocene, northeastern Yucatan peninsula, J. Sedim. *Petrol., 55*, 407420.

Warren, J. (2000). Dolomite: Occurrence, evolution and economically important associations. *Earth-Science Reviews, 52*, 1–81.

Watson, R. T., Zinyowera, M. C., & Moss, R. H. (Eds.). (1998). *The regional impacts of climate change: an assessment of vulnerability*. Cambridge University Press.

Wedephol, K. H. (1968). In Ahren, L. H. (Ed.), *Origin and Distribution of the Elements* (p. 99). Pergimon Press.

Wen, J., Xu, G., Liang, Z., Li, S., Wang, Y., Yang, J., & Nie, Y. (2023). Combing experimental methods and molecular simulations to study self-healing behaviors of polyurethane elastomers containing multiple hydrogen bond networks and flexible blocks. *Physical Chemistry Chemical Physics, 25*(41), 28162–28179. DOI: 10.1039/D3CP02723F PMID: 37818678

WHO. (World Health Organization). (2015). World health statistics 2015.World Health Organization.

Wibowo, Y. G., Ramadan, B. S., Taher, T., & Khairurrijal, K. (2023). Advancements of nanotechnology and nanomaterials in environmental and human protection for combating COVID-19 during and post-pandemic era: A comprehensive scientific review. In *Biomedical materials & devices* (pp. 34–57). Springer., DOI: 10.1007/978-3-031-16891-1_2

Wilson, E. N., Hardie, L. A., & Phillips, O. M. (1990). Dolomitization front geometry, fluid flow patterns, and the origin of massive dolomite: The Triassic Latemar buildup, northern Italy. *American Journal of Science*, *290*(7), 741–796.

Wolf, K. H. (1965). Petrogenesis and paleo-environment of Devonian algal limestones of New South Wales. *Sedimentology*, *4*, 113–178.

World Health Organization (WHO). (2005). *Air Quality Guidelines for Europe. WHO Regional Publications, European Series No. 91*. WHO Regional Office for Europe.

World Health Organization. (2022). *Global report on equity in health technology access*. World Health Organization. https://www.who.int/publications/i/item/9789240065097

Wu, P., Cheng, H., Wang, X., Shi, R., Zhang, C., Arai, M., & Zhao, F. (2021). A self-healing and recyclable polyurethane-urea diels–alder adduct synthesized from carbon dioxide and furfuryl amine. *Green Chemistry*, *23*(1), 552–560. DOI: 10.1039/D0GC03695A

Wu, S., Chuang, W., Ho, J., Wu, H., & Hsu, S. (2023). Self-healing of recombinant spider silk gel and coating. *Polymers*, *15*(8), 1855. DOI: 10.3390/polym15081855 PMID: 37112001

Wu, T., Li, X., Zhang, J., Wang, J., Chen, S., Zhou, Y., & Huang, S. (2023). Thermochromic perovskites for smart solar windows. *Nature Materials*, *22*, 110–118. DOI: 10.1038/s41563-022-01415-1

Xiao, Z., Shao, L., Zhang, N., Wang, J., & Wang, J. (2013). Heavy metal compositions and bio-reactivity of airborne PM_{10} in a valley-shaped city in Northwestern China. *Aerosol and Air Quality Research*, *13*(3), 116–1125. DOI: 10.4209/aaqr.2012.10.0287

Xie, L. S., Skorupskii, G., & Dincă, M. (2020). Electrically conductive metal–organic frameworks. *Chemical Reviews*, *120*(16), 8536–8580. DOI: 10.1021/acs.chemrev.9b00766 PMID: 32275412

Xi, X., Zhang, W., Yang, Z., Zhang, Y., Zhang, H., Zhang, H., & Li, Q. (2016). Smart and flexible supercapacitor based on a porous carbon nanotube film and polyaniline hydrogel. *RSC Advances*, *6*(30), 24946–24951. DOI: 10.1039/C6RA00705H

Xu, F., Weng, D., Li, X., Li, Y., & Sun, J. (2022). Self-healing hydrophilic porous photothermal membranes for durable and highly efficient solar-driven interfacial water evaporation. *CCS Chemistry*, *4*(7), 2396–2408. DOI: 10.31635/ccschem.021.202101111

Xu, S., Zhao, Q., Yin, K., He, G., Zhang, Z., Wang, G., Wen, M., & Zhang, N. (2021). Spatial downscaling of land surface temperature based on a multi-factor geographically weighted machine learning model. *Remote Sensing (Basel)*, *13*(6), 1186. DOI: 10.3390/rs13061186

Yadav, S., Choudhary, N., Bhai, S., Bhojani, G., Chatterjee, S., Ganguly, B., & Paital, A. R. (2021). Recyclable functionalized material for sensitive detection and exceptional sorption of hexavalent chromium and permanganate ions with biosensing applications. *ACS Applied Bio Materials*, *4*(8), 6430–6440. DOI: 10.1021/acsabm.1c00609 PMID: 35006925

Yahyawi, Z., Rbaib, N., & Saadi, A. (2014). Étude descriptive de l'évolution saisonnière des dunes bordières de la plage de Korba (façade orientale, Cap Bon, Tunisie) par la methode de des Modeles Numeriques de Terrain Residuel. *INSTM Bulletin: Marine and Freshwater Sciences*, *41*, 143–158.

Yamashita, Y., Karaki, T., Lee, H.-Y., Wan, H., Kim, H. P., & Jiang, X. (2022). A review of lead perovskite piezoelectric single crystals and their medical transducer applications. *IEEE Transactions on Ultrasonics, Ferroelectrics, and Frequency Control*, *69*(6), 3048–3056. DOI: 10.1109/TUFFC.2022.3160526 PMID: 35298378

Yang, L., Yu, L., Chen, D., Liu, K., Yang, P., & Li, X. (2020). Effects of dolomitization on porosity during various sedimentation-diagenesis processes in carbonate reservoirs. *Minerals (Basel)*, *10*(6), 574.

Yang, Z., Zhou, S., Zu, J., & Inman, D. (2018). High-performance piezoelectric energy harvesters and their applications. *Joule*, *2*(4), 642–697. DOI: 10.1016/j.joule.2018.03.011

Yildizay, H., & Goren, R. (2017). The utilization of Şaphane alunite in geopolymer production. *Journal of the Australian Ceramic Society*, *53*(2), 265–270. DOI: 10.1007/s41779-017-0032-z

Yin, H., Liu, H., Bu, Y., Chen, W., Ding, F., Lin, W., & Zhang, Y. (2024). Regulation of reactivity of calcined ultrafine kaolin products. *Processes (Basel, Switzerland)*, *12*(10), 2268. DOI: 10.3390/pr12102268

Yuan, Z. Y., Banks, C. E., & Li, W. (2025). Introducing smart materials and devices. *Smart Materials and Devices*, *1*, 202511. DOI: 10.1016/j.smad.2025.202511

Yun, T. G., Park, M., Kim, D. H., Kim, D., Cheong, J. Y., Bae, J. G., Han, S. M., & Kim, I. D. (2019). All-transparent stretchable electrochromic supercapacitor wearable patch device. *ACS Nano*, *13*(3), 3141–3150. DOI: 10.1021/acsnano.8b08560 PMID: 30779547

Yu, S., & Jain, P. K. (2019). The chemical potential of plasmonic excitations. *Angewandte Chemie International Edition*, *59*(5), 2085–2088. DOI: 10.1002/anie.201914118 PMID: 31765516

Yuyu, E., Ju, Y., Wang, Z., Chang, Z., Jiang, J., Li, P., & Wang, K. (2022). Tunable and self-healing properties of polysaccharide-based hydrogels through polymer architecture modulation. ACS Sustainable Chemistry &Amp. *ACS Sustainable Chemistry & Engineering*, *10*(42), 14053–14063. DOI: 10.1021/acssuschemeng.2c05125

Zagrarni, M. F. (1999). Sédimentologie, stratigraphie séquentelle et diagenèse des faciès du Crétacé supérieur du Jebel Bireno. Paléogéographie des plates-formes carbonatées du Cénomanien supérieur Coniacien en Tunisie centrale. Unpublished PhD thesis, University of Tunis-El Manar, 220 p.

Zhang, B., Yu, T., Deng, L., Li, Y., Guo, H., Zhou, J., Li, L., & Peng, Y. (2022). Ion-adsorption type rare earth tailings for preparation of alkali-based geopolymer with capacity for heavy metals immobilization. *Cement and Concrete Composites*, *134*, 104768. DOI: 10.1016/j.cemconcomp.2022.104768

Zhang, J., Cheng, M., Ji, D., Liu, Z., Hu, B., Sun, Y., & Wang, Y. S. (2016). Characterization of submicron particles during biomass burning and coal combustion periods in Beijing, China. *The Science of the Total Environment*, *562*, 812–821. DOI: 10.1016/j.scitotenv.2016.04.015 PMID: 27110992

Zhang, N., Hedayat, A., Figueroa, L., Steirer, K. X., Li, L., & Sosa, H. G. B. (2023). Physical, mechanical, cracking, and damage properties of mine tailings-based geopolymer: Experimental and numerical investigations. *Journal of Building Engineering*, *75*, 107075. DOI: 10.1016/j.jobe.2023.107075

Zhang, Y., & Lu, L. W. (2008). Introducing smart structures technology into civil engineering curriculum: Education development at Lehigh University. *Journal of Professional Issues in Engineering Education and Practice*, *134*(1), 41–48. DOI: 10.1061/(ASCE)1052-3928(2008)134:1(41)

Zhang, Y., Wang, M., Zhuo, R., Tang, W., Niu, C., Chen, Q., Yang, W., & Zhao, S. (2025). Study on Self-Healing Technology of Epoxy Resin for Insulator Core Rods. *Journal of Applied Polymer Science*, *142*(21), e56935. DOI: 10.1002/app.56935

Zhang, Y., Wang, Q., Yi, S., Lin, Z., Wang, C., Chen, Z., & Jiang, L. (2024). 4D printing of magnetoactive soft materials for on-demand magnetic actuation transformation. *ACS Applied Materials & Interfaces*, *13*(3), 4174–4184. DOI: 10.1021/acsami.0c19280 PMID: 33398983

Zhang, Z., Wang, H., & Provis, J. L. (2012). Quantitative study of the reactivity of fly ash in geopolymerization by FTIR. *Journal of Sustainable Cement-Based Materials*, *1*(4), 154–166. DOI: 10.1080/21650373.2012.752620

Zhao, D., Zhu, Y., Cheng, W., Chen, W., Wu, Y., & Yu, H. (2020). Cellulose-based flexible functional materials for emerging intelligent electronics. *Advanced Materials*, *33*(28), 2000619. Advance online publication. DOI: 10.1002/adma.202000619 PMID: 32310313

Zhao, K., Zhou, C., Liu, L., & Wu, G. (2025). Self-healing waterborne polyurethane with enhanced mechanical properties based on nano-silica. *Journal of Applied Polymer Science*, *142*(18), e56822. Advance online publication. DOI: 10.1002/app.56822

Zhao, Y., Liu, H., Wang, L., & Li, X. (2021). Multifunctional intelligent materials for environmental remediation: Trends and prospects. *Advanced Functional Materials*, *31*(9), 2007257. DOI: 10.1002/adfm.202007257

Zhou, S., Wang, C., & Ju, J. W. (2022). A numerical chemo-micromechanical damage model of sulfate attack in cementitious materials. *International Journal of Damage Mechanics*, *31*(10), 1613–1638. DOI: 10.1177/10567895221114926

Zhou, Y., Li, L., Han, Z., Li, Q., He, J., & Wang, Q. (2022). Self-healing polymers for electronics and energy devices. *Chemical Reviews*, *123*(2), 558–612. DOI: 10.1021/acs.chemrev.2c00231 PMID: 36260027

Zhou, Y., Wen, J., & Nie, Y. (2024). Microscopic mechanisms of self-healing in polymers revealed by molecular simulations and machine learning. *Macromolecules*, *57*(7), 3258–3270. DOI: 10.1021/acs.macromol.4c00338

Zhu, D., & Miller, R. A. (2004). *Development of Advanced Low Conductivity Thermal Barrier Coatings. July.*

Zhu, M., Li, H., Zhang, Y., Chen, J., Wang, X., & Zhou, Z. (2023). Flexible piezoelectric nanogenerators for energy harvesting. *Nano Energy*, *108*, 108082. DOI: 10.1016/j.nanoen.2023.108082

Zhu, M., Yu, J., Li, Z., & Ding, B. (2022). Self-healing fibrous membranes. *Angewandte Chemie International Edition*, *61*(41), e202208949. Advance online publication. DOI: 10.1002/anie.202208949 PMID: 35862255

Zineb, M., Abdelhak, B., Bilal, B., & Fadila, A. (2024). Characterization and pollution assessment of potentially toxic metals (PTMs) in the sulfide tailings from Boudoukha abandoned mine (Zn–Pb–Cu), NE Algeria. *International Journal of Environmental Research*, *18*(4), 70. DOI: 10.1007/s41742-024-00620-7

About the Contributors

Elhoucine Essefi obtained his PhD form National engineering School of Sfax titled 'Wet Aeolian Sedimentology and Sequence Stratigraphy within the Terrestrial Analogues in Eastern Tunisia: Implications for Wet Aeolian Sedimentology and Sequence Stratigraphy on Mars', in 2013. In 2011, he obtained License in Linguistics and Anglophone Civilization from the Faculty of Arts and Humanities of Sfax. In 2009, he obtained the Master degree from the Faculty of Sciences of Sfax titled Multidisciplinary study of Sidi El Hani Saline Environment: the Geological History and the Climatic Variability. In 2003, he obtained the 'Maitrise' degree in Earth Sciences from the Faculty of Sciences of Sfax. He obtained his Habilitation from the university of Gabes in 2021. From 2022 onward, he occupied a Professor position at University of Sfax. In 2025, he obtained the Internet of Things (License) in Computer Sciences.

. **Suresh C** (Harur, Tamil Nadu, India) is an Associate Professor Aeronautical Engineering at the Global Academy of Technology, Bangalore. With nearly two decades of academic and research experience, he has significantly contributed to the advancement of aerospace education and research in India. He holds a Ph.D. specializing in hail ice impact, projectile impact, underwater explosion, and aircraft structures, with a strong focus on computational and experimental mechanics. His teaching portfolio spans across Aircraft Propulsion, Aircraft Structures, Finite Element Methods, and Emerging Aerospace Technologies, where he emphasizes Outcome-Based Education (OBE), practical learning, and research integration. As the Convenor of GCAAMS2024 (Global Conference on Advances in Aerospace and Materials Science), Dr. Suresh played a key role in fostering collaboration among researchers, industry experts, and academicians. He is a Life Member of the Aeronautical Society of India (AeSI) and has been actively involved in curriculum development, faculty training, and introducing innovative programs such as Emerging Technology Courses, Ability Enhancement Courses, and Soft Skills Development for

Aeronautical Engineers. Dr. Suresh is also driving infrastructure development with initiatives such as the proposal for an Advanced Composite Materials Laboratory, focusing on vacuum infusion, curing, and aerospace applications. His leadership in laboratory design includes the Aircraft Hangar (featuring a Cessna L150 and MIG-21 engine), Aircraft Structures Lab, FEM Lab, and Basic Electrical & Electronics Lab for first-year engineering students. Beyond his academic and research pursuits, Dr. Suresh is deeply committed to student development—ensuring placements, internships, industrial visits, and bridging industry–institute collaboration. His philosophy of teaching blends conceptual clarity, hands-on learning, and critical thinking, preparing students to excel as engineers and innovators. At 39, Dr. Suresh continues to contribute to aerospace engineering education and research with dedication, vision, and passion.

G. Chandra Bose got his undergraduate degree in Mechanical engineering from Bharathidasan University, his Master's degree in Aeronautical Engineering from Madras Institute of Technology, Anna University. Currently, pursuing his Doctorate from the Department of Aeronautical Engineering in Vel Tech Rangarajan Dr. Sagunthala R&D Institute of Science and Technolog, Chennai. He has guided 5 PG & 60 UG scholars on their projects, he has published 8 publications in Journals and international conferences. He is a Life Member of the Institution of Engineers India, the Aeronautical Society of India, Indian Society for Technical Education.

Boopathy Govindarajan earned his undergraduate degree in mechanical engineering from University of Madras, his master's degree in aeronautical engineering from Madras Institute of Technology, and his doctorate from the Department of Aeronautical Engineering, St. Peter's Institute of Higher Education and Research, Chennai. At Vel Tech University, a professorship is currently held in the department of aeronautical engineering. In addition to guiding 6 Ph.D Scholars, guided 8 PG & 90 UG scholars on their significant projects, he has written in 30 international publications. He is a Life Member of the Society of Automotive Engineers INDIA, The Robotics Society and the Aeronautical Society of India.

Hajer Haj Ali obtained her PhD in 2016 from the Faculty of Sciences of Sfax, with a dissertation titled "Study of Fractured Reservoirs: Case of the Bireno Member (Middle Turonian) at Jebel M'rhila and the Abiod Formation (Campanian–Maastrichtian) at Jebel Chérahil, Central Tunisia." In 2010, she earned a Master's degree with the thesis "Relationship Between Fracturing, Dolomitization and Porosity in a Carbonate Reservoir: Case of the Bireno Member (Middle Turonian)." She obtained her Maîtrise degree in Earth Sciences in 2006 from the same institution. She is currently a postdoctoral researcher at the Laboratory of Spectroscopy and

Optical Characterisation of Materials (LaSCOM), Faculty of Sciences of Sfax, University of Sfax, where her research focuses on the development of cementitious materials based on mining wastes and calcined clays from North Tunisia.

Shalini Jaiswal has more than 17 years' experience in academics and research and is the author of more than 65 scientific publications in Scopus\SCI\UGC and other reputed journal\Conference. Dr. Shalini Jaiswal received her D.Phil degree in Organic Chemistry from University of Allahabad, Allahabad India, in 2009. Dr. Shalini Jaiswal has14 month experience as a project fellow on UGC sponsored project entitled Green synthesis of multi- target nucleosides as a potential anti-viral agent. Her research interest includes different area of Chemistry, Nanotechnology, AI, Biotechnology and Environmental science. She is a Life Member of the Indian science congress association. She has reviewed more than 50 papers of various reputed journal and conferences. She is Editorial board member and Reviewer of various reputed journals like TPI (International Journal), IJATES, International journal of advance research, Advanced Scientific Research, IJEEE, Asian Journal of Research in Chemistry (AJRC) etc.Dr. Shalini Jaiswal Worked as Moderator in TOKYO SUMMIT-II (2020) and TOKYO SUMMIT-III (2021), Turkey (2022,2023,2024)

João Labrincha is Full Professor and an expert in ceramics processing. Following his PhD in Materials Science and Engineering (University of Aveiro, 1993) he implemented a research line on Wastes Recycling and Sustainable Use of Resources, teaching disciplines on related topics. He supervised 18 Ph.D and 56 MSc theses and has now 4 new students under progress. João Labrincha participate/d in 25 R&D projects (9 as leader/responsible) and in 16 Contracts financed/in cooperation with industries. João Labrincha has 22 patent applications (two of those as International PTC), and has 410 papers (Science Citation Index). He has >15600 citations and h = 66 (Scopus). He is Associate Editor of Journal of Sustainable Metallurgy and ACerS - International Journal of Ceramic Engineering and Science. He is author/co-author of 8 book chapters and co-editor of 7 books. Between December/2012-2016 he was member of the Operational Group #2 of the European Innovation Partnership on Raw Materials of the European Union. Main Scientific Interests Development of multifunctional eco-materials. Waste-based products (geopolymers, cements and mortars, pigments) for a circular economy.

Rui Novais is an Assistant Professor in the Department of Materials Engineering at the University of Aveiro and a member of the CICECO – Aveiro Institute of Materials Associated Laboratory. Since 2012, he has been working on the development of materials to decarbonize the construction and energy sectors and for environmental remediation applications, with particular emphasis on waste recycling and product sustainability. To date, he has published over 115 articles in scientific journals, of which 97 are in the JCR SCI. He is the first author of a review article published in Progress in Materials Science (IF = 40.0) and co-author of the book Materials and Sustainability (2024). His publications include 8 book chapters and more than 130 presentations (oral and poster) at international conferences. This body of work has accumulated over 4,000 citations and an h-index of 38 (SCOPUS).

Fernando Rocha is Professor of Mineral Resources at the University of Aveiro (Portugal), being the Director of Geobiotec Research Centre. His research is focused on clay minerals, industrial minerals and building materials, with Medical Geology, Coastal and Marine Geology as secondary interests, having published 557 full papers (322 in Web of Science indexed journals; H-index: 38). ResearcherID: A-4532-2009; Scopus Author ID 8626859600; ORCID: 0000-0002-3636-3933 CIENCIA ID: 6F13-70A2-3998

V. Srinivasan earned his undergraduate degree in mechanical engineering from S.K.P Institute of Technology affiliated to Anna University, his master's degree in CAD/CAM from Vel Tech Rangarajan Dr. Sagunthala R & D Institute of Science and Technology and currently pursuing Ph. D in Aeronautical Engineering from Vel Tech Rangarajan Dr. Sagunthala R & D Institute of Science and Technology. Currently doing research in LiDAR (Light detection and Ranging) and other remote sensing and imaging sensors, its integration with UAVs. Obstacle avoidance and autonomous navigation systems in UAV using sensor fusion are his key areas for research.

Index

N

O

P

S

T

V

W

www.ingramcontent.com/pod-product-compliance
Lightning Source LLC
LaVergne TN
LVHW081531300926
849161LV00006B/1385